Juhani Karhumäki   Hermann Maurer
Gheorghe Păun   Grzegorz Rozenberg (Eds.)

Theory Is Forever

Essays Dedicated to Arto Salomaa
on the Occasion of His 70th Birthday

 Springer

Juhani Karhumäki
University of Turku, Department of Mathematics
20014 Turku, Finland
E-mail: karhumak@cs.utu.fi

Hermann Maurer
Graz University of Technology
Institute for Information Systems and Computer Media
Inffeldgasse 16c, 8010 Graz, Austria
E-mail: hmaurer@iicm.edu

Gheorghe Păun
Romanian Academy, Institute of Mathematics
P.O. Box 1-764, 70700 Bucharest, Romania
E-mail: george.paun@imar.ro

Grzegorz Rozenberg
Leiden University, Leiden Institute of Advanced Computer Science (LIACS)
Niels Bohrweg 1, 2333 CA Leiden, The Netherlands
E-mail: rozenber@liacs.nl

The illustration appearing on the cover of this book is the work of Daniel Rozenberg
(DADARA)

Library of Congress Control Number: 2004108213

CR Subject Classification (1998): F.1, F.3, F.4, G.1, G.2

ISSN 0302-9743
ISBN 3-540-22393-2 Springer-Verlag Berlin Heidelberg New York

Springer-Verlag is a part of Springer Science+Business Media

springeronline.com

© Springer-Verlag Berlin Heidelberg 2004
Printed in Germany

Typesetting: Camera-ready by author, data conversion by Boller Mediendesign
Printed on acid-free paper   SPIN: 11019213   06/3142   5 4 3 2 1 0

Lecture Notes in Computer Science 3113

Commenced Publication in 1973
Founding and Former Series Editors:
Gerhard Goos, Juris Hartmanis, and Jan van Leeuwen

Arto Salomaa

Preface

This Festschrift celebrates the 70th birthday of Arto Kustaa Salomaa (born in Turku, Finland on June 6, 1934), one of the most influential researchers in theoretical computer science.

Most of his research concerns theory – he is one of the founding fathers of formal language and automata theory, but he has also made important contributions to cryptography and natural computing. His approach to research in theoretical computer science is exemplary and inspirational for his students, collaborators, and the readers of his papers and books. For him, the role of theory (in computer science) is to discover general rules of information processing that hold within computer science and in the world around us. One should not waste time on research concerning passing artifacts (or fashionable topics of the moment) in computer science – theory should be permanently predictive, insightful, and inspiring. That's why we chose the title "Theory is Forever".

The main source of his influence on theoretical computer science is his publications. Arto is a born writer – his papers and books are always most elegant. He has a unique gift for identifying the real essence of a research problem, and then presenting it in an incisive and eloquent way. He can write about a very involved formal topic and yet avoid a (much too common) overformalization. Many of his writings are genuine jewels and belong to the classics of theoretical computer science. They have inspired generations of students and researchers. Indeed, even computers as well as computer science have learned a lot from Arto's publications – this is nicely illustrated by DADARA on the cover of this volume. His writing talent extends beyond science – he writes beautiful and engaging stories, and his close friends very much enjoy receiving his long, entertaining and informative letters.

There is much other information that could be cited in this preface, such as the fact that he is one of the most celebrated computer scientists (e.g., he holds eight honorary degrees), or that he has been very instrumental in providing the organizational infrastructure for theoretical computer science in Europe (e.g., he is the past President of the European Association for Theoretical Computer Science), or that he is an absolute authority on the Finnish sauna (including both theory and practice). However, all of these accomplishments have been documented already in many places (e.g., in the companion book "Jewels are Forever"[1] published on the occasion of Arto's 65th birthday). Thus we have restricted ourselves to reflections on his research and writings.

We are indebted to all the contributors for their tribute to Arto through this book. We ourselves have benefited enormously through many years of collabo-

---

[1] J. Karhumäki, H. Maurer, G. Păun, G. Rozenberg, Jewels are Forever, Contributions on Theoretical Computer Science in Honor of Arto Salomaa, Springer-Verlag, 1999.

ration with Arto from his guidance and friendship – editing this volume is just a token of our gratitude. We are also indebted to Mrs. Ingeborg Mayer from Springer-Verlag for the pleasant and efficient collaboration in producing this volume. As a matter of fact this collaboration is quite symbolic, as Arto has worked very closely with Springer-Verlag, especially with Mrs. Ingeborg Mayer and Dr. Hans Wössner, on many projects over many years. Finally, our special thanks go to T. Harju, M. Hirvensalo, A. Lepistö, and Kalle Saari for their work on this book.

April 2004

Juhani Karhumäki
Hermann Maurer
Gheorghe Păun
Grzegorz Rozenberg

Table of Contents

Duality for Three: Ternary Symmetry in Process Spaces

Janusz Brzozowski[1] and Radu Negulescu[2]

[1] School of Computer Science, University of Waterloo,
Waterloo, ON, Canada N2L 3G1
brzozo@uwaterloo.ca, http://maveric.uwaterloo.ca
[2] Department of Electrical and Computer Engineering
McGill University, Montreal, Québec, Canada H3A 2A7
radu@macs.ece.mcgill.ca

Abstract. Ternary algebra has been used for detection of hazards in logic circuits since 1948. Process spaces have been introduced in 1995 as abstract models of concurrent processes. Surprisingly, process spaces turned out to be special ternary algebras. We study symmetry in process spaces; this symmetry is analoguous to duality, but holds among three algebras. An important role is played here by the uncertainty partial order, which has been used since 1972 in algebras dealing with ambiguity. We prove that each process space consists of three isomorphic Boolean algebras and elements related to partitions of a set into three blocks.

1 Introduction

The concept of duality is well known in mathematics. In this paper we study a similar concept, but one that applies to three objects instead of two. The road that led to the discovery of these properties deserves to be briefly mentioned, because several diverse topics come together in this work.

The usual tool for the analysis and design of digital circuits is Boolean algebra, based on two values. As early as 1948, however, it was recognized that three values are useful for describing certain phenomena in logic circuits [10]. We provide more information about the use of ternary algebra for hazard detection in Section 2.

Ternary algebra is closely related to ternary logic [11]. This type of logic, allowing a third, ambiguous value in addition to **true** and **false**, was studied by Mukaidono in 1972 [12], who introduced the *uncertainty* partial order, in addition to the usual lattice partial order. This partial order turned out to be very useful; see, for example, [3, 6]. It also plays an important role in the ternary symmetry we are about to describe.

In 1995 Negulescu [13] introduced process spaces as abstract models of concurrent processes. Surprisingly, process spaces turned out to be special types of ternary algebras. It is in process spaces that "ternary duality" exists. Similar properties also hold in so-called *linear logic*, which has been used as another

J. Karhumäki et al. (Eds.): Theory Is Forever (Salomaa Festschrift), LNCS 3113, pp. 1–14, 2004.

framework for representing concurrent processes, and has connections to Petri nets [17]. This topic is outside the scope of the present paper.

The remainder of the paper is structured as follows. Section 2 illustrates hazard detection using ternary algebra. We also recall some basic concepts from lattice theory and summarize the properties of ternary algebras. Process spaces are defined in Section 3. Ternary symmetry is next discussed in Section 4. In Section 5 we show that each process space contains three isomorphic Boolean algebras. Section 6 characterizes elements of a process space that are outside the Boolean algebras, and Section 7 summarizes our results.

We assume that unary operations have precedence over binary operations. For example, $-x + -y$ denotes $(-x) + (-y)$. Sequences of unary operations are written without parentheses; for example, $-/-x$ denotes $-(/(-x))$. Set inclusion is denoted by $\subseteq$ and proper inclusion, by $\subset$. Proofs that are straightforward and involve only elementary set theory are omitted.

2 Ternary Algebras

The logic values are 0 and 1, and a third value, denoted here by Φ, is used to represent an intermediate or uncertain signal. This idea was used by many authors, but we mention here only Eichelberger's 1965 ternary simulation algorithm [8] and its later characterizations [6]. More information about hazard detection can be found in a recent survey [4]. The following example illustrates the use of ternary simulation to detect hazards in logic circuits.

Example 1. Consider the behavior of the circuit of Fig. 1(a) when its input x changes from 0 to 1. Initially, $x = 0$, $y = 1$, and $z = 0$. After the transition, $x = 1$, $y = 0$, and $z = 0$. Thus, z is not supposed to change during this transition. If the inverter has a sufficiently large delay, however, for a short time both inputs to the AND gate may be 1, and there may be a short 1-pulse in z. Such a pulse is undesirable, because it may cause an error in the computation.

In the first part of the ternary simulation, Algorithm A, we change the input to Φ, which indicates that the input is first going through an intermediate, uncertain value. See Fig. 1(b); the first two entries on each line illustrate Algorithm A. The circuit is then analyzed in ternary algebra to determine which gates will undergo changes; the outputs of the changing gates become Φ. In our example, the inverter output becomes uncertain because its input is uncertain. Also, since one input of the AND gate is 1 and the other uncertain, z becomes Φ.

In the second part, Algorithm B, the input is changed to its final binary value, and the circuit is again simulated in ternary algebra. Some gate outputs that became Φ in Algorithm A will become binary, while others remain Φ. In our example, both y and z become 0; see the last two entries in Fig. 1(b). If a gate output has the same (binary) value in the initial state and also at the end of Algorithm B, then that output is not supposed to change during the transition in question. If, however, that output is Φ after Algorithm A is applied, then we have detected a *hazard*, meaning that an undesired pulse may occur. This happens to the output z.   □

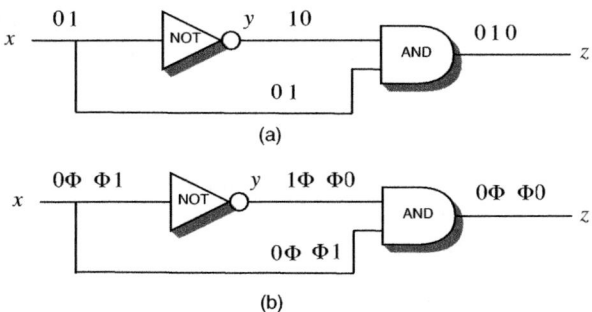

Fig. 1. Circuit with hazard: (a) binary analysis (b) ternary analysis

We now recall some concepts from algebra. For more information about lattices see [1, 7]. We use the following terminology. A *semilattice* [2] is an algebra $(S, \sqcup)$, where S is a set and $\sqcup$ is an idempotent, commutative and associative binary operation on S. We define the partial order $\sqsubseteq_\sqcup$ on S by

$$x \sqsubseteq_\sqcup y \Leftrightarrow x \sqcup y = y.$$

A *bisemilattice* is an algebra $(S, \sqcup, \sqcap)$ in which $(S, \sqcup)$ and $(S, \sqcap)$ are semilattices,

Table 1. Laws of de Morgan Algebras

M1	$x \sqcup x = x$	M1′	$x \sqcap x = x$
M2	$x \sqcup y = y \sqcup x$	M2′	$x \sqcap y = y \sqcap x$
M3	$x \sqcup (y \sqcup z) = (x \sqcup y) \sqcup z$	M3′	$x \sqcap (y \sqcap z) = (x \sqcap y) \sqcap z$
M4	$x \sqcup (x \sqcap y) = x$	M4′	$x \sqcap (x \sqcup y) = x$
M5	$x \sqcup \bot = x$	M5′	$x \sqcap \top = x$
M6	$x \sqcup \top = \top$	M6′	$x \sqcap \bot = \bot$
M7	$x \sqcup (y \sqcap z) = (x \sqcup y) \sqcap (x \sqcup z)$	M7′	$x \sqcap (y \sqcup z) = (x \sqcap y) \sqcup (x \sqcap z)$
M8	$--x = x$		
M9	$-(x \sqcup y) = -x \sqcap -y$	M9′	$-(x \sqcap y) = -x \sqcup -y$

i.e., laws M1–M3, M1′–M3′ of Table 1 hold. A bisemilattice has two partial orders $\sqsubseteq_\sqcup$ and $\sqsubseteq_\sqcap$, the latter defined by

$$x \sqsubseteq_\sqcap y \Leftrightarrow x \sqcap y = x.$$

If a bisemilattice satisfies the absorption laws M4 and M4′, then it is a *lattice*. The two partial orders $\sqsubseteq_\sqcup$ and $\sqsubseteq_\sqcap$ then coincide, and are denoted by $\sqsubseteq$. The converse of $\sqsubseteq$ is denoted by $\sqsupseteq$. The operations $\sqcup$ and $\sqcap$ are the *join* and *meet* of the lattice, respectively. A lattice is *bounded* if it has greatest and least elements

$\top$ (*top*) and $\perp$ (*bottom*) satisfying M5, M6, M5′, M6′. A bounded lattice is represented by $(S, \sqcup, \sqcap, \perp, \top)$. A lattice satisfying the distributive laws M7 and M7′ is *distributive*.

A *de Morgan algebra* is an algebra $(S, \sqcup, \sqcap, -, \perp, \top)$, where $(S, \sqcup, \sqcap, \perp, \top)$ is a bounded distributive lattice, and $-$ is a unary operation, called *quasi-complement*, that satisfies M8 and de Morgan's laws M9 and M9′.

A *Boolean algebra* is a de Morgan algebra $(S, \sqcup, \sqcap, -, \perp, \top)$, which also satisfies the complement laws:

$$x \sqcup -x = \top \qquad\qquad x \sqcap -x = \perp$$

A *ternary algebra* $(S, \sqcup, \sqcap, -, \perp, \Phi, \top)$ is a de Morgan algebra $(S, \sqcup, \sqcap, -, \perp, \top)$ with an additional constant Φ satisfying

> T1 $-\Phi = \Phi$
>
> T2 $(x \sqcup -x) \sqcup \Phi = x \sqcup -x$ T2′ $(x \sqcap -x) \sqcap \Phi = x \sqcap -x$

For more information about ternary algebras the reader is referred to [5, 6, 9, 12]. Here, we mention only the uncertainty partial order and the subset-pair representation of ternary algebras.

Figure 2(a) shows the lattice order $\sqsubseteq$ of the 3-element ternary algebra $\mathbf{T_3} = (\{\perp, \Phi, \top\}, \sqcup, \sqcap, -, \perp, \Phi, \top)$, and Fig. 2(b), its *uncertainty* partial order $\preceq$ [6, 12], where Φ represents the unknown or uncertain value, and $\perp$ and $\top$ are the known or certain values.

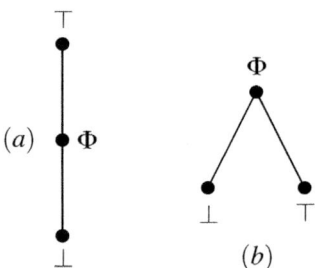

Fig. 2. Partial orders in $\mathbf{T_3}$: (a) $\sqsubseteq$ (b) $\preceq$

For any $x, y \in \mathbf{T_3}$, the least upper bound of $\{x, y\}$ in the partial order $\preceq$ can be expressed as $(x \sqcap y) \sqcup ((x \sqcup y) \sqcap \Phi)$ [6]. We extend this to any ternary algebra $(S, \sqcup, \sqcap, -, \perp, \Phi, \top)$ by defining the binary operation $\vee$ [3] as

$$x \vee y = (x \sqcap y) \sqcup ((x \sqcup y) \sqcap \Phi).$$

It is easily verified that $(S, \vee)$ is a semilattice. The semilattice partial order is

$$x \preceq y \Leftrightarrow x \vee y = y.$$

Let $\mathcal{E}$ be a nonempty set, and $\mathcal{P}$, a collection of ordered pairs (X, X') of subsets of $\mathcal{E}$ such that $X \cup X' = \mathcal{E}$. For $(X, X'), (Y, Y') \in \mathcal{P}$, let

$$(X, X') \sqcup (Y, Y') = (X \cap Y, X' \cup Y'), \tag{1}$$

$$(X, X') \sqcap (Y, Y') = (X \cup Y, X' \cap Y'), \tag{2}$$

$$-(X, X') = (X', X). \tag{3}$$

Let $\bot = (\mathcal{E}, \emptyset)$, $\Phi = (\mathcal{E}, \mathcal{E})$, and $\top = (\emptyset, \mathcal{E})$. Then $(\mathcal{P}, \sqcup, \sqcap, -, \bot, \Phi, \top)$ is a *subset-pair algebra* [5] if $\mathcal{P}$ is closed under $\sqcup$, $\sqcap$, and $-$, and contains the constants $\bot$, Φ, and $\top$. The following result was shown in [5, 9]:

Theorem 1. *Every subset-pair algebra is a ternary algebra, and every ternary algebra is isomorphic to a subset-pair algebra.*

It is easy to verify that

$$(X, X') \sqsubseteq (Y, Y') \Leftrightarrow X \supseteq Y, \text{ and } X' \subseteq Y', \tag{4}$$

$$(X, X') \vee (Y, Y') = (X \cup Y, X' \cup Y'), \tag{5}$$

$$(X, X') \preceq (Y, Y') \Leftrightarrow X \subseteq Y \text{ and } X' \subseteq Y'. \tag{6}$$

3   Process Spaces

The material in this section is based on [14, 15]. The discussion of applications of process spaces is beyond the scope of this paper, and we treat process spaces only as mathematical objects. However, we do give a simple example to motivate the reader.

Let $\mathcal{E}$ be any nonempty set; a *process* x over $\mathcal{E}$ is an ordered pair $x = (X, X')$ of subsets of $\mathcal{E}$ such that $X \cup X' = \mathcal{E}$.

We refer to $\mathcal{E}$ as a set of *executions*. Several different examples of execution sets have been used [14, 15]. For the purposes of this paper, however, we may think of $\mathcal{E}$ as the set of all sequences of *actions* from some action *universe* $\mathcal{U}$; thus $\mathcal{E} = \mathcal{U}^*$. A process $x = (X, X')$ represents a contract between a device and its environment: the device guarantees that only executions from X occur, and the environment guarantees that only executions from X' occur. Thus, $\overline{X} = \mathcal{E} \setminus X$ is the set of executions in which the device violates the contract. Similarly, for executions in $\overline{X'}$, the environment violates the contract. The condition $X \cup X' = \mathcal{E}$, or equivalently $\overline{X} \cap \overline{Y} = \emptyset$, means that the blame for violating the contract can be assigned to either the device or the environment, but not both. The set X is called the set of *accessible* executions of x, and X' is the set of *acceptable* executions.

Example 2. Figure 3 (a) shows a symbol for a buffer, and Fig. 3 (b) shows a sequential machine describing its behavior. The buffer starts in the state marked by an incoming arrow. If it receives a signal on its input a, it moves to a new state. It is expected to respond by producing a signal on its output b and returning to the original state. Thus, the normal operation of the buffer consists of an alternating sequence of a's and b's starting with a. The two states involved in this normal operation are marked g, representing the fact that they are the *goal* states of the process.

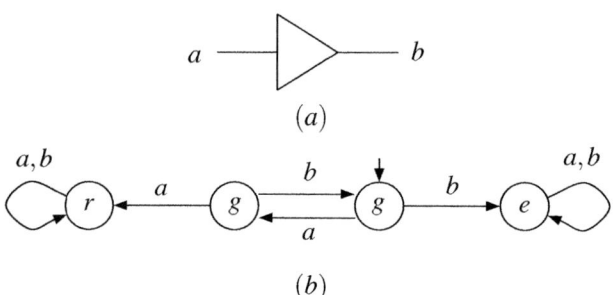

(a)

(b)

Fig. 3. Buffer process: (a) block diagram (b) behavior

It is possible that the environment of the buffer does not behave according to the specified goal, and produces two consecutive a's in the initial state. From the point of view of the buffer, this environment behavior can be rejected as illegal; hence the state diagram moves to a *reject* state marked r, and remains in that state thereafter. It is also possible that the buffer malfunctions by producing b in the initial state. This is a violation of the contract by the buffer, and the process moves to the state labelled e; such executions have been called the *escapes* of the process.

Let L_g be the set of all words taking the machine of Fig. 3 (b) to a state marked g, and let L_e and L_r be defined similarly. One verifies that $L_g = (ab)^*(\epsilon \cup a)$, where ϵ is the empty word, $L_e = (ab)^*b(a \cup b)^*$, and $L_r = (ab)^*aa(a \cup b)^*$. The buffer process is $(X, X') = (L_g \cup L_e, L_g \cup L_r)$. □

The *process space over* $\mathcal{E}$ is denoted by $\mathcal{P}_\mathcal{E}$, and it is the set of all processes over $\mathcal{E}$. Note that each set $\mathcal{E}$ defines a unique process space.

In constructing $\mathcal{P}_\mathcal{E}$ we must put each element of $\mathcal{E}$ in X or X' or both. Hence, if $\mathcal{E}$ has cardinality n, then $\mathcal{P}_\mathcal{E}$ has cardinality 3^n. The smallest process space has three elements. If $\mathcal{E} = \{1\}$, say, then the three processes are: $(\{1\}, \emptyset)$, $(\{1\}, \{1\})$, and $(\emptyset, \{1\})$.

In every process space we identify three special elements: *bottom*, $\perp = (\mathcal{E}, \emptyset)$, *void*, $\Phi = (\mathcal{E}, \mathcal{E})$, and *top*, $\top = (\emptyset, \mathcal{E})$.

Next, we define the main operations on processes. These operations are motivated by applications to concurrent systems, and are related to operations in several theories of concurrency. For more details see [14, 15].

- *Reflection,* defined as in (3). Reflection permutes the roles of the device and the environment. If a process $x = (X, X')$ is the contract seen by the device, then $-x = (X', X)$ represents the same contract as seen by the environment.
- *Refinement,* defined as in (4). If $x = (X, X')$, $y = (Y, Y')$, and $x \sqsubseteq y$, then y is an acceptable substitute for x, because y accesses fewer executions than x, *i.e.,* its device obeys tighter constraints ($Y \subseteq X$), and accepts more executions than x, *i.e.,* its environment has weaker constraints ($Y' \supseteq X'$).
- *Product,* written $\times$, is a binary operation such that

$$(X, X') \times (Y, Y') = (X \cap Y, (X' \cap Y') \cup \overline{(X \cap Y)}). \tag{7}$$

Product models a system formed by two devices operating jointly. The system's accessible executions are those that are accessible to both components. Its set of acceptable executions consists of the executions that are acceptable to both components and those that must be avoided by one of the components.

Refinement is a partial order which induces a lattice over a process space, whose join is given by (1) and meet, by (2). Furthermore, this lattice has $\perp$ and $\top$ as bounds. This, together with reflection, which is defined as in (3), makes an arbitrary process space $(\mathcal{P}_{\mathcal{E}}, \sqcup, \sqcap, -, \perp, \varPhi, \top)$ a subset-pair algebra, and, by Theorem 1, a ternary algebra. Reflection is an involution and it reverses refinement. Thus

$$- - x = x, \tag{8}$$

$$x \sqsubseteq y \Leftrightarrow -x \sqsupseteq -y. \tag{9}$$

One verifies that $(\mathcal{P}_{\mathcal{E}}, \times, \varPhi, \top)$ is a semilattice with identity $\varPhi$ and greatest element $\top$.

An example of a process space is shown in Fig. 4. Here $\mathcal{E} = \{1, 2\}$, and $\mathcal{P}_{\mathcal{E}} = \mathbf{P_9}$ has nine elements. Its partial orders $\sqsubseteq$ and $\preceq$ are shown in the figure. To simplify the notation, we denote $\{1, 2\}$ simply by 12, *etc.*

4   Ternary Symmetry

Process spaces admit a ternary symmetry [14, 16] based on a unary operation, called *rotation* ($/$), and defined by:

$$/x = (\overline{X} \cup \overline{X'}, X), \tag{10}$$

for all $x = (X, X')$.

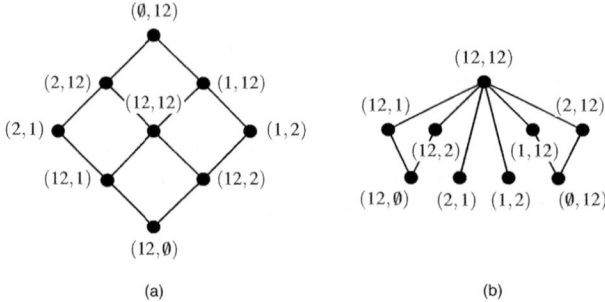

Fig. 4. Partial orders in $\mathbf{P_9}$: (a) $\sqsubseteq$ (b) $\preceq$

Proposition 1. *For any processes x and y we have:*

$$///x = x, \tag{11}$$

$$x \times y = //(/x \sqcap /y), \tag{12}$$

$$-/x = // - x, \quad / - x = -//x, \tag{13}$$

$$/\bot = \Phi, \quad /\Phi = \top, \quad /\top = \bot. \tag{14}$$

Proposition 1 shows that $/$ is bijective, since it is a root of identity, and therefore admits an inverse, namely $//$. Furthermore, Prop. 1 reveals that map $/$ is an isomorphism of the semilattices $(\mathcal{P}_{\mathcal{E}}, \times)$ and $(\mathcal{P}_{\mathcal{E}}, \sqcap)$.

The ternary symmetry brought out by Prop. 1 justifies an alternate representation of processes in a process space as *(set) triplets* [13, 14], defined below. For process $x = (X, X')$, we also write $x = (X_1, X_2, X_3)$ where $X_1 = X \setminus X'$, $X_2 = X \cap X'$, and $X_3 = X' \setminus X$.

Note that the entries in a set triplet "split" $\mathcal{E}$ in the following sense. If $x = (X_1, X_2, X_3)$, then $X_1 \cap X_2 = X_1 \cap X_3 = X_2 \cap X_3 = \emptyset$ and $X_1 \cup X_2 \cup X_3 = \mathcal{E}$. (This "split" is not a "partition" because some of the blocks might be empty.)

We redefine below the main operations on processes in the set triplet representation. These definitions are equivalent to the definition on set pairs given previously.

$$(X_1, X_2, X_3) \times (Y_1, Y_2, Y_3) = ((X_1 \setminus Y_3) \cup (Y_1 \setminus X_3), X_2 \cap Y_2, X_3 \cup Y_3). \tag{15}$$

$$(X_1, X_2, X_3) \sqsubseteq (Y_1, Y_2, Y_3) \Leftrightarrow X_1 \supseteq Y_1 \wedge X_3 \subseteq Y_3. \tag{16}$$

Note that there is no condition on X_2 and Y_2.

$$-(X_1, X_2, X_3) = (X_3, X_2, X_1). \tag{17}$$

$$\bot = (\mathcal{E}, \emptyset, \emptyset), \quad \Phi = (\emptyset, \mathcal{E}, \emptyset), \quad \top = (\emptyset, \emptyset, \mathcal{E}). \tag{18}$$

Rotation has a simpler form in the set triplet representation:

$$/(X_1, X_2, X_3) = (X_3, X_1, X_2). \tag{19}$$

One also verifies that the operation $//$ of *double rotation*, being the composition of two single rotations, satisfies

$$//x = (X', \overline{X} \cup \overline{X'}) \tag{20}$$

and, equivalently,

$$//(X_1, X_2, X_3) = (X_2, X_3, X_1). \tag{21}$$

Several new operations are defined using ternary symmetry. Each definition relates two operations in a way similar to that between $\times$ and $\sqcap$ in Prop. 1. For completeness, we repeat (12).

Definition 1. *For arbitrary processes x and y from process space $\mathcal{P}_{\mathcal{E}}$*

$$x \times y = //(/x \sqcap /y),$$
$$x \otimes y = //(/x \times /y),$$
$$x + y = //(/x \sqcup /y),$$
$$x \oplus y = //(/x + /y).$$

The refinement partial order $\sqsubseteq$ is related to the uncertainty partial order, as shown below. This is remarkable because the notions of refinement and uncertainty as formally defined here were motivated by totally different applications.

Proposition 2. *For arbitrary processes x and y from process space $\mathcal{P}_{\mathcal{E}}$*

$$x \preceq y \Rightarrow /x \sqsubseteq /y.$$

5   Boolean Trios

Let $\mathcal{E}$ be any nonempty set and let $\mathcal{P} = \mathcal{P}_{\mathcal{E}}$ be the process space on $\mathcal{E}$. Let $(\mathcal{P}, \sqcup, \sqcap, -, \bot, \Phi, \top)$ be the process space viewed as a ternary algebra. Let Q_T be the set of all elements comparable to Φ in $\mathcal{P}$.

Theorem 2. *The structure $(Q_T, \sqcup, \sqcap, -, \bot, \Phi, \top)$ is a sub-ternary-algebra of $(\mathcal{P}, \sqcup, \sqcap, -, \bot, \Phi, \top)$.*

Proof. It is easy to verify that Q_T is closed under the two binary operations, and contains the three constants. Since de Morgan's laws hold, we have $x \sqsubseteq y \Leftrightarrow -x \sqsupseteq -y$. In particular, $x \sqsubseteq \Phi \Leftrightarrow -x \sqsupseteq \Phi$, in view of T1. Hence Q_T is also closed under $-$. □

Let Q_M be the set of all the processes of $\mathcal{P}$ that are minimal in the uncertainty partial order $\preceq$. Since $(X, X') \preceq (Y, Y')$ if and only if $X \subseteq Y$ and $X' \subseteq Y'$, a process $x = (X, X')$ is minimal if and only if $X \cap X' = \emptyset$. Otherwise, if there is a common element in X and X', we can remove it from X (or X'), and obtain a smaller process. Thus, if $x \in Q_M$, then x has the form $x = (X, \overline{X})$, for some $X \subseteq \mathcal{E}$. Note that Q_M includes $\top$ and $\bot$.

Let $(\mathcal{P}, \sqcup, \sqcap, -, \bot, \top)$ be the process space $(\mathcal{P}, \sqcup, \sqcap, -, \bot, \Phi, \top)$ viewed as a de Morgan algebra.

Theorem 3. *The structure* $(Q_M, \sqcup, \sqcap, -, \bot, \top)$ *is a sub-de-Morgan-algebra of* $(\mathcal{P}, \sqcup, \sqcap, -, \bot, \top)$. *Furthermore,* $(Q_M, \sqcup, \sqcap, -, \bot, \top)$ *is a Boolean algebra.*

Proof. Let x, y be in Q_M. Then $x = (X, \overline{X})$ and $y = (Y, \overline{Y})$ for some X and Y. Then $x \sqcap y = (X \cup Y, \overline{X} \cap \overline{Y}) = (X \cup Y, \overline{X \cup Y})$, $x \sqcup y = (X \cap Y, \overline{X} \cup \overline{Y}) = (X \cap Y, \overline{X \cap Y})$, *i.e.,* Q_M is closed under both binary operations. Furthermore, $-x = (\overline{X}, X)$, and Q_M is also closed under $-$. We have already noted that Q_M contains $\bot$ and $\top$.

Laws M1–M9, and M1'–M7', M9' hold because they hold in $\mathcal{P}$. Hence we need only to verify the complement laws. We have $x \sqcap -x = (X, \overline{X}) \sqcap (\overline{X}, X) = (X \cup \overline{X}, X \cap \overline{X}) = (\mathcal{E}, \emptyset) = \bot$. Similarly, $x \sqcup -x = \top$. $\square$

We now consider the set of all the processes in Q_T that are below Φ. Let $(\mathcal{P}, \sqcup, \sqcap, -, \bot, \Phi, \top)$ be a process space, and let $Q_L = \{x \in \mathcal{P} \mid x \sqsubseteq \Phi\}$. If $x \in Q_L$, then x has the form $x = (\mathcal{E}, X')$, for some $X' \subseteq \mathcal{E}$. Note that Q_L contains $\bot$ and Φ, and that

$$/x = (\overline{X'}, \mathcal{E}). \tag{22}$$

Secondly, consider the set of all elements above Φ. Let $Q_U = \{x \in \mathcal{P} \mid x \sqsupseteq \Phi\}$. If $x \in Q_U$, then $x = (X, \mathcal{E})$ for some $X \subseteq \mathcal{E}$. Note that Q_L contains Φ and $\top$, and that

$$/x = (\overline{X}, X), \tag{23}$$

for all $x = (\mathcal{E}, X')$ in Q_L.

Finally, recall that elements of Q_M have the form $x = (X, \overline{X})$ and note that

$$/x = (\mathcal{E}, X). \tag{24}$$

If $Q \subseteq \mathcal{E}$, we define $/Q = \{/q \mid q \in Q\}$ and $-Q = \{-q \mid q \in Q\}$.

Proposition 3. $/Q_L = Q_U$, $/Q_U = Q_M$, *and* $/Q_M = Q_L$.

Proof. In view of (22)–(24), we have $/Q_L \subseteq Q_U$, $/Q_U \subseteq Q_M$, and $/Q_M \subseteq Q_L$. Since $///x = x$, it follows that $Q_U = ///Q_U \subseteq //Q_M \subseteq /Q_L$. Thus $/Q_L = Q_U$. The other two equalities follow similarly. $\square$

Proposition 4. *For* $x, y \in Q_L$,

(a) $x \sqcup -/x = \Phi$, $x \sqcap -/x = \bot$,
(b) $/(x \sqcup y) = /x \sqcup /y$, $/(x \sqcap y) = /x \sqcap /y$,
(c) $x \sqcup y = x \otimes y = x + y$, $x \sqcap y = x \times y = x \oplus y$.

Theorem 4. $(Q_L, \sqcup, \sqcap, -/, \bot, \Phi)$ *is a Boolean algebra with join* $\sqcup$, *meet* $\sqcap$, *complement* $-/$, *least element* $\bot$ *and greatest element* Φ.

Proof. We verify that Q_L is closed under $\sqcup, \sqcap$ and $-/$, and contains $\bot$ and Φ. Next we check that the laws of Boolean algebra hold for Q_L. Laws M1–M7, M1'–M7' hold in Q_L, since they hold in $\mathcal{P}$. The complement laws hold by Prop. 4 (a). The involution and de Morgan's laws follow easily using the fact that each element in Q_L is of the form $x = (\mathcal{E}, X')$. $\square$

Proposition 5. *For* $x, y \in Q_U$,

(a) $x \sqcup / - x = \top$, $x \sqcap / - x = \varPhi$,

(b) $/(x \sqcup y) = /x \sqcap /y$, $/(x \sqcap y) = /x \sqcup /y$,

(c) $x \sqcup y = x \times y = x \oplus y$, $x \sqcap y = x + y = x \otimes y$.

Theorem 5. $(Q_U, \sqcup, \sqcap, /-, \varPhi, \top)$ *is a Boolean algebra with join* $\sqcup$, *meet* $\sqcap$, *complement* $/-$, *least element* $\varPhi$ *and greatest element* $\top$. *Moreover, the algebras* $(Q_L, \sqcup, \sqcap, -/, \bot, \varPhi)$ *and* $(Q_U, \sqcup, \sqcap, /-, \varPhi, \top)$ *are isomorphic, an isomorphism being* $/ : Q_L \to Q_U$.

Proof. By Prop. 1, $/$ is a bijection. By Prop. 4 (b), $/$ preserves $\sqcup$ and $\sqcap$. Also $/(-/x) = /-(/x)$, showing that $/$ maps complements correctly. Finally, $/\bot = \varPhi$, and $/\varPhi = \top$.   □

Proposition 6. *For* $x, y \in Q_M$,

(a) $x \sqcap -x = \bot$, $x \sqcup -x = \top$,

(b) $/(x \sqcap y) = /x \sqcup /y$, $/(x \sqcup y) = /x \sqcap /y$,

(c) $x \sqcap y = x \oplus y = x + y$, $x \sqcup y = x \times y = x \otimes y$.

Theorem 6. $(Q_M, \sqcap, \sqcup, -, \top, \bot)$ *is a Boolean algebra with join* $\sqcap$, *meet* $\sqcup$, *complement* $-$, *least element* $\top$ *and greatest element* $\bot$. *Moreover,* $(Q_U, \sqcup, \sqcap, -, \varPhi, \top)$ *and* $(Q_M, \sqcap, \sqcup, -/, \top, \bot)$ *are isomorphic, an isomorphism being* $/ : Q_U \to Q_M$.

Proof. The first claim follows by Theorem 3 and duality in Boolean algebras. Mapping $/$ is a bijection, which behaves like an isomorphism with respect to the binary operations because of Prop. 5 (b). For the unary operation, we have $/(/-x) = //-x = -(/x)$, as required. Finally, $/\varPhi = \top$ and $/\top = \bot$.   □

In a similar fashion we verify the following:

Theorem 7. $(Q_M, \sqcap, \sqcup, -/, \top, \bot)$ *and* $(Q_L, \sqcup, \sqcap, -/, \bot, \varPhi)$ *are isomorphic, an isomorphism being* $/ : Q_M \to Q_L$.

We refer to the three Boolean algebras of a process space as its *Boolean trio*. The Hasse diagram for the partial order $\sqsubseteq$ within the Boolean algebras of the 27-element process space **P₂₇** is shown in Fig. 5. Note that rotation of the Boolean algebras is counterclockwise, whereas rotation of the complement operations in the three algebras is clockwise, since we have $-$, $/-$, and $//- = -/$.

We close this section by showing that $-$ is an isomorphism between Q_L and the dual of Q_U.

Theorem 8. $(Q_L, \sqcup, \sqcap, -/, \bot, \varPhi)$ *and* $(Q_U, \sqcap, \sqcup, /-, \top, \varPhi)$ *are isomorphic, an isomorphism being* $- : Q_L \to Q_U$.

Proof. Since $x \sqsubseteq \varPhi \Leftrightarrow -x \sqsupseteq \varPhi$, we have $Q_U = -Q_L$. Next, $-(x \sqcup y) = -x \sqcap -y$, and $-(x \sqcap y) = -x \sqcup -y$. Also, $-(-/x) = --//x = /x = /--x = /-(-x)$, as required. Finally, $-\bot = \top$, and $-\varPhi = \varPhi$.   □

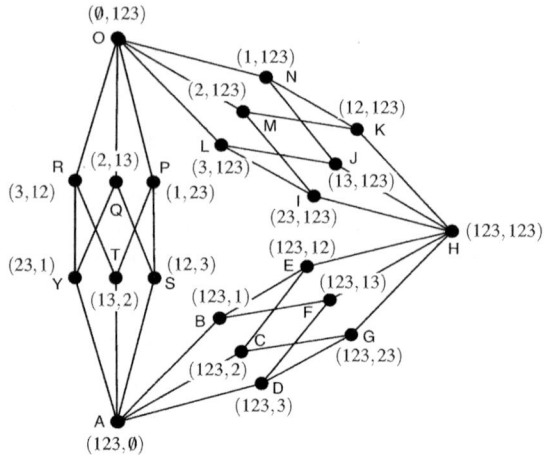

Fig. 5. Boolean trio of $\mathbf{P_{27}}$

By duality of Boolean algebras, algebra $(Q_U, \sqcap, \sqcup, /-, \top, \Phi)$ is isomorphic to $(Q_U, \sqcup, \sqcap, /-, \Phi, \top)$.

Exercise. We now offer the readers an exercise to check their understanding of the concepts presented. To simplify the notation, we introduce the following symbols: $A = (123, \emptyset), B = (123, 1)$, *etc.*, as shown in Fig. 5. The reader who evaluates the following expressions will be richly rewarded (example: $/B = I$):

$/A, \quad //H, \quad -Y, \quad P \sqcup G, \quad //-I;$
$-/G, \quad -/(Q \sqcap R), \quad -Q \sqcup -P, \quad /M, \quad ///-H, \quad -(I \sqcup J), \quad //-H, \quad -/I;$
$/O, \quad Y \sqcup /M, \quad P \oplus R, \quad -/-A.$ □

6   Ordered Tripartitions

We now consider elements $x = (X, X') = (X_1, X_2, X_3)$ that are outside the Boolean trio. Let T be the set of all such elements; these are elements of $\mathcal{P}$ that are incomparable to Φ and are not minimal in the partial order $\preceq$.

Proposition 7. $T = \{(X, X') \mid \emptyset \subset X, X' \subset \mathcal{E}, \text{ and } X \cap X' \neq \emptyset\}$.

We refer to partitions of a set into three blocks as *tripartitions* of the set. An *ordered tripartition* of $\mathcal{E}$ is an ordered triple (X_1, X_2, X_3) of subsets of $\mathcal{E}$, such that $\{X_1, X_2, X_3\}$ is a tripartition of $\mathcal{E}$.

Proposition 8. $T = \{(X_1, X_2, X_3) \mid \{X_1, X_2, X_3\} \text{ is a tripartition of } \mathcal{E}\}$.

A *sextet* is an algebra $(S_6, /, -)$, where S_6 is a set of six elements, and $/$ and $-$ are unary operations satisfying $///x = x$, $--x = x$, and $-/x = //-x$, for all $x \in S_6$. An example of a sextet is shown in Fig. 6(a). Here $S_6 = \{0, 1, \ldots, 5\}$, $/x =$

$x+2 \pmod 6$ and $-x = 5 - x$, where the $-$ on the right-hand side is subtraction of integers. Figure 6(b) shows another example of a sextet. Here, S_6 consists of all the ordered tripartitions generated by the tripartition $\{\{1, 4\}, \{2\}, \{3\}\}$ of $\mathcal{E} = \{1, 2, 3, 4\}$ under the two unary operations $/$ and $-$, the rotation and reflection of triplets.

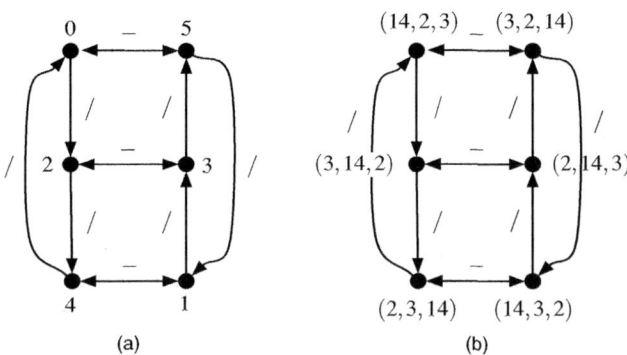

Fig. 6. Illustrating sextets

Proposition 9. *T is a disjoint union of sextets generated by all tripartitions of $\mathcal{E}$.*

If $\mathcal{E}$ has cardinality n, there are 3^n elements in $\mathcal{P}_\mathcal{E}$. In each Boolean algebra of the trio there are 2^n elements, for a total of $3 \times (2^n - 1)$ elements in the trio. Thus there are $3^n - 3 \times (2^n - 1)$ elements in T. For $n = 1$ and $n = 2$, T is empty. For $n = 3$, there are six elements in T. These belong to the sextet generated by the tripartition $\{\{1\}, \{2\}, \{3\}\}$. For $n = 4$, there are 36 elements belonging to the sextets generated by the six tripartitions $\{\{1, 2\}, \{3\}, \{4\}\}, \{\{1, 3\}, \{2\}, \{4\}\}$, $\{\{1, 4\}, \{2\}, \{3\}\}, \{\{1\}, \{2, 3\}, \{4\}\}, \{\{1\}, \{2, 4\}, \{3\}\}, \{\{1\}, \{2\}, \{3, 4\}\}$.

7   Conclusions

We have demonstrated a ternary symmetry similar to duality. We have shown that every process space consists of a trio of Boolean algebras and a disjoint union of sextets generated by all tripartitions of the underlying set.

Acknowledgment

This research was supported by Grants No. OGP0000871 and RGPIN119122 from the Natural Sciences and Engineering Research Council of Canada.

References

1. Balbes, R., Dwinger, P.: Distributive Lattices, University of Missouri Press (1974)
2. Brzozowski, J. A.: De Morgan Bisemilattices. Proc. 30th Int. Symp. on Multiple-Valued Logic, IEEE Comp. Soc. (2000) 173–178
3. Brzozowski, J. A.: Involuted Semilattices and Uncertainty in Ternary Algebras. Int. J. Algebra and Comput. (2004) to appear
4. Brzozowski, J. A., Ésik, Z., Iland, Y.: Algebras for hazard detection. Beyond Two: Theory and Applications of Multiple-Valued Logic, Fitting, M., Orłowska, E., eds., Physica-Verlag, (2003) 3–24
5. Brzozowski, J. A., Lou, J. J., Negulescu, R.: A Characterization of Finite Ternary Algebras. Int. J. Algebra and Comput. **7** (6) (1997) 713–721
6. Brzozowski, J. A., Seger, C-J. H.: Asynchronous Circuits, Springer-Verlag (1995)
7. Davey, B. A., Priestley, H. A.: Introduction to Lattices and Order, Cambridge University Press (1990)
8. Eichelberger, E. B.: Hazard detection in combinational and sequential circuits. IBM J. Res. and Dev. **9** (1965) 90–99
9. Ésik, Z.: A Cayley Theorem for Ternary Algebras. Int. J. Algebra and Comput. **8** (3) (1998) 311–316
10. Goto, M.: Application of Three-Valued Logic to Construct the Theory of Relay Networks (in Japanese). Proc. IEE, IECE, and I. of Illum. E. of Japan (1948)
11. Kleene, S. C.: Introduction to Metamathematics, North-Holland (1952)
12. Mukaidono, M.: On the B-Ternary Logical Function—A Ternary Logic Considering Ambiguity. Trans. IECE, Japan, **55–D** (6) (1972) 355–362. In English in Systems, Computers, Controls **3** (3) (1972) 27–36
13. Negulescu, R.: Process Spaces. Technical Report CS-95-48, Dept. of Comp. Science, University of Waterloo, ON, Canada (1995)
14. Negulescu, R.: Process Spaces and Formal Verification of Asynchronous Circuits. PhD Thesis, Dept. of Comp. Science, University of Waterloo, ON, Canada (1998)
15. Negulescu, R.: Process Spaces. Proc. 11th Int. Conf. on Concurrency Theory (2000) 196–210
16. Negulescu, R.: Generic Transforms on Incomplete Specifications of Asynchronous Interfaces. Proc. 19th Conf. on Math. Found. of Programming Semantics (2003)
17. Troelstra, A. S.: Lectures on Linear Logic, Center for the Study of Language and Information, Stanford University, Stanford, CA (1992)

Mathematical Proofs at a Crossroad?

Cristian S. Calude[1] and Solomon Marcus[2]

[1] University of Auckland, New Zealand
cristian@cs.auckland.ac.nz
[2] Romanian Academy, Romania
solomon.marcus@imar.ro.

Abstract. For more than 2000 years, from Pythagoras and Euclid to Hilbert and Bourbaki, mathematical proofs were essentially based on axiomatic-deductive reasoning. In the last decades, the increasing length and complexity of many mathematical proofs led to the expansion of some empirical, experimental, psychological and social aspects, yesterday only marginal, but now changing radically the very essence of *proof*. In this paper, we try to organize this evolution, to distinguish its different steps and aspects, and to evaluate its advantages and shortcomings. Axiomatic-deductive proofs are not *a posteriori* work, a luxury we can marginalize nor are computer-assisted proofs *bad* mathematics. There is hope for integration!

1 Introduction

From Pythagoras and Euclid to Hilbert and Bourbaki, mathematical proofs were essentially based on axiomatic-deductive reasoning. In the last decades, the increasing length and complexity of many mathematical proofs led to the expansion of some empirical, experimental, psychological and social aspects, yesterday only marginal, but now changing radically the very essence of *proof*. Computer-assisted proofs and the multiplication of the number of authors of a proof became in this way unavoidable.

In this paper, we try to organize this evolution, to distinguish its different steps and aspects and to evaluate its advantages and shortcomings. Various criticisms of this evolution, particularly, Ian Stewart's claim according to which the use of computer programs in a mathematical proof makes it as ugly "as a telephone directory" while purely axiomatic-deductive proofs are "beautiful like Tolstoy's *War and Peace*", will be discussed.

As axiomatic-deductive proofs, computer-assisted proofs may oscillate between ugliness and beauty. The elegance of a computer-program may rival the beauty of a piece of poetry, as the author of the *Art of Computer Programming* convinced us; however, this may not exclude the possibility that a computer-program assisting a proof hides a central idea or obscures the global aspect of the proof. In particular, the program assisting a proof may not be itself "proven correct", as it happened in the proof of the four-color problem, even in the latest, improved 1996 variant.

J. Karhumäki et al. (Eds.): Theory Is Forever (Salomaa Festschrift), LNCS 3113, pp. 15–28, 2004.

Computer-assisted proofs are to usual axiomatic-deductive proofs what (high-school) algebraic approaches are to arithmetic approaches or what analytical approaches are to direct geometric approaches. Arithmetic and intuitive geometry make children's brains more active, but algebra and analytic geometry, leading to routine and general formulas, diminish the intellectual effort and free their brains for new, more difficult problems. Obviously, each of these approaches has advantages and shortcomings, its beauty and ugliness; they are not antithetical, but complementary. Axiomatic-deductive proofs are not *a posteriori* work, a luxury we can marginalize nor are computer-assisted proofs *bad* mathematics. There is hope for integration!

2   Proofs in General

Proofs are used in everyday life and they may have nothing to do with mathematics. There is a whole field of research, at the intersection of logic, linguistics, law, psychology, sociology, literary theory etc., concerning the way people argue: argumentation theory. Sometimes, this is a subject taught to 15 or 16 year-old students.

In the "Oxford American Dictionary" [16] we read:

> *Proof*: 1. a fact or thing that shows or helps to show that something is true or exists; 2. a demonstration of the truth of something, "in proof of my statement"; 3. the process of testing whether something is true or good or valid, "put it to the proof". *To prove*: to give or be proof of; to establish the validity of; to be found to be, "it proved to be a good theory"; to test or stay out. *To argue*: 1. to express disagreement, to exchange angry words; 2. to give reasons for or against something, to debate; 3. to persuade by talking, "argued him into going"; 4. to indicate, "their style of living argues that they are well off". *Argument*: 1. a discussion involving disagreement, a quarrel; 2. a reason put forward; 3. a theme or chain of reasoning.

In all these statements, nothing is said about the means used "to show or help to show that something is true or exists", about the means used "in the process of testing whether something is true or good or valid". In argumentation theory, various ways to argue are discussed, deductive reasoning being only one of them. The literature in this respect goes from classical rhetorics to recent developments such as [28]. People argue by all means. We use suggestions, impressions, emotions, logic, gestures, mimicry, etc.

What is the relation between proof in general and proof in mathematics? It seems that the longer a mathematical proof is, the higher the possibility to contain elements usually belonging to non-mathematical proofs. We have in view emotional, affective, intuitive, social elements related to fatigue, memory gaps, etc. Long proofs are not necessarily computational; the proof of Fermat's theorem and the proof of Bieberbach's conjecture did not use computer programs, but they paid a price for their long lengths.

3   From Proofs to Mathematical Proofs

Why did mathematical proofs, beginning with Thales, Pythagoras and Euclid, till recently, use only deductive reasoning? First of all, deduction, syllogistic reasoning is the most visible aspect of a mathematical proof, but not the only one. Observation, intuition, experiment, visual representations, induction, analogy and examples have their role; some of them belong to the preliminary steps, whose presence is not made explicit, but without which proofs cannot be conceived. As a matter of fact, neither deduction, nor experiment could be completely absent in a proof, be it the way it was conceived in Babylonian mathematics, predominantly empirical, or in Greek mathematics, predominantly logical. The problem is one of *proportion*. In the 1970s, for the first time in the history of mathematics, empirical-experimental tools, under the form of some computer programs, have penetrated massively in mathematics and led to a solution of the four-color problem (4CP), a solution which is still an object of debate and controversy, see Appel and Haken [1], Tymoczko [39], Swart [37], Marcus [27], and A. Calude [10].[3]

Clearly, any proof, be it mathematical or not, is a very heterogeneous process, where different ingredients are involved in various degrees. The increasing role of empirical-experimental factors may recall the Babylonian mathematics, with the significant difference that the deductive component, today impressive, was then very poor. But what is the difference between 'proof' and 'mathematical proof'? The difficulty of this question is related to the fact that proofs which are not typically mathematical may occur in mathematics too, while some mathematical reasonings may occur in non-mathematical contexts. Many combinatorial real-life situations require a mathematical approach, while games like chess require deductive thinking (although chess thinking seems to be much more than deduction). In order to identify the nature of a mathematical proof we should first delimit the idea of a 'mathematical statement', i.e. a statement that requires a mathematical proof. Most statements in everyday life are not of this type. Even most statements of the type 'if ..., then ...' are not mathematical statements. At what moment does mathematics enter the scene? The answer is related to the conceptual status of the involved terms and predicates. Usually, problems raised by non-mathematicians are not yet mathematical problems, they may be farther or nearer to this status. The problem raised to Kepler, about the densest packing, in a container, of some apples of similar dimensions, was very near to a mathematical one and it was easy to find its mathematical version. The task was more difficult for the 4CP, where things like 'map', 'colors', 'neighbor', and 'country' required some delicate analysis until their mathematical models were identified. On the other hand, a question such as 'do you love me?' still remains far from a mathematical modelling process.

---

[3] We have discussed in detail this issue in a previous article [11].

4   Where Does the Job of Mathematicians Begin?

Is the transition from statements in general to mathematical statements the job of mathematicians? Mathematicians are divided in answering this question. Hugo Steinhaus's answer was definitely yes, Paul Erdös's answer was clearly negative. The former liked to see in any piece of reality a potential mathematical problem, the latter liked to deal with problems already formulated in a clear mathematical language. Many intermediate situations are possible, and they give rise to a whole typology of mathematicians. Goethe's remark about mathematicians' habit of translating into their own language what you tell them and making in this way your question completely hermetic refers just to this transition, sometimes of high difficulty.

If in mathematical research both above attitudes are interesting, useful and equally important, in the field of mathematical education of the general public the yes attitude seems more important than the negative one and deserves priority. The social failure of mathematics to be recognized as a cultural enterprise is due, to a large extent, to the insufficient attention paid to its links to other fields of knowledge and creativity. This means that, in general mathematical education, besides the scenario with definitions-axioms-lemmas-theorems-proofs-corollaries-examples-applications we should consider, with at least the same attention, the scenario stressing problems, concepts, examples, ideas, motivations, the historical and cultural context, including links to other fields and ways of thinking. Are these two scenarios incompatible? Not at all. It happens that the second scenario was systematically neglected; but the historical reasons for this mistake will not be discussed here (see more in [29, 30]).

Going back to proof, perhaps the most important task of mathematical education is to explain why, in many circumstances, informal statements of problems and informal proofs are not sufficient; then, how informal statements can be translated into mathematical ones. This task is genuinely related to the explanation of what is the mathematical way of thinking, in all its variants: combinatorial, deductive, inductive, analogical, metaphorical, recursive, algorithmic, probabilistic, infinite, topological, binary, triadic, etc., and, above all, the step-by-step procedure leading to the need to use some means transcending the natural language (artificial symbols of various types and their combinations).

5   Proofs: From Pride to Arrogance

With Euclid's *Elements*, for a long time taken to be a model of rigor, mathematicians became proud of their science, claimed to be the only one giving the feeling of certainty, of complete confidence in its statements and ways of arguing. Despite some mishaps occurring in the 19th century and in the first half of the 20th century, mathematicians continued to trust in axiomatic-deductive rigor, with the improvements brought by Hilbert's ideas on axiomatics and formalization. With Bourbaki's approach, towards the middle of the 20th century, some mathematicians changed pride into arrogance, imposing a ritual excluding any concession to non-formal arguments. 'Mathematics' means 'proof' and 'proof' means 'formal proof', is the new slogan.

Depuis les Grecs, qui dit Mathématique, dit démonstration

is Bourbaki's slogan, while Mac Lane's [25] austere doctrine reads

> *If a result has not yet been given valid proof, it isn't yet mathematics: we should strive to make it such.*

Here, the proof is conceived according to the standards established by Hilbert, for whom a proof is a demonstrative text starting from axioms and where each step is obtained from the preceding ones, by using some pre-established explicit inference rules:

> *The rules should be so clear, that if somebody gives you what they claim is a proof, there is a mechanical procedure that will check whether the proof is correct or not, whether it obeys the rules or not.*

And according to Jaffe and Quinn [20]

> *Modern mathematics is nearly characterized by the use of rigorous proofs. This practice, the result of literally thousands of years of refinement, has brought to mathematics a clarity and reliability unmatched by any other science.*

This is a *linear-growth model* of mathematics (see Stöltzner [36]), a process in two stages. First, informal ideas are guessed and developed, conjectures are made, and outlines of justifications are suggested. Secondly, conjectures and speculations are tested and corrected; they are made reliable by *proving* them. The main goal of proof is to provide reliability to mathematical claims. The act of finding a proof often yields, as a by-product, new insights and possibly unexpected new data. So, by making sure that every step is correct, one can tell once and for all whether a theorem has been proved. Simple! A moment of reflection shows that the case may not be so simple. For example, what if the "agent" (human or computer) checking a proof for correctness makes a mistake (as pointed out by Lakatos [24], agents are fallible)? Obviously, another agent has to check that the agent doing the checking did not make any mistake. Some other agent will need to check that agent, and so on. Eventually either the process continues unendingly (an unrealistic scenario?), or one runs out of agents who could check the proof and, in principle, they could all have made a mistake! Finally, the linear-growth model is built on an asymmetry of proof and conjecture: Posing the latter does not necessarily involve proof.

The Hilbert-Bourbaki model has its own critics, some from outside mathematics such as Lakatos [24]

> *...those who, because of the usual deductive presentation of mathematics, come to believe that the path of discovery is from axiom and/or definitions to proofs and theorems, may completely forget about the possibility and importance of naive guessing*

some from eminent mathematicians as Atiyah [2]:

> [20] *present a sanitized view of mathematics which condemns the subject to an arthritic old age. They see an inexorable increase in standards and are embarrassed by earlier periods of sloppy reasoning. But if mathematics is to rejuvenate itself and break new ground it will have to allow for the exploration of new ideas and techniques which, in their creative phase, are likely to be dubious as in some of the great eras of the past. Perhaps we now have high standards of proof to aim at but, in the early stages of new developments, we must be prepared to act in more buccaneering style.*

Atiyah's point meets Lakatos's [24] views

> *... informal, quasi-empirical mathematics does not grow through a monotonous increase of the number of indubitably established theorems, but through the incessant improvement of guesses by speculation and criticism, by the logic of proof and refutation*

and is consistent with the idea that the linear-growth model tacitly requires a 'quasi-empirical' ontology, as noted by Hirsch in his contribution to the debate reported in [2]:

> *For if we don't assume that mathematical speculations are about 'reality' then the analogy with physics is greatly weakened—and there is no reason to suggest that a speculative mathematical argument is a theory of anything, any more than a poem or novel is 'theoretical'.*

6   Proofs: From Arrogance to Prudence

It is well-known that the doubt appeared in respect to the Hilbert-Bourbaki rigor was caused by Gödel's incompleteness theorem,[4] see, for instance, Kline's *Mathematics, the Loss of Certainty* [21]. It is not by chance that a similar title was used later by Ilya Prigogine in respect to the development of physics. So, arrogance was more and more replaced by prudence. All rigid attitudes, based on binary predicates, no longer correspond to the new reality, and they should be considered 'cum grano salis'. The decisive step in this respect was accomplished by the spread of empirical-experimental factors in the development of proofs.

---

[4] The result has generated a variety of reactions, ranging from pessimism (the final, definite failure of any attempt to formalise all of mathematics) to optimism (a guarantee that mathematics will go on forever) or simple dismissal (as irrelevant for the practice of mathematics). See more in Barrow [4], Chaitin [12] and Rozenberg and Salomaa [34]. The main pragmatical conclusion seems to be that 'mathematical knowledge', whatever this may mean, cannot solely be derived only from some fixed rules. Then, who validates the 'mathematical knowledge'? Wittgenstein's answer was that the *acceptability ultimately comes from the collective opinion of the social group of people practising mathematics.*

7 Assisted Proofs Vs. Long Proofs, or from Prudence to Humility

The first major step was realized in 1976, with the discovery, using a massive computer computation, of a proof of the 4CP. This event should be related to another one: the increasing length of some mathematical proofs. Obviously, the length $l(p(s))$ of the proof $p(s)$ of the statement s should be appreciated in respect to the length $l(s)$ of s. There is a proposal to require the existence of a strictly positive constant k such that, for any reasonable theorem, the ratio $l(p(s))/l(s)$ is situated between $1/k$ and k. But the existence of such a k may remain an eternal challenge.

In the past, theorems with too long a statement were very rare. Early examples of this type can be found in Apollonius's *Conica* written some time after 200 BC. More recent examples include some theorems by Arnaud Denjoy, proved in the first decades of the 20th century, and Jordan's theorem (1870) concerning the way a simple closed curve c separates the plane in two domains whose common frontier is c. A strong trend towards long proofs appears in the second half of the 20th century. We exclude here the artificial situation when theorems with long statements and long proofs can be decomposed into several theorems, with normal lengths. We refer to statements having a clear meaning, whose unity and coherence are lost if they are not maintained in their initial form. The 4CP is just of this type. Kepler's conjecture is of the same type and so are Fermat's theorem, Poincaré's conjecture and Riemann's hypothesis. What about the theorem giving the classification of finite simple groups? In contrast with the preceding examples, in this case the statement of the theorem is very long. It may be interesting to observe that some theorems which are in complete agreement with our intuition, like Jordan's and Kepler's, require long proofs, while some other theorems, in conflict with our intuition, such as the theorem asserting the existence of three domains in the plane having the same frontier, have a short proof. Ultimately, everything depends on the way the mathematical text is segmented in various pieces.

The proof of the theorem giving the typology of the finite simple groups required a total of about fifteen thousand pages, spread in five-hundred separate articles belonging to about three-hundred different authors (see Conder [15]). But Serre [31] is still waiting for experts to check the claim by Aschbacher and Smith to have succeeded filling in the gap in the proof of the classification theorem, a gap already discovered in 1980 by Daniel Gorenstein. The gap concerned that part which deals with 'quasi-thin' groups. Despite this persisting doubt, most parts of the global proof were already published in various prestigious journals. The ambition of rigor was transgressed by the realities of mathematical life. Moreover, while each author had personal control of his own contribution (excepting the mentioned gap), the general belief was that the only person having a global, holistic representation and understanding of this theorem was Daniel Gorenstein, who unfortunately died in 1992. So, the classification theorem is still looking for its validity and understanding.

The story of the classification theorem points out the dramatic fate of some mathematical truths, whose recognition may depend on sociological factors which are no longer under the control of the mathematical community. This situation is not isolated. Think of Fermat's theorem, whose proof (by Wiles) was checked by a small number of specialists in the field, but the fact that here we had several 'Gorensteins', not only one, does not essentially change the situation.

How do exceedingly long proofs compare with assisted proofs? In 1996 Robertson, Sanders, Seymour and Thomas [32] offered a simpler proof of the 4CP. They conclude with the following interesting comment (p. 24):

> We should mention that both our programs use only integer arithmetic, and so we need not be concerned with round–off errors and similar dangers of floating point arithmetic. However, an argument can be made that our "proof" is not a proof in the traditional sense, because it contains steps that can never be verified by humans. In particular, we have not proved the correctness of the compiler we compiled our programs on, nor have we proved the infallibility of the hardware we ran our programs on. These have to be taken on faith, and are conceivably a source of error. However, from a practical point of view, the chance of a computer error that appears consistently in exactly the same way on all runs of our programs on all the compilers under all the operating systems that our programs run on is infinitesimally small compared to the chance of a human error during the same amount of case–checking. Apart from this hypothetical possibility of a computer consistently giving an incorrect answer, the rest of our proof can be verified in the same way as traditional mathematical proofs. We concede, however, that verifying a computer program is much more difficult than checking a mathematical proof of the same length.[5]

Knuth [22] p. 18 confirms the opinion expressed in the last lines of the previous paragraph:

> ... program–writing is substantially more demanding than book–writing. Why is this so? I think the main reason is that a larger attention span is needed when working on a large computer program than when doing other intellectual tasks. ... Another reason is ... that programming demands a significantly higher standard of accuracy. Things don't simply have to make sense to another human being, they must make sense to a computer.

And indeed, Knuth compared his TEX compiler (a document of about 500 pages) with Feit and Thompson's [17] theorem that all simple groups of odd order are cyclic. He lucidly argues that the program might not incorporate as much creativity and "daring" as the proof of the theorem, but they come even when compared on depth of detail, length and paradigms involved. What distinguishes the program from the proof is the "verification": convincing a couple of (human) experts that the proof *works in principle* seems to be easier than making sure that

---

[5] Our emphasis.

the program *really works*. A demonstration that *there exists a way to compile* $T_{E}X$ is not enough! Hence Knuth's warning: "Beware of bugs in the above code: I have only proved it correct, not tried it."

It is just the moment to ask, together with R. Graham: "If no human being can ever hope to check a proof, is it really a proof?" Continuing this question, we may ask: What about the fate of a mathematical theorem whose understanding is in the hands of only a few persons? Let us observe that in both cases discussed above (4CP and the classification theorem) it is not only the global, holistic understanding under question, but also its local validity.

Another example of humility some eminent mathematicians are forced to adopt with respect to yesterday's high exigency of rigor was given recently by one of the most prestigious mathematical journals, situated for a long time at the top of mathematical creativity: *Annals of Mathematics*. We learn from Karl Sigmund [35] that the proof proposed by Thomas Hales in August 1998 and the corresponding joint paper by Hales and Ferguson confirming Kepler's conjecture about the densest possible packing of unit spheres into a container, was accepted for publication in the *Annals of Mathematics*,

> but with an introductory remark by the editors, a disclaimer as it were, stating that they had been unable to verify the correctness of the 250-page manuscript with absolute certainty.

The proof is so long and based to such an extent on massive computations, that the platoon of mathematicians charged with the task of checking it ran out of steam. Robert MacPherson, the *Annals'* editor in charge of the project, stated that "the referees put a level of energy into this that is, in my experience, unprecedented. But they ended up being only 99 percent certain that the proof was correct". However, not only the referees, the author himself, Thomas Hales, 'was exhausted', as Sigmund observes. He was advised to re-write the manuscript: he didn't, but instead he started another project, 'Formal Proof of Kepler' (FPK), a project which puts theorem-verification on equal footing with Knuth's program-verification. Programming a machine to check human reasoning gives a new type of insight which has its own kind of beauty. Here is the bitter-ironical comment by Sigmund:

> After computer-based theorem-proving, this is the next great leap forward: computer-based proof checking. Pushed to the limit, this would seem to entail a self-referential loop. Maybe the purists who insist that a proof is a proof if they can understand it are right after all. On the other hand, computer-based refereeing is such a promising concept, for reviewers, editors, and authors alike, that it seems unthinkable that the community will not succumb to the temptation.

So, what is the perspective? It appears that FPK will require 20 man-years to check every single step of Hales' proof. "If all goes well, we then can be 100 percent certain", concludes Sigmund ([35], p. 67).

Let us recall that Perelman's recent proof of Poincaré's conjecture[6] is still being checked at MIT (Cambridge) and IHES (Paris) and who knows how long this process will be? We enter a period in which mathematical assessment will increase in importance and will use, in its turn, computational means. The job of an increasing number of mathematicians will be to check the work of other mathematicians. We have to learn to reward this very difficult work, to pay it at its correct value.

One could think that the new trend fits the linear–growth model: all experiments, computations and simulations, no matter how clever and powerful, belong to and are to stay at the first stage of mathematical research where informality and guessing are dominant. *This is not the case.* Of course, some automated heuristics will belong only to the first stage. The shift is produced when a large part of the results produced by computing experiments are transferred to the second stage; they no longer only develop the intuition, they no longer only build hypotheses, but *they assist the very process of proof, from discovery to checking, they create a new type of environment in which mathematicians can undertake mathematical research.*

For some a proof including computer programs is like a telephone directory, while a human proof may compete with a beautiful novel. This analogy refers to the exclusive syntactic nature of a computer-based proof (where we learn that the respective proof is valid, but we may not (don't) understand why), contrasting with the attention paid to the semantic aspect, to the understanding process, in the traditional proofs, exclusively made by humans.[7] The criticism implied by this analogy, which is very strong in René Thom's writings, is not always motivated. In fact, an 'elegant' program[8] may help the understanding process of mathematical facts in a completely new way. We confine ourselves to a few examples only:

> ... *if one can program a computer to perform some part of mathematics, then in a very effective sense one does understand that part of mathematics* (G. Tee [38])

> *If I can give an abstract proof of something, I'm reasonably happy. But if I can get a concrete, computational proof and actually produce numbers I'm much happier. I'm rather an addict of doing things on computer, because that gives you an explicit criterion of what's going on. I have a visual way of thinking, and I'm happy if I can see a picture of what I'm working with* (J. Milnor, [7])

> ... *computer-based proofs are often more convincing than many standard proofs based on diagrams which are claimed to commute, arrows which*

---

[6] Mathematicians familiar with Perelman's work expect that it will be difficult to locate any substantial mistakes, cf. Robinson [33].

[7] The conjugate pair rigor-meaning deserves to be reconsidered, cf. Marcus [26].

[8] Knuth's concept of treating a program as a piece of literature, addressed to human beings rather than to a computer; see [23].

are supposed to be the same, and arguments which are left to the reader
(J.-P. Serre [31])

... the computer changes epistemology, it changes the meaning of "to understand." To me, you understand something only if you can program it. (G. Chaitin [14])

It is the right moment to *reject the idea that computer-based proofs are necessarily ugly and opaque not only to being checked for their correctness, but also to being understood in their essence.*

Finally, do axiomatic-deductive proofs remain an *a posteriori* work, a luxury we can marginalize? When asked whether "when you are doing mathematics, can you know that something is true even before you have the proof?", Serre ([31], p. 212) answers: "Of course, this is very common". But he adds: "But one should distinguish between the genuine goal [...] which one feels is surely true, and the auxiliary statements (lemmas, etc.), which may well be intractable (as happened to Wiles in his first attempt) or even downright false [...]."

8 A Possible Readership Crisis and the Globalization of the Proving Process

Another aspect of very long (human or computer-assisted) proofs is the risk of finding no competent reader for them, no professional mathematician ready to spend a long period to check them. This happened with the famous Bieberbach conjecture. In 1916, L. Bieberbach conjectured a necessary condition on an analytic function to map the unit disk injectively to itself. The statement concerns the (normalised) Taylor coefficients a_n of such a function ($a_0 = 0, a_1 = 1$): it then states that $|a_n|$ is at most n, for any positive integer n. Various mathematicians succeeded in proving the required inequality for particular values of n, but not for every n. In March 1984, Louis de Branges (from Purdue University, Lafayette) claimed a proof, but nobody trusted him, because previously he made wrong claims for other open problems. Moreover, nobody in USA agreed to read his 400-pages manuscript to check his proof, representing seven years of hard work. The readership crisis ended when Louis de Branges proposed to the Russian mathematician I. M. Milin that he check the proof; Milin was the author of a conjecture implying Bieberbach's conjecture. De Branges travelled to Leningrad, where after a period of three months of confrontation with a team formed by Milin and two other Russian mathematicians, E. G. Emelianov and G. V. Kuzmina, they all reached the conclusion that various mistakes existing in the proof were all benign. Stimulated by this fact, two German mathematicians, C. F. Gerald and C. Pommerenke (Technical University, Berlin) succeeded in simplifying De Branges's proof.

This example is very significant for the globalization of mathematical research, a result of the globalization of communication and of international cooperation. It is no exaggeration to say that *mathematical proof has now a global dimension.*

9   Experimental Mathematics or the Hope for It

The emergence of powerful mathematical computing environments such as Mathematica, MathLab, or Maple, the increasing availability of powerful (multiprocessor) computers, and the omnipresence of the Internet allowing mathematicians to proceed heuristically and 'quasi-inductively', have created a blend of logical and empirical–experimental arguments which is called "quasi–empirical mathematics" (by Tymoczko [39], Chaitin [13]) or "experimental mathematics" (Borwein, Bailey [8], Borwein, Bailey, Girgensohn [9]). Mathematicians increasingly use symbolic and numerical computation, visualisation tools, simulation and data–mining. New types of proofs motivated by the experimental "ideology" have appeared. For example, the *interactive proof* (see Goldwasser, Micali, Rackoff [18], Blum [5]) or the *holographic proof* (see Babai [3]). And, of course, these new developments have put the classical idea of axiomatic-deductive proof under siege (see [11] for a detailed discussion).

Two programatic 'institutions' are symptomatic for the new trend: the *Centre for Experimental and Constructive Mathematics (CECM)*,[9] and the journal *Experimental Mathematics*.[10] Here are their working 'philosophies':

> *At CECM we are interested in developing methods for exploiting mathematical computation as a tool in the development of mathematical intuition, in hypothesis building, in the generation of symbolically assisted proofs, and in the construction of a flexible computer environment in which researchers and research students can undertake such research. That is, in doing experimental mathematics.* [6]

> Experimental Mathematics *publishes formal results inspired by experimentation, conjectures suggested by experiments, surveys of areas of mathematics from the experimental point of view, descriptions of algorithms and software for mathematical exploration, and general articles of interest to the community.*

For centuries mathematicians have used experiments, some leading to important discoveries: the Gibbs phenomenon in Fourier analysis, the deterministic chaos phenomenon, fractals. Wolfram's extensive computer experiments in theoretical physics paved the way for his discovery of simple programs having extremely complicated behavior [40]. Experimental mathematics—as systematic mathematical experimentation ranging from hypotheses building to assisted proofs and automated proof–checking—will play an increasingly important role and will become part of the mainstream of mathematics. There are many reasons for this trend: they range from logical (the absolute *truth* simply doesn't exist), sociological (correctness is not absolute as mathematics advances by making mistakes and correcting and re–correcting them), economic (powerful computers will be accessible to more and more people), and psychological (results and success

---

[9] www.cecm.sfu.ca.
[10] www.expmath.org.

inspire emulation). The computer is *the* essential, but not the *only* tool. New theoretical concepts will emerge, for example, the systematic search for new axioms. Assisted-proofs are not only useful and correct, but they have their own *beauty* and *elegance*, impossible to find in classical proofs. The experimental trend is not antithetical to the axiomatic-deductive approach, it complements it. Nor is the axiomatic-deductive proof *a posteriori* work, a luxury we can marginalize. There is hope for integration!

Acknowledgment

We are grateful to Greg Chaitin and Garry Tee for useful comments and references.

References

1. K. Appel, W. Haken. *Every Planar Graph is Four Colorable*, Contemporary Mathematics 98, AMS, Providence, 1989.
2. M. Atiyah et al. Responses to 'Theoretical mathematics: Toward a cultural synthesis of mathematics and theoretical physics', *Bulletin of AMS* 30 (1994), 178–211.
3. L. Babai. Probably true theorems, cry wolf? *Notices of AMS* 41 (5) (1994), 453–454.
4. J. Barrow. *Impossibility–The Limits of Science and the Science of Limits*, Oxford University Press, Oxford, 1998.
5. M. Blum. How to prove a theorem so no one else can claim it, *Proceedings of the International Congress of Mathematicians,* Berkeley, California, USA, 1986, 1444–1451.
6. J. M. Borwein. Experimental Mathematics and Integer Relations at `www.ercim.org/publication/Ercim_News/enw50/borwein.html`.
7. J. M. Borwein. `www.cecm.sfu.ca/personal/jborwein/CRM.html`.
8. J. M. Borwein, D. Bailey. *Mathematics by Experiment: Plausible Reasoning in the 21st Century*, A.ÊK.Ê Peters, Natick, MA, 2003.
9. J. M. Borwein, D. Bailey, R. Girgensohn. *Experimentation in Mathematics: Computational Paths to Discovery*, A.ÊK.Ê Peters, Natick, MA, 2004.
10. A. S. Calude. The journey of the four colour theorem through time, *The NZ Math. Magazine* 38, 3 (2001), 27–35.
11. C. S. Calude, E. Calude, S. Marcus. Passages of Proof, Los Alamos preprint archive, `arXiv:math.HO/0305213`, 16 May 2003.
12. G. J. Chaitin. *The Unknowable*, Springer Verlag, Singapore, 1999.
13. G. J. Chaitin. *Exploring Randomness*, Springer Verlag, London, 2001.
14. G. J. Chaitin. *Meta Math!*, E-book at `www.cs.auckland.ac.nz/CDMTCS/chaitin/omega.html`.
15. M. Conder. Pure mathematics: An art? or an experimental science? *NZ Science Review* 51, 3 (1994), 99–102.
16. E. Ehrlich, S. B. Flexner, G. Carruth, J. M. Hawkins. *Oxford American Dictionary*, Avon Publishers of Bard, Camelot, Discus and Flare Books, New York, 1982.
17. W. Feit, J. G. Thomson. Solvability of groups of odd order, *Pacific J. Math.* 13 (1963), 775–1029.

18. S. Goldwasser, S. Micali, C. Rackoff. The knowledge complexity of interactive proof–systems, *SIAM J. Comput.*, 18(1) (1989), 186–208.

19. R. Hersh. *What Is Mathematics, Really?*, Vintage, London, 1997.

20. A. Jaffe and F. Quinn. Theoretical mathematics: Toward a cultural synthesis of mathematics and theoretical physics, *Bulletin of AMS* 29 (1993), 178–211.

21. M. Kline. *Mathematics: The Loss of Certainty*, Oxford University Press, Oxford, 1982.

22. D. E. Knuth. Theory and practice, *EATCS Bull.* 27 (1985), 14–21.

23. D. E. Knuth. *Literate Programming*, CSLI Lecture Notes, no. 27, Stanford, California, 1992.

24. I. Lakatos. *Proofs and Refutations. The Logic of Mathematical Discovery*, John Worrall and Elie Zahar (eds.), Cambridge University Press, Cambridge, 1966.

25. S. Mac Lane. Despite physicists, proof is essential in mathematics, *Synthese* 111 (1997), 147–154.

26. S. Marcus. No system can be improved in all respects, in G. Altmann, W. Koch (eds.) *Systems; New Paradigms for the Human Sciences*, Walter de Gruyter, Berlin, 1998, 143–164.

27. S. Marcus. *Ways of Thinking*, Scientific and Encyclopedic Publ. House, Bucharest, 1987. (in Romanian)

28. C. Perelman, L. Olbrechts-Tyteca. *Traité de l'Argumentation. La Nouvelle Rhetorique*, Éditions de l'Université de Bruxelles, Bruxelles, 1988.

29. G. Pólya. *How to Solve It*, Princeton University Press, Princeton, 1957. (2nd edition)

30. G. Pólya. *Mathematics and Plausible Reasoning*, Volume 1: *Induction and Analogy in Mathematics*, Volume 2: *Patterns of Plausible Inference*, Princeton University Press, Princeton, 1990. (reprint edition)

31. M. Raussen, C. Skau. Interview with Jean-Pierre Serre, *Notices of AMS*, 51, 2 (2004), 210–214.

32. N. Robertson, D. Sanders, P. Seymour, R. Thomas. A new proof of the four-colour theorem, *Electronic Research Announcements of AMS* 2,1 (1996), 17–25.

33. S. Robinson. Russian reports he has solved a celebrated math problem, *The New York Times*, April 15 (2003), p.ÊD3.

34. G. Rozenberg, A. Salomaa. *Cornerstones of Undecidability*, Prentice-Hall, New York, 1994.

35. K. Sigmund. Review of George G. Szpiro. "Kepler's Conjecture", Wiley, 2003, *Mathematical Intelligencer*, 26, 1 (2004), 66–67.

36. M. Stöltzner. What Lakatos could teach the mathematical physicist, in G. Kampis, L. Kvasz, M. Stöltzner (eds.). *Appraising Lakatos. Mathematics, Methodology and the Man*, Kluwer, Dordrecht, 2002, 157–188.

37. E. R. Swart. The philosophical implications of the four-colour problem, *American Math. Monthly* 87, 9 (1980), 697–702.

38. G. J. Tee. Computers and mathematics, *The NZ Math. Magazine* 24, 3 (1987), 3–9.

39. T. Tymoczko. The four-colour problem and its philosophical significance, *J. Philosophy* 2,2 (1979), 57–83.

40. S. Wolfram. *A New Kind of Science*, Wolfram Media, 2002.

41. *Experimental Mathematics*: Statement of Philosophy, www.expmath.org/expmath /philosophy.html.

Rational Relations as Rational Series

Christian Choffrut

LIAFA, UMR 7089, Université Paris 7
2 Pl. Jussieu, Paris Cedex 75251
France
cc@liafa.jussieu.fr

Abstract. A rational relation is a rational subset of the direct product of two free monoids: $R \subseteq A^* \times B^*$. Consider R as a function of A^* into the family of subsets of B^* by posing for all $u \in A^*$, $R(u) = \{v \in B^* \mid (u, v) \in R\}$. Assume $R(u)$ is a finite set for all $u \in A^*$. We study how the cardinality of $R(u)$ behaves as the length of u tends to infinity and we show that there exists an infinite hierachy of growth functions.

Keywords: free monoid, rational relation, rational series.

1 Introduction

It is a elementary result in mathematics that the n-th term of a sequence of reals satisfying a linear recurrence equation

$$u_n = a_1 u_{n-1} + a_2 u_{n-2} + \ldots + a_k u_{n-k}$$

is asymptotically equivalent to a linear combination of expressions of the form $P(n)\lambda^{n-k}$ where λ is a root of the characteristic polynomial of the recurrence, k its multiplicity and $P(n)$ a polynomial of degree $k-1$, cf. [4, Theorem 6.8] or [6, Lemma II.9.7]. Not less known is the fact that the u_n's are the coefficients of a rational series in one variable, or equivalently of the infinite expansion of the quotient of two polynomials on the field of the reals. The natural extension to rational series in a finite number of non-commuting variables is completely solved in [7] where it is shown that when the growth function of the coefficients is subexponential, it is polynomial with positive integer exponent. For exponential growth, some indication can be found in [10].

The purpose of this paper is concerned with a less classical extension. We still consider rational series in non-commuting variables but the coefficients belong to the family of rational subsets of a free monoid and we study the asymptotic behaviour (in some precise way which is specified later) of the coefficients.

2 Preliminaries

We refer the reader to the textbooks [1,3,6] for all definitions which are not recalled here, such as the notions of semiring, finite automaton, and the like.

J. Karhumäki et al. (Eds.): Theory Is Forever (Salomaa Festschrift), LNCS 3113, pp. 29–34, 2004.

2.1   Rational Series

We denote by A a finite set of *letters* (the *alphabet*) and by A^* the free monoid it generates. An element u of A^* is called a *word* or a *string*. Its *length* $|u|$ is the number of letters occurring in u. The *empty* word has length 0 and is denoted by 1.

Given a semiring $\mathbb{K}$, we denote by $\mathbb{K}\langle\!\langle A^*\rangle\!\rangle$ the monoid algebra of A^* over the semiring $\mathbb{K}$. We view its elements as formal sums (i.e., *series*) of terms of the form ku where $k \in \mathbb{K}$ and $u \in A^*$. This algebra is provided with the usual operations of sum, product and star restricted to the elements whose constant term is zero. The family $\mathrm{Rat}_{\mathbb{K}}(A^*)$ of rational series over the semiring $\mathbb{K}$ is the smallest family of series containing the series reduced to the constant 0 and the terms ka for all $a \in A$ and closed under the operations of sum, product and restricted star. By Kleene's Theorem we know that any rational subset is recognized by a finite automaton with multiplicities [6, 5]. Here, we are concerned with the case when $\mathbb{K}$ is the semiring of rational subsets of the free monoid B^* for some finite alphabet B, denoted by $\mathrm{Rat}_{\mathcal{B}}B^*$ where $\mathcal{B}$ is the Boolean semiring $\{0, 1\}$, or more succinctly $\mathrm{Rat}B^*$.

2.2   Rational Relations

We recall that a relation $R \subseteq A^* \times B^*$ is *rational* if it is a rational subset of the product monoid $A^* \times B^*$, i.e., if it belongs to the smallest family of subsets of the monoid $A^* \times B^*$, containing the singletons and closed under the operations of *subset union, product* $(X \cdot Y = \{x \cdot y \mid x \in X, y \in Y\}$ where the product is meant componentwise) and *star* $(X^* = \bigcup_{n\geq 0} X^n)$.

The connection with the rational series is stated in the following basic result, cf., [5, Theorem I 1.7.].

Proposition 1. *A relation $R \subseteq A^* \times B^*$ is rational if and only if the series $R(u) = \{v \in B^* \mid (u, v) \in R\}$ is rational over the semiring $\mathrm{Rat}B^*$.*

This Proposition breaks the symmetry between the *input* alphabet A and the *output* alphabet B. The set $\{v \in B^* \mid (u, v) \in R\}$ is called the *image* of $u \in A^*$. The *domain* of R, denoted by $\mathrm{Dom}(R)$, is the subset of strings $u \in A^*$ whose image is non-empty. Assuming all input strings have finite image, we say that the rational relation R has *asymptotic growth function* $g : \mathbb{N} \longrightarrow \mathbb{N}$, if the following two conditions are satisfied ($\|X\|$ denotes the cardinality of X)

1) $\|R(w)\| = O(g(|w|)$ holds for all $w \in \mathrm{Dom}R$

2) for some infinite (length-) increasing sequence of words $(w_n)_{n>0}$ we have $\|R(w_n)\| = \theta(g(|w_n|)$

Observe that the hypothesis that all input words have finite image is not a strong requirement, since it can be easily shown that the restriction of a rational relation to the subset of input strings with finite image, i.e., the relation $R_{<\infty} = \{(u, v) \in A^* \times B^* \mid (u, v) \in R$ and $\|R(u)\| < \infty\}$ is a rational subset of $A^* \times B^*$. The problem is first studied by Schützenberger in [8].

2.3   Finite Transducer

Rational relations can be computed by a construct which is a natural extension of a finite automaton.

A *transducer* is a quadruple $\mathcal{T} = (Q, Q_-, Q_+, \mu)$ where Q is the finite set of *states*, $Q_- \subseteq Q$ (resp. $Q_+ \subseteq Q$) the set of *initial* (resp. *final*) *states* and $\mu : A^* \to \mathrm{Rat}(B^*)^{Q \times Q}$ is a *linear representation* of the monoid A^* into the multiplicative monoid of square matrices in the semiring $\mathrm{Rat}(B^*)$. Observe that the input and output alphabets A and B are understood in the definition of a transducer.

A linear representation is traditionally pictured as a finite labeled graph. E. g., consider the representation $\mu : \{a, b\}^* \to \mathrm{Rat}\{b\}^{*3 \times 3}$ defined by

$$\mu(a) = \begin{pmatrix} 1 & \emptyset & \emptyset \\ \emptyset & b & \emptyset \\ \emptyset & \emptyset & 1 \end{pmatrix} \qquad \mu(b) = \begin{pmatrix} 1 & 1 & \emptyset \\ \emptyset & \emptyset & 1 \\ \emptyset & \emptyset & 1 \end{pmatrix}$$

Identify the rows and colu;ns with the integers $1, 2, 3$ and choose $Q_- = \{1\}$ and $Q_+ = \{3\}$. It should be clear how to pass from the matrix representation to the following graph representation (by abuse of notation, the sets consisting of a unique word are identified with this word, i.e., we write b, instead of the more rigorous $\{b\}$).

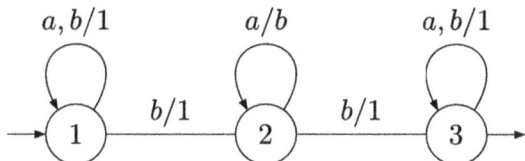

The relation *computed* by $\mathcal{T}$, denoted $||\mathcal{T}||$, is the relation: $\{(u, v) \in A^* \times B^* \mid \exists q_- \in Q_-, q_+ \in Q_+, v \in \mu_{q_-,q_+}(u)\}$. In the above example, the relation computed by the transducer is $\{(u, a^n) \in A^* \times a^* \mid u \in A^* b a^n b A^*\}$.

Since we are concerned with rational relations for which every input has finite image, all entries of the matrices $\mu(a)$ for $a \in A$ are finite subsets of B^*.

3   An Infinite Hierachy

The rational relations with finite but unbounded image, i.e., for which there is no integer N such that $||R(u)|| < N$ holds for all $u \in \mathrm{Dom} R$, are characterized in [9] by two local conditions on the structure of the tansducer. A relatively direct consequence is that if the growth is subexponential, then it is bounded by $O(n^k)$ for some integer k. We shall prove in this note a more precise result by showing that for each integer k, there exists a rational relation whose growth function is in $\theta(n^{\frac{k}{2}})$.

In order to exhibit an infinite hierarchy, consider the alphabets $A = \{a_1, \ldots, a_k\}$ and $B = \{b, c\}$. Set $W_i = A^* - a_i A^* - A^* a_{i+1}$ for $i = 1, \ldots, k-1$ with the convention $W_0 = A^* - A^* a_1$ and $W_k = A^* - a_k A^*$. Define the relation

$$R(w) = \{b^{n_1} c b^{n_2} c \ldots c b^{n_k} \mid w \in W_0 a_1^{n_1} W_1 a_2^{n_2} \ldots a_k^{n_k} W_k\} \qquad (1)$$

for all $w \in (a_1^+ a_2^+ \ldots a_k^+)^*$ (where as usual, we set $X^+ = X X^*$). Observe that the relation is indeed rational. We draw a transducer computing the relation when $k = 2$ and let the reader guess why we do not draw it for values of k greater than 2.

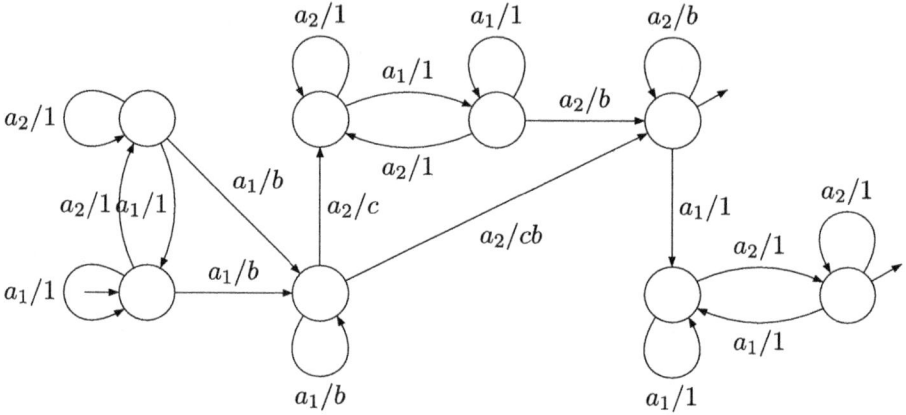

Proposition 2. *We have* $\|R(w)\| = O(|w|^{\frac{k}{2}})$ *for all* $w \in \mathrm{Dom} R$. *Furthermore, for some infinite (length-) increasing sequence of words* $(w_n)_{n>0}$ *we have* $\|R(w_n)\| = \theta(|w_n|^{\frac{k}{2}})$.

Proof. Let us first prove the last claim. Consider the words $w_n = \prod_{1 \leq i \leq n} a_1^i a_2^i \cdot \ldots \cdot a_k^i$. Then we have $R(w_n) = \{b^{n_1} c b^{n_2} \ldots c b^{n_k} \mid 1 \leq n_1 \leq n_2 \ldots \leq n_k \leq n\}$. A simple computation leads to $|w_n| = \theta(\frac{k}{2} n^2)$ and $\|R(w_n)\| = \binom{kn}{k}$ which completes the verification.

Let us now turn to the main claim and set

$$K = \limsup\{r \mid \frac{\|R(w)\|}{|w|^r} = O(1) \text{ for all } w \in (a_1^+ a_2^+ \ldots a_k^+)^+\}$$

The previous claim shows that $K \geq \frac{k}{2}$. In order to prove the equality let us make a few observations. To that order, consider the standard decomposition of an arbitrary word of the domain of R.

$$w = \left(\prod_{1 \leq j \leq n} \prod_{1 \leq i \leq k} a_i^{r_{ij}} \right) = a_1^{r_{11}} a_2^{r_{21}} \ldots a_k^{r_{k1}} \ldots a_1^{r_{1n}} a_2^{r_{2n}} \ldots a_k^{r_{kn}}$$

Call *spectrum* of w the function which assigns to each $1 \leq i \leq k$, the number $\sigma_w(i)$ of different exponents $r_{i,j}$, with $1 \leq j \leq n$. Let N be the maximum of the

$\sigma_w(i)$'s when i from 1 to k. Then the number of elements in the image of w is in $O(N^k)$. It now suffices to give a lower bound for the length of w.

Observation 1: We may assume that $\{r_{ij} \mid 1 \leq j \leq n\} = \{1, \ldots, \sigma_w(i)\}$ holds for $1 \leq i \leq k$.

Indeed, for a fixed $1 \leq i \leq k$, the bijection which to each exponent r_{ij} associates its rank among the exponents of the letter a_i, does not increase the length of the word and does not modify the cardinality of its image (e. g., the sequence $4, 2, 5, 4, 7$ of exponents would be normalized as $2, 1, 3, 2, 4$).

The second observation is obvious.

Observation 2: We may assume that for each $1 \leq i \leq k$ there exists $n - \sigma_w(i) + 1$ occurrences of exponent 1 and one occurrence of each exponent $2, \ldots, \sigma_w(i)$. In particular, if i is the value of the index which achieves the maximum $\sigma_w(i)$ (equal to N), the length $|w|_{a_i}$ in the letter a_i is at least equal to $\theta(N^2)$.

As a consequence, by observation 2 the length $|w|$ of a word w whose spectrum satisfies $\max\{\sigma_w(i) \mid 1 \leq i \leq k\} = N$, is not less than $\theta(N^2)$. Since the cardinality of $R(w)$ under this hypothesis is in $O(N^k)$, the proof is completed.

□

Actually we may relax the condition on the number of generators of the free monoid and establish the same result on the binary alphabet $A = \{a, b\}$. For each integer k, consider the family $\mathcal{R}_k$ of rational relations with subexponential growth

$$R \subseteq A^* \times A^* \text{ is in } \mathcal{R}_k \text{ if and only if } \|R(w)\| = O(|w|^{\frac{k}{2}}) \tag{2}$$

Theorem 1. *The hierachy 2 is strict.*

Proof. Consider the relation in (1) and define $E = \{(ab^i a, a_i) \mid i = 1, \ldots, k\}^*$. Then the composition $E \circ R$ is a rational relation with growth function $\theta(n^{\frac{k}{2}})$.

□

4   Further Developments

There are plenty of possible variations of the problem of the asymptotic growth of rational series. We may wish to study more general semirings for the coefficients (provided we can assciate a numerical value, such as the cardinality, as in this note) or study rational series over more general than free monoids, or more ambitiously, extend both the coefficient semiring and the monoid simultaneously. For example, consider an $\mathbb{N}$-rational series over the direct product of two free monoids A^* and B^*. Given such a series s, denote by $s_{(u,v)}$ the coefficient associated with the pair $(u, v) \in A^* \times B^*$: $s = \sum_{u \in A^*, v \in B^*} s_{(u,v)}(u, v)$. The growth function of the coefficients of the series is the function $g : \mathbb{N} \rightarrow \mathbb{N}$ defined as $g(n) = \max\{s_{(u,v)} \mid |u| + |v| = n\}$. The example worked out by Wich in order to exhibit a logarithmic degree of ambiguity for linear context-free languages can be directly interpreted in terms of such series. Indeed, consider the following three unambiguous rational series of the direct product $\{a,b\}^* \times \{a,b\}^*$.

$$G = \{(a^{i_1}ba^{i_2}b \ldots a^{i_n}b, a^{2i_1}ba^{2i_2}b) \ldots a^{2i_n}b \mid n > 0, i_1, i_2, \ldots i_n \geq 0\}$$
$$M = \{a^nb, 1) \mid n \geq 0\}$$
$$R = \{(a^{2i_1}ba^{2i_2}b \ldots a^{2i_n}b, a^{i_1}ba^{i_2}b \ldots a^{i_n}b) \mid n > 0, i_1, i_2, \ldots i_n \geq 0\}$$

By considering pairs of the form

$$(aba^4b \ldots a^{2^k}b, a^2ba^8b \ldots a^{2^{k-1}}b)$$

for some integer k, it is shown that the product GMR has logarithmic growth, see [11] for details.

More surprisingly, the simplest case of $\mathbb{N}$-rational series in k commuting variables is not settled, at least as far as we know. The result in [2] proves in a special case that the asymptotic growth can be of the form $\frac{\lambda^n}{\sqrt{n}}$ for some real λ.

5   Open Problems

PROBLEM 1. Prove or disprove that every rational relation whose growth is subexponential has an asymptotic growth function of the form $|w|^{\frac{k}{2}}$ for some integer $k \geq 0$.

PROBLEM 2. Does there exist an algorithm for computing the exponent of the asymptotic growth function of a subexponential rational relation?

References

1. J. Berstel. *Transductions and context-free languages*. B. G. Teubner, 1979.
2. A. Bertoni, C. Choffrut, M. Goldwurm, and V. Lonati. On the number of occurrences of a symbol in words of regular languages. *Theoret. Comput. Sci.*, 302(1-3):431–456, 2003.
3. S. Eilenberg. *Automata, Languages and Machines*, volume A. Academic Press, 1974.
4. P. Henrici. *Elements of numerical analysis*. John Wiley, 1964.
5. J. Sakarovitch. *Eléments de théorie des automates*. Vuibert Informatique, 2003.
6. A. Salomaa and M. Soittola. *Automata-Theoretic Aspects of Formal Power Series*, volume Texts and Monographs in Compuer Science. Springer Verlag, 1978.
7. M.-P. Schützenberger. Finite counting automata. *Information and Control*, 5:91–107, 1962.
8. M.-P. Schützenberger. Sur les relations rationnelles fonctionnelles entre monoïdes libres. *Theoret. Comput. Sci.*, 9:243–259, 1976.
9. A. Weber. On the valuedness of finite transducers. *Acta Informatica*, 27:749–780, 1990.
10. A. Weber and H. Seidl. On the degree of ambiguity of finite automata. *Theoret. Comput. Sci.*, 88:325–349, 1991.
11. K. Wich. Sublinear ambiguity. In *Proceedings of the Conference MFCS'2000*, number 1893 in LNCS, pages 690–698. Springer-Verlag, 2000.

Networks of Standard Watson-Crick $D0L$ Systems with Incomplete Information Communication

Erzsébet Csuhaj-Varjú*

Computer and Automation Research Institute
Hungarian Academy of Sciences
Kende utca 13-17, 1111 Budapest, Hungary
csuhaj@sztaki.hu

Abstract. Watson-Crick $D0L$ systems ($WD0L$ systems) are variants of $D0L$ systems with controlled derivations, inspired by the phenomenon of Watson-Crick complementarity of the familiar double helix of DNA. These systems are defined over a DNA-like alphabet, i.e. each letter has a complementary letter and this relation is symmetric. Depending on a special condition, called the trigger, a parallel rewriting step is applied either to the string or to its complementary string. A network of Watson-Crick $D0L$ systems (an $NWD0L$ system) is a finite set of $WD0L$ systems over a common DNA-like alphabet which act on their own strings in parallel and after each derivation step send copies some of the generated words to the other nodes. In [2] it was shown that the so-called standard $NWD0L$ systems form a class of computationally complete devices, that is, any recursively enumerable language can be determined by a network of standard Watson-Crick $D0L$ systems. In this paper we prove that the computational power of these constructs does not change in the case of a certain type of incomplete information communication, namely where the communicated word is a non-empty prefix of the generated word. An analogous statement can be given for the case where the communicated word is a non-empty suffix of the string.

1 Introduction

Watson-Crick complementarity, motivated by the well-known characteristics of the familiar double helix of DNA, is a fundamental concept in DNA computing. According to this phenomenon, two DNA strands form a double strand if they are complement of each other. A notion, called a Watson-Crick $D0L$ system (a $WD0L$ system), where the paradigm of complementarity is considered in the operational sense, was introduced and proposed for further investigations in [8, 9].

---

* Research supported in part by the Hungarian Scientific Research Fund "OTKA" Grant no. T 042529 and by Project "MolCoNet - A thematic network on Molecular Computing, European Commission, Information Society Technologies Programme, IST 2001-32008".

J. Karhumäki et al. (Eds.): Theory Is Forever (Salomaa Festschrift), LNCS 3113, pp. 35–48, 2004.

A Watson-Crick $D0L$ system is a $D0L$ system over a so-called DNA-like alphabet Σ and a mapping ϕ, called the trigger for complementarity transition. In a DNA-like alphabet each letter has a complementary letter and this relation is symmetric. The letters of a DNA-like alphabet are called purines and pyrimidines, a terminology extended from the DNA-alphabet of the four nucleotides A, C, T, and G which are the sequence elements forming DNA strands. The complementary letter of each purine is a pyrimidine and the complementary letter of each pyrimidine is a purine. The trigger is a logical-valued mapping over the set of strings over the DNA-like alphabet with the following property: the ϕ-value of the axiom is 0, and whenever the ϕ-value of a string is 1, then the ϕ-value of its complementary word must be 0. (The complement of a string is obtained by replacing each letter with its complementary letter.) The derivation in the Watson-Crick $D0L$ system is defined as follows: when the new string is computed by applying the morphism of the $D0L$ system, then it is checked according to the trigger. If the ϕ-value of the obtained string is 0 (the string is a so-called good word), then the derivation continues in the usual manner. If the obtained string is a so-called bad one, that is, its ϕ-value is equal to 1, then the string is changed for its complement and the derivation continues with this complementary string.

The idea behind the concept is the following: in the course of a computational or a developmental process things can go wrong to such extent that it is advisable to continue with the complementary string, which is always available [12]. Watson-Crick complementarity is viewed as an operation: together with or instead of a word we consider its complementary word.

Particularly important variants of Watson-Crick $D0L$ systems are the so-called standard Watson-Crick $D0L$ systems ($SW D0L$ systems). The controlled derivation in a standard Watson-Crick $D0L$ system is defined as follows: after rewriting the string by applying rules of the $D0L$ system in parallel, the number of occurrences of purines and that of pyrimidines in the obtained string are counted. If in the new string there are more occurrences of pyrimidines than that of purines, then each letter in the string is replaced by its complementary letter and the derivation continues from this string, otherwise the derivation continues in the usual manner. Thus, in this case the trigger is defined through the number of occurrences of purines and pyrimidines in the string.

Watson-Crick $D0L$ systems have been studied in details during the years. The interested reader can find further information on the computational power and different properties of these systems in [1, 16–18, 12–15, 7].

Another research direction was initiated in [3] where networks of Watson-Crick $D0L$ systems ($NW D0L$ systems) were introduced and their behaviour was studied. A network of Watson-Crick $D0L$ systems (an $NW D0L$ system) is a finite set of $W D0L$ systems over a common DNA-like alphabet which act on their own strings in parallel and after each derivation step communicate copies some of the generated words to the other nodes. The condition for communication is determined by the trigger for turning to the complement.

In [3] *NWD0L* systems with two main variants of protocols were studied: in the first case (protocol (a)), after a parallel rewriting step the nodes keep the good strings and the corrected strings (complements of the bad strings) and communicate a copy of each good string they obtained to each other node. In the second case (protocol (b)), the nodes, again, keep both the good and the corrected strings but communicate the copies of the corrected strings. The two protocols realize diferent philosophies: in the first case the nodes inform each other about their correct activities, in the second case they give information on the correction of their failures.

The research was continued in [4], where three results were established about the power of so-called standard networks of Watson-Crick *D0L* systems (or *NSWD0L* systems). Two of them show how it is possible to solve in linear time well-known NP-complete problems, namely, the Hamiltonian Path Problem and the Satisfiability Problem. The third one shows how in the very simple case of four-letter DNA alphabets we can obtain weird (not even *Z*-rational) patterns of the population growth of the strings in the network.

Network architectures are in the focus of interest in present computer science. One of the main areas of investigations is to study how powerful computational tools can be obtained by using networks of simple computing devices functioning with simple communication protocols. In [2] it was shown that any recursively enumerable language can be obtained as the language of an extended *NSWD0L* system using protocol (a). The language of an extended *NSWD0L* system is the set of words which are over a special sub-alphabet of the system (the terminal alphabet) and which appear at a dedicated node, the master node, at a derivation step during the functioning of the system.

In this paper we deal with networks of standard *WD0L* systems with a certain type of incomplete information communication. We study the computational power of *NSWD0L* systems where the node sends to each other node a good non-empty prefix of every good non-empty word obtained by parallel rewriting (this can be the whole good word itself) and keeps the obtained good words and the complements of the bad words. A node is allowed to send different prefixes of the same string to different nodes, but it is allowed to send only one prefix of the string to a certain node. It also might happen that the same word is communicated to a node as the chosen prefix of two different words and/or from two different nodes, but after the communication a communicated good string will be present at the destination node always only in one copy.

We prove that in this case extended networks of standard Watson-Crick *D0L* systems form a class of computationally complete devices, i.e. any recursively enumerable language can be obtained by these constructs. An analogous statement can be given for the case where the communicated strings are good non-empty suffixes of the string.

2   Preliminaries and Basic Notions

Throughout the paper we assume that the reader is familiar with the basic notions of formal language theory. For further details and unexplained notions consult [6], [10], and [11].

The set of non-empty words over an alphabet Σ is denoted by Σ^+; if the empty string, λ, is included, then we use notation Σ^*. A set of strings $L \subseteq \Sigma^*$ is said to be a language over alphabet Σ. For a string $w \in L$ and for a set $U \subseteq \Sigma$, we denote by $|w|_U$ the number of occurrences of letters of U in w.

A string u is said to be a prefix (a suffix) of a string $w \in \Sigma^*$ if $w = uz$ ($w = zu$) holds for $u, z \in \Sigma^*$; u is called a proper prefix (a proper suffix) if $u \neq w$ and $u \neq \lambda$ holds. In the sequel, we shall denote by $pref(w)$ and $suf(w)$ the set of prefixes and the set of suffixes of a string w, respectively.

Now we recall the basic notions concerning standard Watson-Crick $D0L$ systems [8].

By a DNA-like alphabet we mean an alphabet Σ with $2n$ letters, $n \geq 1$, where Σ is of the form $\Sigma = \{a_1, \ldots, a_n, \bar{a}_1, \ldots, \bar{a}_n\}$. Letters a_i and $\bar{a}_i$, $1 \leq i \leq n$, are said to be complementary letters. $\Sigma_1 = \{a_1, \ldots, a_n\}$ is said to be the sub-alphabet of purines of Σ and $\Sigma_2 = \{\bar{a}_1, \ldots, \bar{a}_n\}$ is called the sub-alphabet of pyrimidines.

A string $w \in \Sigma^*$ is said to be good (or correct) if $|w|_{\Sigma_1} \geq |w|_{\Sigma_2}$ holds, otherwise the string is called bad (or not correct). The empty word is a good word.

We denote by h_w the letter to letter endomorphism of a DNA-like alphabet Σ mapping each letter to its complementary letter.

A standard Watson-Crick $D0L$ system (an $SWD0L$ system, for short) is a triple $H = (\Sigma, P, w_0)$, where Σ is a DNA-like alphabet, the alphabet of the system, P is a set of pure context-free rules over Σ, the set of rewriting rules of the system, and w_0 is a non-empty good (correct) word over Σ, the axiom of H. Furthermore, P is complete and deterministic, that is, P has for each letter b in Σ exactly one rule of the form $b \to u$, with $u \in \Sigma^*$.

The direct derivation step in H is defined as follows: for two strings $x, y \in \Sigma^*$ we say that x directly derives y in H, denoted by $x \Longrightarrow_H y$, if $x = x_1 \ldots x_m$, $y = z_1 \ldots z_m$, $m \geq 1$, and $z_i = y_i$ if $y_1 \ldots y_m$ is a good word and $z_i = h_w(y_i)$ otherwise, where $x_i \to y_i \in P$, $1 \leq i \leq m$. The empty word, λ, derives directly itself. The parallel rewriting of each x_i onto y_i, $1 \leq i \leq m$, is denoted by $x_1 \ldots x_m \Longrightarrow_P y_1 \ldots y_m$.

Thus, if after applying a parallel rewriting to the string the obtained new string has less occurrences of purines than that of pyrimidines, then the new string must turn to its complement and the derivation continues from this complementary word, otherwise the derivation continues in the usual manner.

Now we recall the basic notions concerning networks of standard Watson-Crick $D0L$ systems [3, 2].

By a network of standard Watson-Crick $D0L$ systems (an $NSWD0L$ system, for short) with m components, where $m \geq 1$, we mean an $m + 1$-tuple

$$\Gamma = (\Sigma, (P_1, w_1), \dots, (P_m, w_m)),$$

where

- Σ is a DNA-like alphabet, the alphabet of the system,
- P_i is a complete deterministic set of pure context-free rules over Σ, the set of rules of the i-th component (or the i-th node) of Γ, $1 \leq i \leq m$, and
- w_i is a good (correct) non-empty word over Σ, the axiom of the i-th component, $1 \leq i \leq m$.

The first component, (P_1, w_1), is said to be the master node. (We note that, for our convenience, any other node can be distinguished as the master node, this does not mean any change in the meaning of the definition).

$NSW D0L$ systems function by changing their states according to parallel derivation steps performed in the $W D0L$ manner and a communication protocol.

By a state of an $NSW D0L$ system $\Gamma = (\Sigma, (P_1, w_1), \dots, (P_m, w_m))$, $m \geq 1$, we mean an m-tuple $(L_1, \dots, L_m)$, where L_i is a set of good words over Σ, $1 \leq i \leq m$.

The initial state of the system is $(\{w_1\}, \dots \{w_m\})$.

Modifying the notion of protocol (a), introduced in [3], we define the functioning of an $NSW D0L$ sytem which uses communication protocol (x, a), where $x \in \{pref, suf\}$. In this case, after the parallel rewriting step, the node sends a good non-empty prefix (a good non-empty suffix) of every obtained good non-empty string to each other node and keeps the obtained good words and the complements of the generated bad words. Notice that the communicated word can be the good word itself. Observe that a node is allowed to send different prefixes of the same string to different nodes, but it is allowed to send only one prefix of the string to a certain node. It also might happen that the same word is communicated to a node as the chosen prefix of two different words and/or from two different nodes, but after communication a communicated good string will be present at the destination node always only in one copy.

Let $\Gamma = (\Sigma, (P_1, w_1), \dots, (P_m, w_m))$, $m \geq 1$, be an $NSW D0L$ system and let $s_1 = (L_1, \dots, L_m)$ and $s_2 = (L'_1, \dots, L'_m)$ be two states of Γ.

We say that s_1 directly derives s_2 according to protocol (x, a), where $x \in \{pref, suf\}$, written as $s_1 \overset{(x,a)}{\Longrightarrow}_\Gamma s_2$, if the following condition holds: for each i, $1 \leq i \leq m$,

$$L'_i = A'_i \cup B'_i \bigcup_{j=1, j \neq i}^{m} C'_j,$$

where

$A'_i = \{z \mid z = h_w(y), x \Longrightarrow_{P_i} y, x \in L_i, \ y \text{ is a bad string}\},$
$B'_i = \{y \mid x \Longrightarrow_{P_i} y, x \in L_i, \ y \text{ is a good string}\},$

and C_j is a set of elements obtained from the elements of

$B'_j = \{v_{j_1}, \dots, v_{j_{r_j}}\} = \{y \mid x \Longrightarrow_{P_j} y, x \in L_j, \text{ is a good string}\}$

as follows:

$$C'_j = \{v'_{j_1}\} \cup \cdots \cup \{v'_{j_r}\},$$

where $v'_{j_k} \in x(v_{j_k})$, for $x \in \{pref, suf\}$, v'_{j_k} is a non-empty good word and $v_{j_k} \in B'_j$, $1 \le k \le r_j$. If B'_j is the empty set, then C'_j is empty as well.

The transitive and reflexive closure of $\overset{(x,a)}{\Longrightarrow}_\Gamma$ is denoted by $\overset{(x,a)^*}{\Longrightarrow}_\Gamma$.

The language of an $NSWD0L$ system Γ using protocol (x,a) for $x \in \{pref, suf\}$ $\Gamma = (\Sigma, (P_1, w_1), \ldots, (P_m, w_m))$, $m \ge 1$, is

$$L_{(x,a)}(\Gamma) = \{u_1 \in L_1 \mid (\{w_1\}, \ldots, \{w_m\}) \overset{(x,a)^*}{\Longrightarrow}_\Gamma (L_1, \ldots, L_m)\}.$$

That is, the language of Γ is the set of strings which appear at the master node at some derivation step of the functioning of the system, including the axiom.

By an extended $NSWD0L$ system (an $ENSWD0L$ system, for short) we mean an $m + 2$-tuple $\Gamma = (\Sigma, T, (P_1, w_1), \ldots, (P_m, w_m))$, $m \ge 1$, where $T \subseteq \Sigma$ and all other components of Γ are defined in the same way as in the case of $NSWD0L$ systems.

The language of an extended $NSWD0L$ system Γ using protocol (x,a) for $x \in \{pref, suf\}$ is defined by

$$L_{(x,a)}(\Gamma) = \{u_1 \in (T^* \cap L_1) \mid (\{w_1\}, \ldots, \{w_m\}) \overset{(x,a)^*}{\Longrightarrow}_\Gamma (L_1, \ldots, L_m)\}.$$

3   Computational Power of $ENSWD0L$ Systems

In the following we show that any recursively enumerable language can be obtained as the language of an extended $NSWD0L$ system using communication protocol $(pref, a)$. Since the language of any extended $NSWD0L$ system is a recursively enumerable language, the statement implies that $ENSWD0L$ systems are as powerful as Turing machines. Analogous statement can be given for the case of protocol (suf, a), by modifying the proof of the above statement. The idea of the proof is to simulate the generation of the words of the recursively enumerable language of an Extended Post Correspondence (EPC) by an $ENSWD0L$ system.

Let $T = \{a_1, \ldots, a_n\}$ be an alphabet, where $n \ge 1$. An Extended Post Correspondence (an EPC, for short) is a pair $P = (\{(u_1, v_1), \ldots, (u_r, v_r)\}, (z_{a_1}, \ldots, z_{a_n}))$, where $u_j, v_j, z_{a_i} \in \{0,1\}^*$, $1 \le j \le r$, $1 \le i \le n$.
The language represented by P in T, written as $L(P)$, is

$$L(P) = \{x_1 \ldots x_m \in T^* \mid \text{there are indices } s_1, \ldots, s_t \in \{1, \ldots, r\}, \ t \ge 1,$$
$$\text{such that } u_{s_1} \ldots u_{s_t} = v_{s_1} \ldots v_{s_t} z_{x_1} \ldots z_{x_m}\}.$$

It is known that for each recursively enumerable language L there exists an Extended Post Correspondence P such that $L = L(P)$ [5].

We note that the above definition remains correct and the statement remains true if we suppose that the words u_i, v_i, z_{a_j} are given over $\{1, 2\}$ instead of $\{0, 1\}$. We shall use this observation to make the construction simpler, so in the sequel we shall consider this version of the EPC and the above statement. Thus, we can consider the words $u_{s_1} \ldots u_{s_t}$ and $v_{s_1} \ldots v_{s_t} z_{x_1} \ldots z_{x_m}$ as numbers in the base three notation and therefore we can speak about their values.

According to the above theorem, a word $w = x_1 \ldots x_m$, $x_i \in T$, $1 \le i \le m$, is in L if and only if there exist indices $s_1, \ldots, s_t \in \{1, \ldots, r\}$ such that the two words $u_{s_1} \ldots u_{s_t}$ and $v_{s_1} \ldots v_{s_t} z_{x_1} \ldots z_{x_m}$ have the same value as numbers in the base three notation.

It is easy to see that we can determine the words of L as follows: We start the generation with a string of the form $u_{s_1} v_{s_1}$, $s_1 \in \{1, \ldots, r\}$. Then we add u-s and v-s to the string in the correct manner to obtain a string of the form $\alpha\beta$ with $\alpha = u_{s_1} \ldots u_{s_t}$ and $\beta = v_{s_1} \ldots v_{s_t}$, for $t \ge 1$. Then, in the second phase of the generation we add x-s and z-s to the string in a correct manner to obtain $x_1 \ldots x_m u_{s_1} \ldots u_{s_t} v_{s_1} \ldots v_{s_t} z_{x_1} \ldots z_{x_m}$. In the final phase we check whether $\alpha = u_{s_1} \ldots u_{s_t}$ and $\beta' = v_{s_1} \ldots v_{s_t} z_{x_1} \ldots z_{x_m}$ are equal or not, and if they are equal, then we eliminate both substrings from the string. If the empty word is in L, then after the first phase of the above procedure, we continue with the final generation phase. The reader can observe that the words of L can also be obtained if in the previous procedure we represent α, β, and β' with strings with exactly as many occurrences of a certain letter, say, A and B, respectively, as the value of α, β, and β', respectively, according to the base three notation. Thus, we can simulate the appending of a pair (u_j, v_j) or (a_i, z_{a_i}) to the string in generation by modifying the number of occurrences of letters A and B in the word. This observation will be used in our construction.

We shall use the following notation in the sequel: for a word $u \in \{1, 2\}^*$, we denote by $val(u)$ the value of u as a number in the base three notation and by $dig(u)$ the length of u (the number of digits in u).

Theorem 1. *For every recursively enumerable language L there exists an $ENSW D0L$ system Γ such that $L_{(pref,a)}\Gamma) = L$.*

Proof. Let L be a recursively enumerable language with $L \subseteq T^*$, where $T = \{a_1, \ldots, a_n\}$, $n \ge 1$, and let L be represented by an EPC

$$P = (\{(u_1, v_1), \ldots, (u_r, v_r)\}, (z_{a_1}, \ldots, z_{a_n})),$$

where $u_j, v_j, z_{a_i} \in \{1, 2\}^*$, $1 \le j \le r$, $1 \le i \le n$. We construct an $ENSW D0L$ system Γ such that $L_{(pref,a)}(\Gamma) = L(P)$ and Γ, functioning with protocol $(pref, a)$, simulates the generation of words of L according to P.

For each pair (u_j, v_j), $1 \le j \le r$, and for each pair (a_i, z_{a_i}), $1 \le i \le n$, Γ will have a dedicated node which simulates the effect of appending the pair to the string in generation in a correct manner. Furthermore, Γ will also have a node dedicated for deciding whether or not the two substrings representing the auxiliary substrings α and β' (see the short explanation before the theorem) are equal. The nodes of Γ will also able to check whether or not a string of a certain form which arrives at the node from another node is a good proper prefix of the word that served as the source of the communication at the other node. If a string of this form is a good proper prefix of the original string to be communicated, then in the course of the further derivation steps this string will have an occurrence of the trap symbol at the first position. Both the trap symbol, F, and its complementary symbol, $\overline{F}$, cannot be cancelled from a string at any node. Thus, neither the string with the trap symbol, nor any word originating from this string (by rewriting at any node or by communication to any node) can take part in a derivation of a terminal word. For the sake of easier reading, we also use the short term " the node for the pair (u, v) or (a, z_a)" in the sequel instead of the long version "the node dedicated for simulating the effect of adding the pair (u, v) or (a, z_a) to the string in generation."

Now we define Γ. To help the legibility, we provide the reader only with the necessary details.

Let

$$\Gamma = (\Sigma, T, (P_e, w_e),$$
$$(P_{(u_1,v_1)}, w_{(u_1,v_1)}), \ldots, (P_{(u_r,v_r)}, w_{(u_r,v_r)}),$$
$$(P_{(a_1,z_{a_1})}, w_{(a_1,z_{a_1})}), \ldots, (P_{(a_n,z_{a_n})}, w_{(a_n,z_{a_n})}))),$$

where n and r are given by EPC P.

Let

$$\Sigma = \{X, \overline{X} \mid X \in \{A_p, B_p, \$_{p,A}, \$_{p,B}, \$_{0,p}, \$_{1,p}\}, 1 \le p \le 3\}\} \cup$$
$$\{X, \overline{X} \mid X \in \{Y_j, A_{1,j}, B_{1,j}, \$_{1,A,j}, \$_{1,B,j}, \$_{0,1,j}, \$_{1,1,j}\}, 1 \le j \le r\} \cup$$
$$\{X, \overline{X} \mid X \in \{Z_i, A_{2,i}, B_{2,i}, \$_{2,A,i}, \$_{2,B,i}, \$_{0,2,i}, \$_{1,2,i}\}, 1 \le i \le n\} \cup$$
$$\{X_i, \overline{X_i} \mid X \in \{a, b, c, d, f\}, 1 \le i \le n\} \cup$$
$$\{X, \overline{X} \mid X \in \{A_4, B_4, A_5, B_5, Z, E, F\}\}.$$

We note that F is the so-called trap symbol, and each node contains the rule $F \to F$ and $\overline{F} \to F$. The axioms are defined as follows: $w_{(u_j,v_j)} = E$, for $1 \le j \le r$, $w_{(a_i,z_{a_i})} = F$, for $1 \le i \le n$, and $w_e = F$. The master node is (P_e, w_e).

In the following we define the rule sets of the nodes, with some explanations concerning their functioning.

The rule set $P_{(u_j,v_j)}$ of the node dedicated for simulating the effect of appending the pair $(u_j, v_j,)$ $1 \le j \le r$, to the string consists of the following rules:

$$A_1 \to A_{1,j}\overline{A_{1,j}}, \qquad B_1 \to B_{1,j}\overline{B_{1,j}}, \qquad \$_{1,A} \to \$_{1,A,j}\overline{\$_{1,A,j}},$$

$$\$_{1,B} \to \$_{1,B,j}, \qquad \$_{0,1} \to \overline{Y_j}\$_{0,1,j}\overline{\$_{0,1,j}}, \quad \$_{1,1} \to \$_{1,1,j}\overline{\$_{1,1,j}},$$

$$A_{1,j} \to A_1^{3^{dig(u_j)}}, \qquad B_{1,j} \to B_1^{3^{dig(v_j)}}, \qquad \$_{1,A,j} \to A_1^{val(u_j)}\$_{1,A},$$

$$\$_{1,B,j} \to B_1^{val(v_j)}\$_{1,B}, \ \$_{0,1,j} \to \$_{0,1}, \qquad \$_{1,1,j} \to \$_{1,1}.$$

$P_{(u_j,v_j)}$ also contains $Y_j \to F$, and $\overline{X} \to \lambda$, for $X \in \{Y_j, A_{1,j}, B_{1,j}, S_{1,A,j}, S_{0,1,j}, S_{1,1,j}\}$, and $X \to F$ for any other letter X of Σ different from E and the letters with the above listed rules. The node also has the rule $E \rightarrow \$_{0,1}\$_{1,1}A_1^{k_j}\$_{1,A}B_1^{l_j}\$_{1,B}$, where k_j is equal to the value of u_j, and l_j is equal to the value of v_j.

We give some explanations to the functioning of this node. The rules of the node are constructed in such way that a string which appears at the node under the functioning of the system can lead to a terminal word of Γ only if it represents $\alpha\beta$ in the first phase of the generation of a word in $L(P)$ according to EPC P. The strings which are of different forms either already have an occurrence of the trap symbol at the first position or will obtain it in the course of the following derivation (rewriting steps and communication). Then neither this string with F or any other string which originates from it can lead to a terminal word in Γ.

Suppose that a string found at this node is a non-empty prefix v' of a string of the form

$$v = \$_{0,1}\$_{1,1}A_1^{k_1}\$_{1,A}B_1^{l_1}\$_{1,B},$$

where k_1 is equal to the value of $u_{s_1} \ldots u_{s_t}$ and l_1 is equal to the value of $v_{s_1} \ldots v_{s_t}$, $s_1, \ldots, s_t \in \{1, \ldots, r\}$, for some $t \geq 1$. Then, by applying the rules of the node, the obtained string will be the corresponding prefix v''' of the string

$$v'' = \overline{Y_j}\$_{0,1,j}\overline{\$_{0,1,j}}\$_{1,1,j}\overline{\$_{1,1,j}}(A_{1,j}\overline{A_{1,j}})^{k_1}\$_{1,A,j}\overline{\$_{1,A,j}}(B_{1,j}\overline{B_{1,j}})^{l_1}\$_{1,B,j}.$$

This string, v''', is a good string if and only if v' contains $\$_{1,B}$, that is, either v was communicated from another node, or v was obtained at this node by the previous parallel rewriting step. Otherwise, v''' turns to its complement and in the next derivation step the new string will obtain an occurrence of the trap symbol at the first position. (Observe that v'' represents a string of the form $\alpha\beta$, see the explanation before the theorem.) Then neither this new string with F, nor any other string which originates from this word can take part in the derivation of a terminal word of Γ in the further steps of the derivation. Similarly, if v'' is a good string and it is communicated to another node, then it will change for a string of trap symbols at the destination node. (Notice that in this case v'' has only one good non-empty prefix, namely itself.) The same holds for the strings of the form like v'' which arrive from another node $(P_{(u_h,v_h)}, w_{(u_h,v_h)})$, with $h \neq j$, $1 \leq h \leq r$. We shall see below, that strings arriving from node (P_e, w_e) or from a node $(P_{(a_i,z_{a_i})}, w_{(a_i,z_{a_i})})$, for $1 \leq i \leq n$, either already have an occurrence of F at the first position or will be rewritten onto a string with the trap symbol at the first position in the next derivation step. Thus, neither they, nor any word originating from these strings can take part in the generation of a terminal word of Γ in the course of the further derivation steps. Thus, suppose that $v''' = v''$. Then, by applying the rules of the node to v', we obtain the string

$$\$_{0,1}\$_{1,1}A_1^{k_2}\$_{1,A}B^{l_2}\$_{1,B},$$

where $k_2 = k_1 \cdot 3^{dig(u_j)} + val(u_j)$ and $l_2 = l_1 \cdot 3^{dig(v_j)} + val(v_j)$. Thus, the rewriting simulates the effect of appending the pair (u_j, v_j) to the string $u_{s_1} \ldots u_{s_t} v_{s_1} \ldots v_{s_t}$ in the correct manner, to represent $u_{s_1} \ldots u_{s_t} u_j v_{s_1} \ldots v_{s_t} v_j$.

The rule set $P_{(a_i,z_{a_i})}$ of the node dedicated for simulating the effect of appending the pair (a_i, z_{a_i}), $1 \leq i \leq n$, to the string contains the following rules:

$$A_1 \to A_{2,i}\overline{A_{2,i}}, \qquad B_1 \to B_{2,i}\overline{B_{2,i}}, \qquad \$_{1,A} \to \$_{2,A,i}\overline{\$_{2,A,i}},$$

$$\$_{1,B} \to \$_{2,B,i}, \qquad \$_{0,1} \to \overline{Z_i}\$_{0,2,i}\overline{\$_{0,2,i}}, \; \$_{1,1} \to \$_{1,2,i}\overline{\$_{1,2,i}},$$

$$A_2 \to A_{2,i}\overline{A_{2,i}}, \qquad B_2 \to B_{2,i}\overline{B_{2,i}}, \qquad \$_{2,A} \to \$_{2,A,i}\overline{\$_{2,A,i}},$$

$$\$_{2,B} \to \$_{2,B,i}, \qquad \$_{0,2} \to \overline{Z_i}\$_{0,2,i}\overline{\$_{0,2,i}}, \; \$_{1,2} \to \$_{1,2,i}\overline{\$_{1,2,i}},$$

$$A_{2,i} \to A_2, \qquad B_{2,i} \to B_2^{3^{dig(z_{a_i})}}, \qquad \$_{2,A,i} \to \$_{2,A},$$

$$\$_{2,B,i} \to B_2^{val(z_{a_i})}\$_{2,B}, \; \$_{0,i,2} \to \$_{0,2}, \qquad \$_{1,i,2} \to b_i\$_{1,2}.$$

Moreover, it also contains productions $b_h \to b_{h,i}\overline{b_{h,i}}$, $b_{h,i} \to b_h$, $\overline{b_{h,i}} \to \lambda$ for $1 \le h \le n$, and $Z_i \to F$, $\overline{X} \to \lambda$ for $X \in \{Z_i, A_{2,i}, B_{2,i}, \$_{2,A,i}, \$_{0,2,i}, \$_{0,2,i}\}$ and $X \to F$ for any other letter X of Σ different from the letters with productions listed above. Letter b_i represents $a_i \in T$, $1 \le i \le n$. Let $T_b = \{b_i \mid 1 \le i \le n\}$.

Again, we give some explanations to the functioning of this node. Analogously to the previous case, the productions of this node are constructed in such way that a string which appears at this node can lead to a terminal word in Γ only if it represents either $\alpha\beta$ or $u\alpha\beta'$ in the first, respectively in the second phase of the generation of a word in $L(P)$. The strings which are of different forms either already have an occurrence of the trap symbol at the first position or will obtain it in the course of the following derivation (rewriting and communication) and neither the string with F nor any other string originating from it can lead to a terminal word in Γ. Suppose that a string found at this node is a good non-empty prefix v' of a string v of the form

$$v = \$_{0,p}u'\$_{1,p}A_p^{k_1}\$_{p,A}B_p^{l_1}\$_{p,B},$$

where $u' \in T_b^*$, $p \in \{1,2\}$, and $u' = \lambda$ for $p = 1$. String u' is obtained from $u \in T^*$ by replacing any occurrence of a_i in u with b_i, for $1 \le i \le n$.

Furthermore, k_1 is equal to the value of $u_{s_1} \ldots u_{s_t}$ and l_1 is equal to the value of $v_{s_1} \ldots v_{s_t}z_u$, where $s, \ldots, s_t \in \{1, \ldots, r\}$, $t \ge 1$, and z_u is the sequence of z-s corresponding to u for $u \ne \lambda$ and $z_u = \lambda$ for $u = \lambda$. Then, similarly to the case of the nodes $P_{(u_j,v_j)}$, $1 \le j \le r$, we can show that in two derivation steps we obtain from v' either the string of the form

$$\$_{0,2}u'b_i\$_{1,2}A_2^{k_2}\$_{2,A}B_2^{l_2}\$_{2,B},$$

where $k_2 = k_1$ and $l_2 = l_1 \cdot 3^{dig(z_{a_i})} + val(z_{a_i})$, or a string is obtained which has an occurrence of F at the first position. Then neither this latter string, nor any string originating from this one (by rewriting or communication) can lead to a terminal word of Γ. Indeed, if v' is a proper prefix of v, then v' will derive in two derivation steps a string with the trap symbol at the first position, since if $\$_{2,B}$ and thus $\$_{2,B,i}$ does not occur in the string, symbol $\overline{Z_i}$ turns to its complement and then Z_i is rewritten to F. Thus, starting from v, we can simulate the effect of appending the pair (a_i, z_{a_i}) to the string $uu_{s_1} \ldots u_{s_t} v_{s_1} \ldots v_{s_t} z_u$, obtaining a string which represents $ua_i u_{s_1} \ldots u_{s_t} v_{s_1} \ldots v_{s_t} z_u z_{a_i}$.

As in the case of node $P_{(u_j,v_j),w_{u_j,v_j}}$, any string which arrives from another node and is not of the form v, above, either already has an occurrence of the trap symbol at the first position or will obtain it in the course of the following derivation and then neither the string nor any string originating from this string will take part in a derivation in Γ which leads to a terminal word.

Finally, we list the rules in the rule set P_e of the node dedicated for deciding whether the generated string satisfies EPC P or not, that is, whether the corresponding two strings, α and β', mentioned in the explanation before the theorem, are equal or not. This is done by using the possibility of turning to the complement. To help the reader in understanding how the decision is done, we list the rules together with a derivation.

We note that analogously to the case of the other nodes, the rules of P_e are defined in such way that only those strings appearing at this node lead to a terminal word which represent strings of the form $x_1 \ldots x_m u_{s_1} \ldots u_{s_t} v_{s_1} \ldots v_{s_t} z_{x_1} \ldots z_{x_m}$, with $x_i \in T$, $1 \le i \le m$, where $\alpha = \beta'$ for $\alpha = u_{s_1} \ldots u_{s_t}$ and $\beta' = v_{s_1} \ldots v_{s_t} z_{x_1}$. $\ldots \cdot z_{x_m}$. (See the explanation before the theorem).

Let

$$\$_{0,p} u' \$_{1,p} A_p^k \$_{p,A} B_p^l \$_{p,B}$$

be a string at node (P_e, w_e), where $u' \in T_b^*$, and $p \in \{1,2\}$. We note that $u' = \lambda$ for $p = 1$. Then, at the first step, with rules

$$A_p \to A_3 \overline{A_3}, \quad B_p \to B_3 \overline{B_3}, \qquad \$_{p,A} \to \$_{3,A} \overline{\$_{3,A}},$$

$$\$_{p,B} \to \$_{3,B}, \quad \$_{0,p} \to \overline{Z} \$_{0,3} \overline{\$_{0,3}}, \quad \$_{1,p} \to \$_{1,3} \overline{\$_{1,3}},$$

and $b_i \to c_i \overline{c_i}$, for $1 \le i \le n$, the string is rewritten to

$$\overline{Z} \$_{0,3} \overline{\$_{0,3}} u'' \$_{1,3} \overline{\$_{1,3}} (A_3 \overline{A_3})^k \$_{3,A} \overline{\$_{3,A}} (B_3 \overline{B_3})^l \$_{3,B},$$

where $u'' = \lambda$ for $u' = \lambda$ (and $p = 1$, above) and $u'' = c_{i_1} \overline{c_{i_1}} \ldots c_{i_m} \overline{c_{i_m}}$ for $u' = b_{i_1} \ldots b_{i_m}$, with $b_{i_j} \in T_b$, for $1 \le j \le m$.

Then, either the string is a good string and then the generation continues with this string, otherwise the string turns to its complement, obtaining letter Z at the first position.

The rule set P_e also contains rules $Z \to F$, $c_i \to d_i$, $\overline{c_i} \to \overline{d_i}$, where $1 \le i \le n$, $A_3 \to \overline{A_4}$, $B_3 \to B_4$, $A_4 \to F$, $\overline{B_4} \to F$, and $X \to \lambda$ for $X \in \{\overline{Z}, \overline{A_3}, \overline{B_3}, \$_{3,A}, \overline{\$_{3,A}}, \$_{3,B}, \overline{\$_{3,B}}, \$_{1,3}, \overline{\$_{1,3}}, \$_{0,3}, \overline{\$_{0,3}}\}$. Thus, in the next derivation step either a string with F at the first position or a string of the form

$$v'' = d_{i_1} \overline{d_{i_1}} \ldots d_{i_m} \overline{d_{i_m} A_4}^k B_4^l$$

is obtained. The string with F at the first position and any other string originating from it will never lead to a terminal word. Suppose that the derivation continues with v''.

Then, the derivation at (P_e, w_e) will lead to a string over T only if $k \leq l$, otherwise the string turns to its complement and at the next derivation step occurrences of the trap symbol F will be introduced, and thus neither the string nor any other string originating from it can lead to a terminal word.

Suppose that the derivation leads to a terminal string at node (P_e, w_e). Then, having productions $d_i \to f_i$, $\overline{d_i} \to \overline{f_i}$, $1 \leq i \leq n$, $\overline{A}_4 \to A_5$, $B_4 \to \overline{B}_5$, $\overline{A}_5 \to F$, $B_5 \to F$ in P_e, we obtain a string of the form

$$v''' = f_{i_1} \overline{f}_{i_1} \cdots f_{i_m} \overline{f_{i_m}} A_5^k \overline{B}_5^l.$$

Again, the derivation will lead to a terminal string at this node only if $k \geq l$, otherwise, at the next derivation step occurrences of the trap symbol will be introduced.

Suppose that a derivation to a terminal word continues at node (P_e, w_e). Having rules $f_i \to a_i$, $\overline{f}_i \to \lambda$, $a_i \to a_i$, $1 \leq i \leq n$, and $A_5 \to \lambda$, $\overline{B}_5 \to \lambda$ we obtain string $a_{i_1} \cdots a_{i_m}$. For any other letter X in Σ, not listed with productions above, the node has the rule $X \to F$.

Notice that the derivation results in the empty word if and only if $\lambda \in L(P)$ holds.

Now we should prove that Γ derives all words of L but not more.

Suppose that $x_1 \ldots x_m \in L$, $x_i \in T$, $1 \leq i \leq m$, that is, there are indices $s_1, \ldots, s_t \in \{1, \ldots, r\}$ such that $u_{s_1} \ldots u_{s_t} = v_{s_1} \ldots v_{s_t} z_{x_1} \ldots z_{x_m}$ holds. Then $x_1 \ldots x_m$ can be obtained in Γ as follows: First E, the axiom of the node for simulating the effect of adding the pair (u_{s_1}, v_{s_1}), the axiom of node for (u_{s_1}, v_{s_1}), for short, is rewritten to the string representing $u_{s_1} v_{s_1}$ in the coded form, and then, by communication the string is forwarded to the node for (u_{s_2}, v_{s_2}). Then, the communicated string is rewritten in two derivation steps at this node and it is forwarded to the next node for (u, v) in the order. We continue this procedure while the string representing $u_{s_1} \ldots u_{s_t} v_{s_1} \ldots v_{s_t}$ is generated at node for $(u_{s_t} v_{s_t})$. Then, the string is communicated to the node for (x_1, z_{x_1}), where it is rewritten in two derivation steps and then it is communicated to the next node in the order, a node for some pair (x, z_x). Continuing this procedure, we finish this part of the generation at node for (x_m, z_{x_m}) with a string representing $x_1 \ldots x_m u_{s_1} \ldots u_{s_t} v_{s_1} \ldots v_{s_t} z_{x_1} \ldots z_{x_m}$. Then the string is forwarded to node (P_e, w_e), where in some steps its substring representing $\alpha \beta' = u_{s_1} \ldots u_{s_t} v_{s_1} \ldots v_{s_t} z_{x_1} \ldots z_{x_m}$ is eliminated and the corresponding letters from T are introduced. Thus, $x_1 \ldots x_m$ is an element of $L(\Gamma)$. The procedure for computing $\lambda \in L(P)$, if $\lambda \in L(P)$, is analogous.

We should prove that Γ does not generate a word not in L. By the definition of the rule sets of the nodes, we can see that for each string generated at the node or communicated to the node, the node for the pair (u_j, v_j), $1 \leq j \leq r$, either produces a new string representing a word of one of the forms $u_j v_j$ or $u_{s_1} \ldots u_{s_t} u_j v_{s_1} \ldots v_{s_t} v_j$, $s_1 \ldots, s_t \in \{1, \ldots, r\}$, $t \geq 1$, or it produces a new string which contains the trap symbol F at the first position which does not make possible to generate a terminal word. Then this string and any other string originating from this one is irrelevant from the point of view of generation of terminal words of Γ. Analogously, for each string generated at the node or communicated to the node, it holds that the node for (a_i, z_{a_i}), $1 \leq i \leq n$, either produces a string representing a string of the form $ua_i u_{s_1} \ldots u_{s_t} v_{s_1} \ldots v_{s_t} z_u z_{a_i}$, $u \in T^*$, z_u is the sequence of z-s which corresponds to u, or it generates a string with an occurrence of the trap symbol, F. The latter case leads to strings irrelevant from the point of view of generation of words of Γ. But, only those strings have no occurrence of the trap symbol at the first position at the above two types of nodes which represent strings that correspond to the respective generation phases of words of L according to EPC P. Similarly to the above cases, the master node, (P_e, w_e), either produces a terminal string (or the empty word) from a string it has generated or it received by communication, or the node generates a string with an occurrence of the trap symbol. Thus, any terminal word (including the empty word) which can be generated by Γ can be generated according to P but not more. Hence the result.

By standard techniques it can be shown that any language of an extended $NSW D0L$ system is a recursively enumerable language. Thus we can state the following theorem:

Theorem 2. *The class of languages of $ENSW D0L$ systems is equal to the class of recursively enumerable languages.*

Modifying the proof of Theorem 1, an analogous statement can be given for the case of communication protocol (suf, a). The idea of the proof is to change the role of the endmarker symbols $\$_{0,p}$ and $\$_{p,B}$, for $p = 1, 2, 3$, in the procedure of checking whether the communicated string is a proper subword of the original string or not. We give this statement without the proof, the details are left to the reader.

Theorem 3. *For every recursively enumerable language L there exists an $ENSW D0L$ system Γ such that $L_{(suf,a)}\Gamma) = L$.*

4   Final Remarks

In this paper we examined the computational power of $ENSW D0L$ systems with a certain type of incomplete information communication, namely where the nodes communicate good non-empty prefixes (suffixes) of the good strings they obtained by rewriting. It is an interesting open question how large computational power can be obtained if some other way of incomplete communication is chosen. For example, it would be interesting to study the case where the node communicates an arbitrary non-empty good subword of the good words obtained by parallel rewriting or the case where the node splits a copy of the word to be communicated into as many pieces as the number of the other nodes in the network and these splitted subwords are distributed among the different nodes. We plan to return to these topics in the future.

References

1. J. Csima, E. Csuhaj Varjú and A. Salomaa, Power and size of extended Watson-Crick L systems. Theoretical Computer Science 290 (2003), 1665-1678.
2. E. Csuhaj-Varjú, Computing by networks of Watson-Crick $D0L$ systems. In: Proc. Algebraic Systems, Formal Languages and Computation. (M. Ito, ed.), RIMS Kokyuroku 1166, August 2000, Research Institute for Mathematical Sciences, Kyoto University, Kyoto, 43-51.
3. E. Csuhaj-Varjú and A. Salomaa, Networks of Watson-Crick $D0L$ systems. In: Proc. of the International Conference "Words, Languages & Combinatorics", Kyoto, Japan, 14-18, 2000. (M. Ito and T. Imaoka, eds.), World Scientific, Singapore, 2003, 134-150.
4. E. Csuhaj-Varjú and A. Salomaa, The Power of Networks of Watson-Crick $D0L$ Systems. In: Aspects of Molecular Computing. Essays Dedicated to Tom Head on the Occasion of His 70th Birthday. (N. Jonoska, Gh. Păun, and G. Rozenberg, eds.), Lecture Notes in Computer Science 2950, Springer, 2004, 106-118.
5. V. Geffert, Context-free-like forms for phrase-structure grammars. Proc. MFCS'88, LNCS 324, Springer Verlag, 1988, 309-317.
6. Handbook of Formal Languages. Vol. I-III. (G. Rozenberg and A. Salomaa, eds.), Springer Verlag, Berlin-Heidelberg-New York, 1997.
7. J. Honkala and A. Salomaa, Watson-Crick $D0L$ systems with regular triggers. Theoretical Computer Science 259 (2001), 689–698.
8. V. Mihalache and A. Salomaa, Watson-Crick $D0L$ systems. EATCS Bulletin 62 (1997), 160-175.
9. V. Mihalache and A. Salomaa, Language-theoretic aspects of DNA complementarity. Theoretical Computer Science 250 (2001), 163-178.
10. G. Păun, G. Rozenberg and A. Salomaa, DNA Computing. New Computing Paradigms. Springer-Verlag, Berlin, Heidelberg, New York, 1998.
11. G. Rozenberg and A. Salomaa, The Mathematical Theory of L systems. Academic Press, New York, London, 1980.
12. A. Salomaa, Turing, Watson-Crick and Lindenmayer. Aspects of DNA Complementarity. In: Unconventional Models of Computation. (C.S. Calude, J. Casti, and M. J. Dinneen, eds.), Springer Verlag, Singapore, Berlin, Heidelberg, New York, 1998, 94-107.
13. A. Salomaa, Watson-Crick Walks and Roads on $D0L$ Graphs. Acta Cybernetica 14 (1) (1999), 179-192.
14. A. Salomaa, Iterated morphisms with complementarity on the DNA alphabet. In: Words, Semigroups, Transductions. (M. Ito, G. Paun and S. Yu, eds.), World Scientific, Singapore, 2001, 405-420.
15. A. Salomaa, Uni-transitional Watson-Crick $D0L$ systems. Theoretical Computer Science 281 (2002), 537-553.
16. A. Salomaa and P. Sosík, Watson-Crick $D0L$ systems: the power of one transition. Theoretical Computer Science 301 (2003), 187-200.
17. P. Sosík, $D0L$ systems + Watson-Crick complement = universal computation. In: Machines, Computations and Universality. (M. Margenstern and Y. Rogozhin, eds.), Lecture Notes in Computer Science 2055, Springer, 2001, 308-320.
18. P. Sosík, Watson-Crick $D0L$ systems: generative power and undecidable problems. Theoretical Computer Science 306 (2003), 101-112.

On the Size of Components of Probabilistic Cooperating Distributed Grammar Systems[*]

Erzsébet Csuhaj-Varjú[1] and Jürgen Dassow[2]

[1] Computer and Automation Research Institute, Hungarian Academy of Sciences
Kende utca 13-17, H–1111 Budapest, Hungary
csuhaj@sztaki.hu
[2] Otto-von-Guericke-Universiät Magdeburg, Fakultät für Informatik
PSF 4120, D–39016 Magdeburg, Germany
dassow@iws.cs,uni-magdeburg.de

Abstract. Probabilistic cooperating distributed grammar systems introduced in [1] are systems of probabilistic grammars in the sense of [9], i.e., a probability is associated with any transition from one rule to another rule and with any transition from one probabilistic grammar to another probabilistic grammar; a probabilistic grammar stops, if the chosen rule cannot be applied; and the generated language contains only words where the product of the transitions is larger than a certain cut-point). We study the families obtained with cut-point 0 by restricting the number of rules in a probabilistic component. We show that at most two productions in any component are sufficient to generate any recursively enumerable language. If one restricts to probabilistic components with one production in any component, then one obtains the family of deterministic ET0L systems.

1 Introduction

Cooperating distributed grammar systems have been introduced in [3] as a formal language theoretic approach to the blackboard architecture known from the distributed problem solving. Essentially, such a system consists of some context-free grammars (called the components) which work on a common sentential form and where the conditions for a grammar to start and/or to stop are prescribed in a protocol or derivation mode. The most investigated derivation mode is the so-called t-mode, where a grammar has to work as long as it can apply some of its productions, and if a component has finished its derivation, then any other enabled component can start. It has been shown in [3] that cooperating distributed grammar system have the same generative power as ET0L systems known from

[*] This research was supported in part under grant no. D-35/2000 and HUN009/00 by the Intergovernmental S&T Cooperation Programme of the Office of Research and Development Division of the Hungarian Ministry of Education and its German partner, the Federal Ministry of Education and Research (BMBF).

J. Karhumäki et al. (Eds.): Theory Is Forever (Salomaa Festschrift), LNCS 3113, pp. 49–59, 2004.
© Springer-Verlag Berlin Heidelberg 2004

the theory of developmental or Lindenmayer systems (see [7] and [8]). Furthermore, in [6] it has been proved that any ET0L language can be generated by a cooperating distributed grammar system where any component has at most five productions.

Obviously, instead of context-free grammars one can also use another type of grammars as basic grammars. For instance, in [3] and [11] some variants of Lindenmayer systems have been taken as basic grammars. In [1], probabilistic grammars introduced by A. SALOMAA in [9] have been used as components, i.e., with any transition from one rule to another rule a probability is associated and the generated language contains only words where the product of the transition is larger than a certain cut-point. Moreover, in the case of grammar systems a probability is associated with a transition from one probabilistic grammar to another one, and in the t-mode of derivation a probabilistic grammar stops, if after the application of a rule one chooses a rule which cannot be applied. In [1] it has been shown that any language generated by a probabilistic cooperating distributed grammar system with cut-point $c > 0$ is a finite language and that any recursively enumerable language can be generated by a probabilistic cooperating distributed grammar system with cut-point 0.

In [2] a theorem analogous to the result of [6] mentioned above has been given: For any recursively enumerable language L, there is a probabilistic cooperating distributed grammar system Γ such that any probabilistic component of Γ contains at most six productions and L is the language generated by Γ with cut-point 0. In this paper we improve this result. We show that at most two productions in the probabilistic components are sufficient to generate (with cut-point 0) any recursively enumerable language. If one restricts to probabilistic cooperating distributed grammar systems with one production in any component, then one obtains the same generative power as the power of deterministic ET0L systems.

2   Definitions

An n-dimensional vector $(a_1, a_2, \ldots, a_n)$ is called *probabilistic*, if $0 \le a_i \le 1$ for $1 \le i \le n$ and $\sum_{i=1}^{n} a_i = 1$. The cardinality of a (finite) set M is denoted by $\#(M)$. The set of non-empty words over an alphabet V is denoted by V^+; if the empty string, denoted by λ, is included, then we use the notation V^*.

We recall the notions of matrix grammar, Indian parallel programmed grammars and extended deterministic tabled Lindenmayer systems. For further details we refer to [5], [10], [8] and [7].

A matrix grammar is a construct $G = (N, T, S, M, F)$ where N and T are the disjoint alphabets of nonterminals and terminals, respectively, $S \in N$, $M = \{m_1, m_2, \ldots m_n\}$ is a finite set of finite sequences of context-free productions, i.e., for $1 \le i \le n$, $m_i = (A_{i1} \to w_{i1}, A_{i2} \to w_{i2}, \ldots, A_{ir_i} \to w_{ir_i})$ with $r_i \ge 1$, $A_{ij} \in N$ and $w_{ij} \in (N \cup T)^*$ for $1 \le i \le n, 1 \le j \le r_i$ and F is a finite subset of $\{A_{ij} \to w_{ij} \mid 1 \le i \le n, 1 \le j \le r_i\}$. The sequences m_i, $1 \le i \le n$, are called matrices.

Let x and y be two words of $(N \cup T)^*$. We say that x directly derives y by an application of $m_i \in M$, written as $x \Longrightarrow_{m_i} y$ if there are words $x_1, x_2 \ldots, x_{r_i+1}$ such that $x = x_1$, $y = x_{r_i+1}$ and, for $1 \leq j \leq r_i$, one of the following conditions is satisfied:

i) $x_j = x_j' A_{ij} x_j''$ for some $x_j', x_j'' \in (N \cup T)^*$ and $x_{j+1} = x_j' w_{ij} x_j''$ or

ii) A_{ij} does not occur in x_j, $A_{ij} \to w_{ij} \in F$ and $x_{j+1} = x_j$.

The language $L(G)$ generated by a matrix grammar $G = (N, T, S, M, F)$ consists of all words $z \in T^*$ such that there is derivation

$$S \Longrightarrow_{m_{i_1}} z_1 \Longrightarrow_{m_{i_2}} z_2 \Longrightarrow_{m_{i_3}} \cdots \Longrightarrow_{m_{i_s}} z_s = z$$

for $s \geq 1$ and some matrices $m_{i_1}, m_{i_2}, \ldots, m_{i_s} \in M$.

It is well-known (see [5] for a proof) that the family of languages generated by matrix grammars coincides with the family $\mathcal{L}(RE)$ of recursively enumerable languages.

We say that a matrix grammar $G = (N, T, S, M, F)$ is in normal form if

- $N = N_1 \cup N_2 \cup \{S, Z\}$ with $N_1 \cap N_2 = \emptyset$, $S, Z \notin N_1 \cup N_2$,
- all matrices of M have one of the following forms
    (1) $(S \to x)$ with $x \in L(G)$, $|x| \leq 1$
    (2) $(S \to AX)$ with $A \in N_1$, $X \in N_2$,
    (3) $(A \to w, X \to Y)$ with $A \in N_1$, $w \in (N_1 \cup T)^*$, $X, Y \in N_2$, $X \neq Y$
    (4) $(A \to Z, X \to Y)$ with $A \in N_1$, $X, Y \in N_2$
    (5) $(A \to w, X \to Y)$ with $A \in N_1$, $w \in T^*$, $X \in N_2$, $a \in T$,
- F consists of all rules of the form $A \to Z$ with $A \in N_1$.

For $X \in N_2$, we say that m is an X-matrix, if m contains a rule with right hand side X. By $n_G(X)$ we denote the number of X-matrices.

Lemma 1. *For any recursively enumerable language L, there is a matrix grammar G in normal form such that $L = L(G)$.*

Proof. The statement is shown in Theorem 1.3.7 of [5] for matrix grammars in accurate binary normal form, which is obtained from our normal form by deletion of the conditions $X \neq Y$ for matrices of type (3). However, it can be seen from the proof in [5] that our additional condition is satisfied.

An *Indian parallel programmed* grammar is a construct $G = (N, T, S, P)$ where any rule $p \in P$ has the form $p = (A \to w, \sigma, \varphi)$ where $A \to w$ is a context-free production with $A \in N$ and $w \in (N \cup T)^*$ and σ and φ are subsets of P called the success field and failure field, respectively. The language $L(G)$ generated by G consists of all words $z \in T^*$ which can be obtained by a derivation of the form

$$S = z_0 \Longrightarrow_{p_1} z_1 \Longrightarrow_{p_2} z_2 \Longrightarrow_{p_3} \cdots \Longrightarrow_{p_n} z_n = z$$

where $n \geq 1$, and for $1 \leq i \leq n - 1$, one of the following conditions are satisfied:

i) $p_i = (A_i \rightarrow w_i, \sigma_i, \varphi_i)$,
   $z_{i-1} = x_1 A_i x_2 A_i \ldots x_{r-1} A_i x_r$ with $x_i \in ((N \cup T) \setminus \{A_i\})^*$
   $z_i = x_1 w_i x_2 w_i \ldots x_{r-1} w_i x_r$, and
   $p_{i+1} \in \sigma_i$,

or

ii) $p_i = (A_i \rightarrow w_i, \sigma_i, \varphi_i)$,
    A_i does not occur in z_{i-1},
    $z_i = z_{i-1}$, and
    $p_{i+1} \in \varphi_i$.

An *extended deterministic tabled Lindenmayer system* (for short EDT0L system) is an $n + 3$-tuple $G = (V, T, h_1, h_2, \ldots, h_r, w)$, where V is an alphabet, the set T (of terminals) is a subset of V, $w \in V^+$ is the axiom and, for $1 \leq i \leq r$, $h_i : V \rightarrow V^*$ is a morphism.

For two strings $x = x_1 x_2 \ldots x_n$ with $n \geq 1$, $x_i \in V$ for $1 \leq i \leq n$ and $y \in V^*$ we say that x directly derives y, if there is a morphism h_j, $1 \leq j \leq r$, such that $y = h_j(x) = h_j(x_1) h_j(x_2) \ldots h_j(x_n)$. The language $L(G)$ generated by an EDT0L system is defined as the set of all words over T which can be obtained from w by a sequence of direct derivation steps, i.e.,

$$L(G) = \{ h_{i_1}(h_{i_2}(\ldots (h_{i_s}(w)) \ldots)) \mid 1 \leq i_j \leq r \text{ for } 1 \leq j \leq s \} \cap T^*.$$

By $\mathcal{L}(EDT0L)$ we denote the family of languages generated by EDT0L systems.

We now define the central concept of this paper, the probabilistic cooperating distributed grammar systems.

A *probabilistic cooperating distributed grammar system* is a construct

$$\Gamma = (N, T, S, (P_1, \delta_1, \phi_1, \phi_1'), (P_2, \delta_2, \phi_2, \phi_2'), \ldots, (P_n, \delta_n, \phi_n, \phi_n'), \delta)$$

where

- N and T are disjoint alphabets of nonterminals and terminals, respectively,
- S called the axiom is an element of N,
- n is a positive integer,
- for $1 \leq i \leq n$,
  - $P_i = \{p_{i1}, p_{i2}, \ldots, p_{ik_i}\}$ is a finite set of context-free productions (i.e., each p_{ij} is of the form $A \rightarrow w$ with $A \in N$ and $w \in (N \cup T)^*$) where the given order of the k_i elements of P_i is fixed,
  - δ_i is a k_i-dimensional probabilistic vector, whose j-th component gives the probability to start a derivation, which uses only rules of P_i, with the j-th rule p_{ij} of P_i,
  - ϕ_i is a k_i-dimensional vector, whose j-th component is a k_i-dimensional probabilistic vector $\phi_i(j) = (\phi_{ij1}, \phi_{ij2}, \ldots, \phi_{ijk_i})$ whose k-th component ϕ_{ijk} gives the probability that after an application of p_{ij} we apply p_{ik} as the next rule,

- ϕ_i' is a n-dimensional probabilistic vector, whose j-th component $\phi_i'(j)$ gives the probability that after an application of the component P_i we continue with the component P_j,

- δ is an n-dimensional probabilistic vector, whose i-th component $\delta(i)$ gives the probability that the derivation starts with the i-th component P_i.

The constructs $(P_i, \delta_i, \phi_i, \phi_i')$, $1 \le i \le n$, are called the components of Γ. Sometimes we also say that P_i is a component.

Let

$$D : x \Longrightarrow_{p_{j_1}} x_1 \Longrightarrow_{p_{j_2}} x_2 \Longrightarrow_{p_{j_3}} \cdots \Longrightarrow_{p_{j_s}} x_s = y$$

be a derivation which only uses rules of P_i. Shortly, we write

$$D : x \Longrightarrow_{P_i}^* y .$$

We say that D is a t-derivation with respect to P_i, if

i) y is a word over the terminal alphabet T, or
ii) the production $p_{j_{s+1}} \in P_i$ chosen to be applied in the next step cannot be applied.

With a t-derivation D we associate in case i) or ii) the values

$$v(D) = \delta_i(j_1) \cdot \phi_{ij_1 j_2} \cdot \phi_{ij_2 j_3} \cdots \cdot \phi_{ij_{s-1} j_s}$$

or

$$v(D) = \delta_i(j_1) \cdot \phi_{ij_1 j_2} \cdot \phi_{ij_2 j_3} \cdots \cdot \phi_{ij_{s-1} j_s} \cdot \phi_{ij_s j_{s+1}} ,$$

respectively.

Let

$$D' : S \Longrightarrow_{P_{i_1}}^* z_1 \Longrightarrow_{P_{i_2}}^* z_1 \Longrightarrow_{P_{i_3}}^* \cdots \Longrightarrow_{P_{i_{r-1}}}^* z_{r-1} \Longrightarrow_{P_{i_r}}^* z_r = z \qquad (1)$$

be a derivation such that any subderivation

$$D_j : z_{j-1} \Longrightarrow_{P_{i_j}}^* z_j$$

is a t-derivation with respect to P_{i_j}. With D' we associate the value

$$v(D') = \delta(i) \cdot \prod_{j=1}^{r-1} \phi_{i_j}'(i_{j+1}) \cdot \prod_{i=1}^{r} v(D_i)$$

(the first factor gives the probability to start with the component P_{i_1}, the second factor takes into consideration the transitions from one component to another one, whereas the third factor measures the derivations D_j). The language $L(\Gamma, c)$ with cut-point c consists of all words $z \in T^*$ which can be obtained by a derivation D' of the form (1) such that $v(D') > c$.

Our definitions given above differ slightly from the definitions presented in [2]. Especially, we have a more accurate value associated with a derivation. However, it is easy to see that both definitions are equivalent.

In [1] it has been shown that any language generated by a probabilistic cooperating distributed grammar system with cut-point $c > 0$ is a finite language, and that any recursively enumerable language can be generated by a probabilistic cooperating distributed grammar system with cut-point 0. The latter statement can be seen easily since a probabilistic cooperating distributed grammar system with cut-point 0 can be transformed in a programmed grammar where the success field of a rule $p_{ij} \in P_i$ consists of all rules $p_{ik} \in P_i$ such that $\phi_{ijk} > 0$ and the failure field of p_{ij} consists of all rules $p_{rs} \in P_r$, $1 \leq r \leq n$, with $\phi'_i(r) > 0$ and $\delta_r(s) > 0$.

In this paper we shall discuss only languages with cut-point 0. Therefore we use the notation $L(\Gamma)$ instead of $L(\Gamma, 0)$.

By $\mathcal{L}(PCD_rCF)$ we denote the family of all languages generated by probabilistic cooperating distributed grammar system Γ with cut-point 0 where each component of Γ contains at most r productions.

The following lemma immediately follows from the definitions.

Lemma 2. *For any* $r \geq 1$, $\mathcal{L}(PCD_rCF) \subseteq \mathcal{L}(PCD_{r+1}CF)$.

3   Results

We start with an investigation of probabilistic grammar systems where each component contains only one rule.

Lemma 3. *For any probabilistic cooperating distributed grammar system* Γ *where each component contains exactly one rule, there is a Indian parallel programmed grammar* G *such that* $L(G) = L(\Gamma)$.

Proof. Let $\Gamma = (N, T, S, (\{A_1 \rightarrow w_1\}, \delta_1, \phi_1, \phi'_1), \ldots, (\{A_n \rightarrow w_n\}, \delta_n, \phi_n, \phi'_n), \delta)$ be an arbitrary probabilistic cooperating distributed grammar system where each component contains exactly one rule. Then $\delta_i = \phi_i(1) = (1)$ for $1 \leq i \leq n$.

Let us assume that we have to apply $(\{A_i \rightarrow w_i\}, (1), (1), \phi'_i)$ to the sentential form w. If A_i occurs in w and in w_i, then we have to apply the rule $A_i \rightarrow w_i$ ad infinitum and do not terminate. Thus we do not change the generated language if we substitute $(\{A_i \rightarrow w_i\}, (1), (1), \phi'_i)$ by $(\{A_i \rightarrow F\}, (1), (1), \phi'_i)$ where F is an additional symbol (because we are not able to terminate the letter F). Therefore, without loss of generality, we can assume that A_i does not occur in w_i for $1 \leq i \leq n$.

The application of $(\{A_i \rightarrow w_i\}, (1), (1), \phi'_i)$ to $w = x_1 A_i x_2 A_i \ldots x_k A_i x_{k+1}$ with $x_j \in ((N \cup T) \setminus \{A_i\})^*$ leads to $w' = x_1 w_i x_2 w_i \ldots x_k w_i x_{k+1}$, i.e. we have performed a derivation step as in an Indian parallel mode (see condition i) of the definition of the derivation step in a Indian parallel programmed grammar).

We now construct the Indian parallel programmed grammar

$$G = (N \cup \{S'\}, T, S', \{p_1, p_2, \ldots, p_n\}$$

with

$$p_0 = (S' \to S, \{p_j \mid \delta(j) > 0\}, \{p_j \mid \delta(j) > 0\}),$$
$$p_i = (A_i \to w_i, \{p_j \mid \phi_i'(j) > 0\}, \{p_j \mid \phi_i(j) > 0\}) \text{ for } 1 \le i \le n.$$

Any derivation in G starts with an application of p_0 and leads to S to which all rules can be applied which correspond to components of Γ whose start probability is greater than 0. Moreover, in the sequel we have $w \Longrightarrow w'$ by application of the component P_i in Γ if and only if we have $w \Longrightarrow w'$ by application of p_i in G. Therefore $L(G) = L(\Gamma)$ follows.

Lemma 4. $\mathcal{L}(EDT0L) \subseteq \mathcal{L}(PCD_1CF)$.

Proof. Let $L \in \mathcal{L}(EDT0L)$. By [8], Chapter V, Theorem 1.3 (since the construction in its proof gives a EDT0L system, if we start with an EDT0L system, this theorem holds for EDT0L systems, too), there is an EDT0L system $G = (V, T, \{h_1, h_2\}, w)$ (with only two homomorphisms) such that $L = L(G)$. Let

$$V = \{a_1, a_2, \dots, a_m\}, \ V' = \{a_i' \mid 1 \le i \le m\} \text{ and } V'' = \{a_i'' \mid 1 \le i \le m\}.$$

Moreover, if $w = b_1 b_2 \dots b_n$, $b_i \in V$ for $1 \le i \le n$, then we set $w' = b_1' b_2' \dots b_n'$ and $w'' = b_1'' b_2'' \dots b_n''$. Furthermore, we define the homomorphism $h : V' \to T \cup \{F\}$ by $h(a') = a$ for $a \in T$ and $h(a) = F$ in the remaining cases.

We now construct the probabilistic CD grammar system

$$\Gamma = (V' \cup V'' \cup \{S, F\}, T, S, (P_1, \delta_1, \phi_1, \phi_1'), \dots, (P_{4m+1}, \delta_{4m+1}, \phi_{4m+1}, \phi_{4m+1}'), \delta)$$

where

$$\delta = (0, 0, \dots, 0, 1),$$
$$P_{4m+1} = \{S \to w'\}, \ \delta_{4m+1} = \phi_{4m+1}(1) = (1),$$
$$\phi_{4m+1}'(j) = \begin{cases} 1/2 & j \in \{1, 3m+1\} \\ 0 & \text{otherwise} \end{cases},$$
$$P_m = \{a_m' \to a_m''\}, \ \delta_m = \phi_m(1) = (1),$$
$$\phi_m'(j) = \begin{cases} 1/2 & j \in \{m+1, 2m+1\} \\ 0 & \text{otherwise} \end{cases},$$
$$P_{2m} = \{a_m'' \to h_1(a_m)'\}, \ \delta_{2m} = \phi_{2m}(1) = (1),$$
$$\phi_{2m}'(j) = \begin{cases} 1/2 & j \in \{1, 3m+1\} \\ 0 & \text{otherwise} \end{cases},$$
$$P_{3m} = \{a_m'' \to h_2(a_m)'\}, \ \delta_{3m} = \phi_{3m}(1) = (1),$$
$$\phi_{3m}'(j) = \begin{cases} 1/2 & j \in \{1, 3m+1\} \\ 0 & \text{otherwise} \end{cases},$$
$$P_{4m} = \{a_m'' \to h(a_m')\}, \ \delta_{4m} = \phi_{4m}(1) = (1),$$
$$\phi_{4m}' = (0, 0, \dots, 0, 1),$$

and for $1 \le i \le m - 1$,

$$P_i = \{a_i' \to a_i''\}, \ \delta_i = \phi_i(1) = (1),$$

$$\phi_i'(j) = \begin{cases} 1 & j = i + 1 \\ 0 & \text{otherwise} \end{cases},$$

$$P_{m+i} = \{a_i'' \to h_1(a_i)'\}, \ \delta_{m+i} = \phi_{m+i}(1) = (1),$$

$$\phi_{m+i}'(j) = \begin{cases} 1 & j = m + i + 1 \\ 0 & \text{otherwise} \end{cases},$$

$$P_{2m+i} = \{a_i'' \to h_2(a_i)'\}, \ \delta_{2m+i} = \phi_{2m+i}(1) = (1),$$

$$\phi_{2m+i}'(j) = \begin{cases} 1 & j = 2m + i + 1 \\ 0 & \text{otherwise} \end{cases},$$

$$P_{3m+i} = \{a_i' \to h(a_i')\}, \ \delta_{3m+i} = \phi_{3m+i}(1) = (1),$$

$$\phi_{3m+i}'(j) = \begin{cases} 1 & j = m + i + 1 \\ 0 & \text{otherwise} \end{cases}.$$

By our construction, we have to start with the component P_{4m+1} which leads to w' and we have to continue with P_1 or P_{3m+1}.

Let us now assume that we have a sentential form v' for some $v \in V^*$ and that we can apply the components P_1 or P_{3m+1}.

Using P_{3m+1} we substitute all occurrences of a_1' by $h(a_1')$ and have to continue with P_{3m+2} which corresponds to a substitution of all occurrences of a_2' by $h(a_2')$ and so on. After using P_{4m} we have replaced all letters of v' and obtain $h(v')$. We get a word containing an F (and the derivation cannot be terminated) or the terminal word v.

Using P_1 we replace by the use of the component $P_1, P_2, \ldots, P_m$ in succession all occurrences of primed letters by the corresponding two-primed version, i.e. we get v''. Moreover, we have to continue with P_{m+1} or P_{2m+1}. In the former case we apply in succession the components $P_{m+1}, P_{m+2}, \ldots, P_{2m}$ and replace each letter a_i'' by $h_1(a_i)'$. Thus we obtain $h_1(v)'$. In the latter case we obtain $h_2(v)'$. Therefore we have simulated a derivation step according to the EDT0L system G.

By these remarks it is obvious that $L = L(G) = L(\Gamma)$

Now we turn to probabilistic grammar systems with at most two rules in a component.

Lemma 5. $\mathcal{L}(RE) \subseteq \mathcal{L}(PCD_2CF)$.

Proof. Let L be a recursively enumerable language. By Lemma 1, there is a matrix grammar $G = (N, T, S, M, F)$ in normal form such that $L = L(G)$. Let us assume that, for $1 \le i \le 5$, there are k_i matrices of type (i) in M. We set $l_0 = 0$ and $l_i = k_1 + k_2 + \cdots + k_i$ for $1 \le i \le 5$. We number the matrices of type (i) from $l_{i-1} + 1$ to l_i.

We define the probabilistic cooperating distributed grammar system

$$\Gamma = (N, T, S, (P_1, \delta_1, \phi_1, \phi_1'), (P_2, \delta_2, \phi_2, \phi_2'), \ldots, (P_{l_5+1}, \delta_{l_5+1}, \phi_{l_5+1}, \phi_{l_5+1}'), \delta)$$

where the component P_i is associated with the i-th matrix m_i, by

$$\delta(i) = \begin{cases} 1/l_2 & \text{for } 1 \leq i \leq l_2, \\ 0 & \text{for } l_2 + 1 \leq l_5 \end{cases}$$

(we start with a component associated with a matrix of type (1) or (2)),

$$P_{l_5+1} = \{S \rightarrow S\}, \; \delta_{l_5+1} = (1), \; \phi_{l_5+1}(1) = (1), \; \phi_{l_5+1}'(j) = \begin{cases} 1 & \text{for } j = l_5 + 1 \\ 0 & \text{otherwise} \end{cases}$$

(if this component has to be applied to a sentential form containing S, then we have to replace $S \rightarrow S$ ad infinitum; if S is not present in the sentential form, we have to apply without changes this component again and again, i.e., if we have to apply this component, we cannot terminate),

$$P_i = \{S \rightarrow x\}, \; \delta_i = (1), \; \phi_i(1) = (1), \; \phi_i'(j) = \begin{cases} 1 & \text{for } j = 1 \\ 0 & \text{otherwise} \end{cases}$$

for $1 \leq i \leq l_1$ and $m_i = (S \rightarrow x)$ (if we start with a component corresponding to a matrix of type (1), we generate a terminal word and stop the derivation),

$$P_i = \{S \rightarrow AX\}, \; \delta_i = (1), \; \phi_i(1) = (1),$$
$$\phi_i'(j) = \begin{cases} 1/n_G(X) & \text{if } m_j \text{ is an } X\text{-matrix} \\ 0 & \text{otherwise} \end{cases}$$

for $l_1 + 1 \leq i \leq l_2$ and $m_i = (S \rightarrow AX)$ (if we start with a component corresponding to a matrix of type (2), we generate AX which is a simulation of an application of m_i and continue with a X-matrix of type (3),(4) or (5) as in the matrix grammar),

$$P_i = \{A \rightarrow w, X \rightarrow Y\}, \; \delta_i = (1, 0), \; \phi_i(1) = \phi_i(2) = (0, 1),$$
$$\phi_i'(j) = \begin{cases} 1/n_G(Y) & \text{if } m_j \text{ is a } Y\text{-matrix} \\ 0 & \text{otherwise} \end{cases}$$

for $l_2 + 1 \leq i \leq l_3$ and $m_i = (A \rightarrow w, X \rightarrow Y)$ (if we apply a component corresponding to a matrix of type (3) to a word zX, $z \in (N_1 \cup T)^*$, we substitute one occurrence of A in z by w and the only occurrence of X by Y thus simulating an application of the matrix and continue with a Y-matrix of type (3),(4) or (5) as in the matrix grammar; if A does not occur in z, then we immediately pass without changing the sentential form to the following component which corresponds to a matrix of (3), (4) or (5), i.e., we have the same situation as before the application of P_i),

$$P_i = \{A \to F, X \to Y\}, \ \delta_i = (0,1), \ \phi_i(1) = \phi_i(2) = (1,0),$$

$$\phi_i'(j) = \begin{cases} 1/n_G(Y) & \text{if } m_j \text{ is a } Y\text{-matrix} \\ 0 & \text{otherwise} \end{cases}$$

for $l_3 + 1 \le i \le l_4$ and $m_i = (A \to F, X \to Y)$ (if we apply a component corresponding to a matrix of type (4) to a word zX, $z \in (N_1 \cup T)^*$, we first substitute X by Y; if A is present in z, we replace all occurrences of A by F such that the derivation cannot be terminated since there is no rule for F; if A is not present in z, we continue with a Y-matrix of type (3), (4) or (5) as in the matrix grammar; therefore in terminating derivations we have simulated a derivation step according to G),

$$P_i = \{A \to w, X \to a\}, \ \delta_i = (1,0), \ \phi_i(1) = \phi_i(2) = (0,1),$$

$$\phi_i(j) = \begin{cases} 1 & \text{for } j = l_5 + 1 \\ 0 & \text{otherwise} \end{cases}$$

for $l_4 + 1 \le i \le l_5$ and $m_i = (A \to w, X \to a)$ (if we apply a component corresponding to a matrix of type (5), we again simulate the application of the matrix; moreover, we have to terminate since otherwise we have to continue with the last component which results in a non-terminating infinite derivation by the remark added to this component).

It is easy to see that all sentential forms are terminal words or of the form zX with $z \in (N_1 \cup T)^*$. Thus by the above explanations any derivation of Γ simulates a derivation of G. Thus $L(\Gamma) \subseteq L(G)$.

Moreover, it is easy to see that any derivation of G can be simulated in Γ. Thus $L(G) \subseteq L(\Gamma)$, too. Hence $L(\Gamma) = L(G) = L$.

We now combine our results to obtain a hierarchy with respect to the number of rules in the components. We shall obtain a hierarchy with two levels only.

Theorem 1. *For any $r \ge 2$,*

$$\mathcal{L}(EDT0L) = \mathcal{L}(PCD_1CF) \subset= \mathcal{L}(PCD_rCF) = \mathcal{L}(RE).$$

Proof. By [4], Lemma 4, any language generated by an Indian parallel programmed grammar is contained in $\mathcal{L}(EDT0L)$. If we combine this result with Lemma 3, then we obtain $\mathcal{L}(PCD_1CF) \subseteq \mathcal{L}(EDT0L)$. Together with Lemma 4 we get $\mathcal{L}(PCD_1CF) = \mathcal{L}(EDT0L)$.

By Lemma 5, Lemma 2, and the result from [1] that any probabilistic cooperating distributed grammar system generates a recursively enumerable language, we have

$$\mathcal{L}(RE) \subseteq \mathcal{L}(PCD_2CF) \subseteq \mathcal{L}(PCD_rCF) \subseteq \mathcal{L}(RE)$$

for $r \ge 2$. This implies the remaining equalities $\mathcal{L}(PCD_rCF) = \mathcal{L}(RE)$ for $r \ge 2$.

References

1. K. Arthi and K. Krithivasan, Probabilistic cooperating distributed grammar systems. Submitted.
2. K. Arthi, K. Krithivasan and E. Csuhaj-Varjú, On the number of rules in components of cooperating distributed grammar systems with probabilities. *Journal of Automata, Languages and Combinatorics* **7** (2002) 433-446.
3. E. Csuhaj-Varjú and J. Dassow, Cooperating distributed grammar systems. *EIK* **26** (1990) 49-63.
4. J. Dassow, On some extensions of Indian parallel context-free grammars. *Acta Cybernetica* **4** (1980) 303–310.
5. J. Dassow and Gh. Păun, *Regulated Rewriting in Formal Language Theory.* Springer-Verlag, Berlin, 1989.
6. J. Dassow, Gh. Păun and St. Skalla, On the size of components of cooperating grammars. In: *Results and Trends in Computer Science* (Eds.: J. Karhumäki, H. Maurer, G.Rozenberg), LNCS 812, 1994, 325-343.
7. G. T. Herman and G. Rozenberg, *Developmental Systems and Languages.* North-Holland, Amsterdam, 1975.
8. G. Rozenberg and A. Salomaa, *The Mathematical Theory of L Systems.* Academic Press, New York, 1980.
9. A. Salomaa, Probabilistic and weighted grammars. *Inform. Control* **15** (1969) 529–544.
10. A. Salomaa, *Formal Languages.* Academic Press, New York, 1973.
11. D. Wätjen, On cooperating/distributed limited 0L systems. *J. Inform. Proc. Cyb. EIK* **29** (1993) 129–142.

Remarks on Sublanguages Consisting of Primitive Words of Slender Regular and Context-Free Languages

Pál Dömösi[1*], Carlos Martín-Vide[2], and Victor Mitrana[2,3]

[1] Institute of Mathematics and Informatics, Debrecen University
Debrecen, Egyetem tér 1., H-4032, Hungary
domosi@math.klte.hu

[2] Research Group on Mathematical Linguistics, Universitat Rovira i Virgili,
PL. Imperial Tarraco, 1, 43005 Tarragona, Spain
cmv@astor.urv.es

[3] Faculty of Mathematics and Computer Science, University of Bucharest
Str. Academiei 14, 70109, Bucharest, Romania
vmi@fll.urv.es

Abstract. In this note we investigate the languages obtained by intersecting slender regular or context-free languages with the set of all primitive words over the common alphabet. We prove that these languages are also regular and, respectively, context-free. The statement does not hold anymore for either regular or context-free languages. Moreover, the set of all non-primitive words of a slender context-free language is still context-free. Some possible directions for further research are finally discussed.

1 Introduction

Combinatorial properties of words and languages play an important role in mathematics and theoretical computer science (algebraic coding, combinatorial theory of words, etc.), see, e.g., [2],[5],[8],[16].

A word is called primitive if it cannot be expressed as the power of another word. There has been conjectured [1] that the set of all primitive words over a given alphabet is not context-free. However, this language satisfies different necessary conditions for context-free languages (see [1] for further details). Hopefully, this conjecture requires new methods based on the structure of context-free languages and perhaps will lead to sharper necessary conditions for languages to be context-free.

A language is slender if the number of its words of any length is bounded by a constant. It was proved, first in [6], and later, independently, in [13] and [10], that slender regular and USL-languages coincide. A similar characterization of

---

* This work was supported by grant from Dirección General de Universidades, Secretaría de Estado de Educatión y Universidades, Ministerio de Educación, Cultura y Deporte (SAB2001-0081), España.

J. Karhumäki et al. (Eds.): Theory Is Forever (Salomaa Festschrift), LNCS 3113, pp. 60–67, 2004.

slender context-free languages was reported in [7] and later, independently, in [4] and [11]. It was showed that every slender context-free language is UPL and vice versa, statement conjectured in [10].

It is known that the intersection of a regular language with the set of primitive words over the common alphabet is not necessarily regular. Since the set of all primitive words over an alphabet with at least two letters is not regular, it suffices to take the regular language consisting of all words over such an alphabet. We prove that if the regular language is slender then the above intersection is always regular. It immediately follows that the set of all non-primitive words of a slender regular language is regular too. Similar results hold for slender context-free languages as well. Furthermore, we prove that, similar to the case of regular languages, the set of all primitive words of a context-free language is not necessarily context-free.

This note is organized as follows: in the next section we fix the basic notions and notations and recall several results which will be used in later reasonings. The third section is dedicated to the sets of all primitive words of slender regular languages. The main result of this section states that these languages are always regular. As an immediate consequence, the set of all non-primitive words of a slender regular language is regular. A similar investigation is done in the forth section for slender context-free languages. The obtained results are similar, namely the sets of all primitive and non-primitive words of a slender context-free language are both context-free. The paper end by a short section dedicated to some open problems.

2   Preliminaries

We give some basic notions in formal language theory; for all unexplained notions the reader is referred to [12].

A *word* (over Σ) is a finite sequence of elements of some finite non-empty set Σ. We call the set Σ an *alphabet,* the elements of Σ *letters.* If u and v are words over an alphabet Σ, then their *catenation* uv is also a word over Σ. Especially, for every word u over Σ, $u\lambda = \lambda u = u$, where λ denotes the *empty word.* Given a word u, we define $u^0 = \lambda$, $u^n = u^{n-1}u$, $n > 0$, $u^* = \{u^n : n \geq 0\}$ and $u^+ = u^* \setminus \{\lambda\}$.

The *length* $|w|$ of a word w is the number of letters in w, where each letter is counted as many times as it occurs. Thus $|\lambda| = 0$. By the *free monoid* Σ^* *generated by* Σ we mean the set of all words (including the *empty word* λ) having catenation as multiplication. We set $\Sigma^+ = \Sigma^* \setminus \{\lambda\}$, where the subsemigroup Σ^+ of Σ^* is said to be the *free semigroup generated by* Σ. Subsets of Σ^* are referred to as *languages* over Σ.

A *primitive word* over an alphabet Σ is a nonempty word not of the form w^m for any nonempty word $w \in \Sigma^+$ and integer $m \geq 2$. The set of all primitive words over Σ will be denoted by $Q(\Sigma)$, or simply by Q if Σ is understood. Q has received special interest: Q and $\Sigma^+ \setminus Q$ play an important role in the algebraic theory of codes and formal languages (see [8] and [16]).

We denote by $card(H)$ the *cardinality* of the finite set H. A language $L \subseteq \Sigma^*$ is said to be *k-slender* if $card(\{w \in L : |w| = n\}) \leq k$, for every $n \geq 0$. A language is *slender* if it is k-slender for some positive integer k. A 1-slender language is called a *thin* language. A language $L \subseteq \Sigma^*$ is said to be a *union of single loops* (or, in short, USL) if for some positive integer k and words $u_i, w_i \in \Sigma^*$, $w_i \in \Sigma^+$, $1 \leq i \leq k$,

$$(*) \quad L = \bigcup_{i=1}^{k} u_i v_i^* w_i.$$

A language $L \subseteq \Sigma^*$ is called a *union of paired loops* (or UPL, in short) if for some positive k and words $u_i, w_i, y_i \in \Sigma^*$, $v_i, x_i \in \Sigma^+$, $1 \leq i \leq k$,

$$(**) \quad L = \bigcup_{i=1}^{k} \{u_i v_i^n w_i x_i^n y_i : n \geq 0\}.$$

For a USL (or UPL) language L the smallest k such that $(*)$ (or $(**)$) holds is referred to as the USL-index (or UPL-index) of L. A USL language L is said to be a *disjoint union of single loops* (DUSL, in short) if the sets in the union $(*)$ are pairwise disjoint. In this case the smallest k such that $(*)$ holds and the k sets are pairwise disjoint is referred to as the DUSL-index of L. The notions of a *disjoint union of paired loops* (DUPL) and DUPL-index are defined analogously considering the relation $(**)$.

For slender regular languages, we have the following characterization, first proved in [6], and later, independently, in [13] and [10] ([14] and [15] are an extended abstract and a revised form, respectively, of [13]).

Theorem 1. *For a given language L, the following conditions are equivalent:*
  (i) L is regular and slender.
  (ii) L is USL.
  (iii) L is DUSL.
Moreover, if L is regular and slender, then the USL- and DUSL-indices of L are effectively computable.

The following result is taken from [10].

Theorem 2. *Every UPL language is DUPL, slender, linear and unambiguous.*

The next characterization of slender context-free languages was proved in [7] and later, independently, in [4] and [11]. It was also conjectured in [10].

Theorem 3. *Every slender context-free language is UPL.*

Any cyclic permutation of a primitive (non-primitive) word remains primitive (non-primitive) as formally stated in [16, 17].

Theorem 4. *Let $i \geq 1$ and $uv \in \{p^i : p \in Q\}$. Then $vu \in \{p^i : p \in Q\}$, too. In other words, the sets $\{p^i : p \in Q\}$ ($i \geq 1$) are closed under cyclic permutations of words.*

We shall use the following result from [18].

Theorem 5. *Let $f, g \in Q, f \neq g$. Then $fg^n \in Q$ or $fg^{n+1} \in Q$ for all $n \geq 2$.*

Let $u \neq \lambda$ and let f be a primitive word with an integer $k \geq 1$ having $u = f^k$. We write $\sqrt{u} = f$ and call f the *primitive root* of the word u. The uniqueness of primitive root was proved in [9] (see also [16]).

Theorem 6. *If $u \neq \lambda$, then there exists a unique primitive word f and a unique integer $k \geq 1$ such that $u = f^k$.*

The next statement, useful in what follows, is also from [9].

Theorem 7. *Let $f, g \in Q, f \neq g$. Then $f^m g^n \in Q$ for all $m \geq 2, n \geq 2$.*

The following result reported in [2, 3] will also be applied in the sequel. (For a weaker version of this statement see also [16].)

Theorem 8. *Let u and v be two nonempty words, and, $p, q \geq 0$ integers. If u^p and v^q contain a common prefix or suffix of length $|u| + |v| - gcd(|u|, |v|)$ (where $gcd(|u|, |v|)$ denotes the greatest common divisor of $|u|$ and $|v|$) then $u = w^m$ and $v = w^n$, for some word w and positive integers m, n.*

Finally, we need one more result taken from [18].

Theorem 9. *Let $p, q \in Q, p \neq q$. Then $card(p^+ q^+ \setminus Q) \leq 1$.*

3   Intersecting Slender Regular Languages with Q

We start with some preliminary results. First it is easy to note that Theorem 9 can be extended, in a certain sense, to arbitrary words instead of primitive ones. Assume $\Sigma = \{a, b\}, p = a^2, q = ba^2b$. Then, of course, $p, pq \notin Q$. Theorem 7 implies $pq^n \in Q, n \geq 2$, hence $card(pq^+ \setminus Q) = 2$ In general, we have the following result.

Lemma 1. *Let $u, v \in \Sigma^+$ such that $\sqrt{u} \neq \sqrt{v}$. Then $card(uv^* \setminus Q) \leq 2$.*

Proof. By Theorem 9, $(\sqrt{u})^+ (\sqrt{v})^+ \setminus Q$ has at most one element. Therefore, $uv^* \setminus Q$ has at most one element if $u \in Q$. Assume $u \in \Sigma^+ \setminus Q$ and let $u = (\sqrt{u})^s$ for some $s > 1$. Then, by Theorem 7, $uv^n \in Q$ whenever $n \geq 2$. Therefore, uv^* has at most two non-primitive words.   □

Next we prove the following statement.

Lemma 2. *Let $u, w \in \Sigma^*$ and $v \in \Sigma^+$.*
   (i) If $uw = \lambda$, then $uv^*w \setminus Q = \{\lambda\}$.
   (ii If $uw \neq \lambda$ and $\sqrt{wu} \neq \sqrt{v}$, then $card(uv^*w \setminus Q) \leq 2$.

Proof. Using Theorem 4, it is enough to prove that $card(wuv^* \setminus Q) \leq 2$ whenever $uw, v \in \Sigma^+$ such that $\sqrt{wu} \neq \sqrt{v}$. But this is a direct consequence of Lemma 1.

$\square$

Now we can state the main result of this section.

Theorem 10. *The family of slender regular languages is closed under intersection with the set of all primitive words.*

Proof. Let L be a slender regular language; by Theorem 1 L is a DUSL, hence $L = \bigcup_{i=1}^{k} u_i v_i^* w_i$ for some positive integer k and words $u_i, v_i, w_i, 1 \leq i \leq k$, such that $u_i v_i^* w_i \cap u_j v_j^* w_j = \emptyset$ for all $1 \leq i \neq j \leq k$. If $\sqrt{w_i u_i} = \sqrt{v_i}$ or $u_i w_i = \lambda$ for some i, then all words in the set $u_i v_i^+ w_i$ are non-primitive. If $\sqrt{w_i u_i} \neq \sqrt{v_i}$, then each set $u_i v_i^* w_i$ contains at most two non-primitive words. Therefore,

$$L \cap Q = F \cup \bigcup_{i \in I} (u_i v_i^* w_i \setminus R_i),$$

where $I = \{i : 1 \leq i \leq k, \sqrt{w_i u_i} \neq \sqrt{v_i}\}$, $F = \{u_i w_i : 1 \leq i \leq k, i \notin I, u_i w_i \in Q\}$, and R_i, $i \in I$ are finite sets containing at most two words. By the closure properties of regular languages it follows that $L \cap Q$ is regular. The slenderness of $L \cap Q$ is obvious.

$\square$

Since the class of regular languages is closed under set difference, by Theorem 10 we also have:

Corollary 1. *The class of slender regular languages is closed under set difference with the language of primitive words.*

4   Intersecting Slender Context-Free Languages with Q

Now we start a similar investigation to that from the previous section for the class of slender context-free languages. Again, we first need some preliminary results.

Lemma 3. *Let $u, w, y \in \Sigma^*, v, x \in \Sigma^+$. If $\{k : \sqrt{yuv^k w} = \sqrt{x}\}$ is a finite set, then $\{uv^n wx^n y : n \geq 0\} \setminus Q$ is finite as well.*

Proof. Let us first consider the case $uwy = \lambda$. Clearly, $\sqrt{v} \neq \sqrt{x}$, otherwise the set $\{k : \sqrt{yuv^k w} = \sqrt{x}\}$ would be infinite. Then the statement follows from Theorem 7.

Assume now that $uwy \neq \lambda$ and let k_0 be the maximal k such that $\sqrt{yuv^k w} = \sqrt{x}$, therefore $\sqrt{yuv^n w} \neq \sqrt{x}$ for any $n \geq k_0$, Let $n > \max(k_0, 3)$.

If $yuv^n w \notin Q$, then by Theorem 7 we infer that $yuv^n wx^n \in Q$, hence, by Theorem 4, $uv^n wx^n y \in Q$.

If $yuv^n w \in Q$, then by Theorem 5 and the choice of n, $yuv^n wx^n \in Q$ holds.

$\square$

Lemma 4. *Let $u, w, y \in \Sigma^*, v, x \in \Sigma^+$ such that the set $\{k : \sqrt{yuv^k w} = \sqrt{d}\}$ is infinite. Then $\{uv^n wx^n y : n \geq 1\} \cap Q = \emptyset$.*

Proof. Case 1. $uwy = \lambda$. Then, since $\{k : \sqrt{yuv^k w} = \sqrt{x}\}$ is infinite, there exist infinitely many $k \geq 1$ with $\sqrt{v^k} = \sqrt{x}$. On the other hand, for every $k \geq 1$, we have $\sqrt{v^k} = \sqrt{x}$ if and only if $\sqrt{v} = \sqrt{x}$. But this implies $v^k x^k \notin Q, k \geq 1$.

Case 2. $uwy \neq \lambda$. First we prove that $\sqrt{wyu} = \sqrt{v}$. Indeed, assume $\sqrt{wyu} \neq \sqrt{v}$. If $wyu \notin Q$, then by Theorem 7, $wyuv^n \in Q, n \geq 2$. If $wyu \in Q$, then by Theorem 5, $wyuv^n \in Q, n \geq 3$. Therefore, by Theorem 4, $yuv^n w \in Q$, $n \geq 3$. But then for every $s, t \geq 3$, we obtain $\sqrt{yuv^s w} = \sqrt{yuv^t w}$ if and only if $s = t$. Therefore, if $\sqrt{yuv^k w} = \sqrt{x}$ then $\sqrt{yuv^{k+\ell} w} \neq \sqrt{x}$, for any $\ell \geq 1$, which implies that $\{k : \sqrt{yuv^k w} = \sqrt{x}\}$ is finite, a contradiction. Thus, we have $\sqrt{wyu} = \sqrt{v}$ (with $yuw \neq \lambda$). Furthermore, $\sqrt{yuv^s w} = \sqrt{yuv^t w}$, for all $s, t \geq 1$.

On the other hand, since $\{k : \sqrt{yuv^k w} = \sqrt{x}\}$ is infinite, there exist infinitely many $k \geq 1$ with $\sqrt{yuv^k w} = \sqrt{x}$. Hence, using $\sqrt{yuv^s w} = \sqrt{yuv^t w}$ for all $s, t \geq 1$, we obtain $\sqrt{yuv^k w} = \sqrt{x}$ for all $k \geq 1$. Thus, we get $\{uv^n wx^n w : n \geq 1\} \cap Q = \emptyset$ as we stated. $\square$

As a consequence, we have the following result similar to Theorem 10 and Corollary 1. Note that unlike the family of regular languages, the family of context-free languages is not closed under set difference.

Theorem 11. *The class of slender context-free languages is closed under intersection and set difference with the language of primitive words.*

Proof. Let L be a slender context-free language; by Theorems 3 and 2 L is a DUPL. Consequently, $L = \bigcup_{i=1}^{k} u_i v_i^* w_i x_i^* y_i$ for some positive integer k and words $u_i, v_i, w_i, x_i, y_i, 1 \leq i \leq k$, such that $u_i v_i^* w_i x_i^* y_i \cap u_j v_j^* w_j x_j^* y_j = \emptyset$ for all $1 \leq i \neq j \leq k$.

By Lemma 3, if $\{p : \sqrt{y_i u_i v_i^p w_i} = \sqrt{x_i}\}$ is a finite set, then $u_i v_i^* w_i x_i^* y_i$ contains a finite set of non-primitive words.

By Lemma 4, if $\{p : \sqrt{y_i u_i v_i^p w_i} = \sqrt{x_i}\}$ is an infinite set, then $u_i v_i^* w_i x_i^* y_i$ contains a primitive word only, provided that $u_i w_i y_i \in Q$, or no primitive word, otherwise.

In conclusion

$$L \cap Q = F \cup \bigcup_{i \in I}(u_i v_i^* w_i x_i^* y_i \setminus R_i),$$

where $I = \{i : 1 \leq i \leq k, \{p : \sqrt{y_i u_i v_i^p w_i} = \sqrt{x_i}\}$ is a finite set $\}$, $F = \{u_i w_i y_i : 1 \leq i \leq k, i \notin I, u_i w_i y_i \in Q\}$, and R_i, $i \in I$ are finite sets. By the closure properties of context-free languages it follows that $L \cap Q$ is context-free.

Analogously,

$$L \setminus Q = (\bigcup_{i \in I} R_i) \cup \bigcup_{i \notin I}(u_i v_i^* w_i x_i^* y_i \setminus F),$$

where I, F and R_i are the same sets as above. Obviously, $L \setminus Q$ is also context-free.

In both cases, the languages are slender since they are sublanguages of a slender language.   □

This result does not hold anymore for arbitrary context-free languages. Indeed, let us take the well-known context-free language $L = \{ww^R : w \in \{a,b\}^+\}$, where w^R s the *mirror image* of the word w. We use the pumping lemma for showing that $L \cap Q$ is not context-free. Clearly, $x = a^n b a^n a^n b a^n$ lies in $L \cap Q$ for arbitrarily large n. However, any attempt to pump two subwords of x satisfying the requirements of pumping lemma leads to a word which cannot be at the same time in L and primitive. We can state this as:

Theorem 12. *The family of context-free languages is not closed under intersection with the language of primitive words.*

5   Final Remarks

We finish this note with a brief discussion on possible directions, which appears of interest to us, for further research. There are a lot of subclasses of regular and context-free languages: locally-testable, poly-slender, Parikh-slender, dense, complete, periodic, quasi-periodic, etc. A natural continuation is to investigate which of these classes are closed under the intersection with the language of primitive words. Alternatively, in some cases, it appears attractive to study when the intersection of languages in these classes and the language of primitive words leads to a regular or context-free language.

References

1. P. Dömösi, S. Horvath, M. Ito, Formal languages and primitive words, *Publ. Math. Debrecen* **42** (1993) 315–321.
2. N.J. Fine, H.S. Wilf, Uniqueness theorems for periodic functions, *Proceedings of the American Mathematical Society* **16** (1965) 109-114.
3. M. Harrison, *Introduction to Formal Language Theory*, Addison-Wesley, Reading, Mass. 1978.
4. L. Ilie, On a conjecture about slender context-free languages, *Theoret. Comput. Sci.* **132** (1994) 427-434.
5. B. Imreh, M. Ito, On some special classes of regular languages. In: *Jewels are Forever* (J. Karhumäki, H. Maurer, G. Păun, G. Rozenberg, eds.) Springer-Verlag 1999, 25-34.
6. M. Kunze, H. J. Shyr, G. Thierrin, h-bounded and semidiscrete languages, *Inform. Control* **51** (1981) 147–187.
7. M. Latteux, G. Thierrin, Semidiscrete context-free languages, *Internat. J. Comput. Math.* **14** (1983) 3–18.
8. M. Lothaire, *Combinatorics on Words*, Addison-Wesley 1983.
9. R.C. Lyndon, M.P. Schützenberger, The equation $a^M = b^N c^P$ in a free group, *Michigan Math. J.* **9** (1962) 289–298.
10. G. Păun, A. Salomaa, Thin and slender languages, *Discrete Appl. Math.* **61** (1995) 257-270.

11. D. Raz, Length considerations in context-free languages, *Theoret. Comput. Sci.* **183** (1997) 21-32.

12. A. Salomaa, *Formal Languages*, Academic Press NY, 1973.

13. J. Shallit, Numeration systems, linear recurrences, and regular sets. *Research Report CS-91-32*, July, 1991, Computer Science Department, University of Waterloo, Canada.

14. J. Shallit, Numeration systems, linear recurrences, and regular sets. (Extended abstract.) In: *Proc. ICALP'92, LNCS 623*, Springer-Verlag, Berlin, 1992, 89–100.

15. J. Shallit, Numeration systems, linear recurrences, and regular sets. *Information and Computation* **113** (1994) 331–347.

16. H.J. Shyr, *Free Monoids and Languages*, Ho Min Book Company, Taiwan, 1991.

17. H.J. Shyr, G. Thierrin, Disjunctive languages and codes, *FCT'77, LNCS 56*, Springer-Verlag, Berlin, 1977, 171–176.

18. H.J. Shyr, S.S. Yu, Non-primitive words in the language $p^+ q^+$, *Soochow J.Math.* **20** (1994) 535–546.

A Semiring-Semimodule Generalization of ω-Context-Free Languages[*]

Zoltán Ésik[1] and Werner Kuich[2]

[1] University of Szeged
ze@inf.u-szeged.hu
[2] Technische Universität Wien
kuich@tuwien.ac.at

Abstract. We develop an algebraic theory on semiring-semimodule pairs for ω-context-free languages. We define ω-algebraic systems and characterize their solutions of order k by behaviors of algebraic finite automata. These solutions are then set in correspondence to ω-context-free languages.

1 Introduction

The purpose of our paper is to give an algebraic approach independent of any alphabets and languages for ω-context-free languages. The paper continues the research of Ésik, Kuich [5–7] and uses again pairs consisting of a semiring and a semimodule, where the semiring models a language with finite words and the semimodule models a language with ω-words.

The paper consists of this and two more sections. We assume the reader of this paper to be familiar with the definitions of Ésik, Kuich [5–7]. But to increase readibility, we repeat the necessary definitions concerning semiring-semimodule pairs and quemirings in this section. In Section 2, ω-algebraic systems and ω-algebraic power series are considered. The solutions of order k of these ω-algebraic systems are characterized by behaviors of algebraic finite automata. The ω-algebraic systems and ω-algebraic power series are then connected in Section 3 to ω-context-free grammars and ω-context-free languages, respectively.

Suppose that S is a semiring and V is a commutative monoid written additively. We call V a (left) S-semimodule if V is equipped with a (left) action

$$S \times V \to V$$
$$(s, v) \mapsto sv$$

subject to the following rules:

$$s(s'v) = (ss')v, \quad (s + s')v = sv + s'v, \quad s(v + v') = sv + sv',$$
$$1v = v, \quad 0v = 0, \quad s0 = 0,$$

[*] Partially supported by Aktion Österreich-Ungarn, Wissenschafts- und Erziehungskooperation, Projekt 53ÖU1. Additionally, the first author was supported, in part, by the National Foundation of Hungary for Scientific Research, grant T 35163.

J. Karhumäki et al. (Eds.): Theory Is Forever (Salomaa Festschrift), LNCS 3113, pp. 68–80, 2004.

for all $s, s' \in S$ and $v, v' \in V$. When V is an S-semimodule, we call (S, V) a *semiring-semimodule pair*.

Suppose that (S, V) is a semiring-semimodule pair such that S is a starsemiring and S and V are equipped with an omega operation $^\omega : S \to V$. Then we call (S, V) a *starsemiring-omegasemimodule pair*.

Ésik, Kuich [5] define a *complete semiring-semimodule pair* to be a semiring-semimodule pair (S, V) such that S is a complete semiring and V is a complete monoid, and an *infinite product operation* $\prod$ is defined, mapping infinite sequences over S to V. Moreover, the infinite sums and products have to satisfy certain conditions assuring that computations with these obey the usual laws.

Suppose that (S, V) is complete. Then we define

$$s^* = \sum_{i \geq 0} s^i, \qquad s^\omega = \prod_{i \geq 1} s,$$

for all $s \in S$. This turns (S, V) into a starsemiring-omegasemimodule pair.

Following Bloom, Ésik [2] we define a matrix operation $^\omega : S^{n \times n} \to V^{n \times 1}$ on a starsemiring-omegasemimodule pair (S, V) as follows. When $n = 0$, M^ω is the unique element of V^0, and when $n = 1$, so that $M = (a)$, for some $a \in S$, $M^\omega = (a^\omega)$. Assume now that $n > 1$ and decompose M into blocks a, b, c, d with a of dimension 1×1 and d of dimension $(n-1) \times (n-1)$: $M = \begin{pmatrix} a & b \\ c & d \end{pmatrix}$. Then

$$M^\omega = \begin{pmatrix} (a + bd^*c)^\omega + (a + bd^*c)^* bd^\omega \\ (d + ca^*b)^\omega + (d + ca^*b)^* ca^\omega \end{pmatrix}.$$

Moreover, we define matrix operations $^{\omega_k} : S^{n \times n} \to V^{n \times 1}$, $0 \leq k \leq n$, as follows. Assume that $M \in S^{n \times n}$ is decomposed into blocks a, b, c, d with a of dimension $k \times k$ and d of dimension $(n-k) \times (n-k)$: $M = \begin{pmatrix} a & b \\ c & d \end{pmatrix}$. Then

$$M^{\omega_k} = \begin{pmatrix} (a + bd^*c)^\omega \\ d^*c(a + bd^*c)^\omega \end{pmatrix}.$$ Observe that $M^{\omega_0} = 0$ and $M^{\omega_n} = M^\omega$.

Suppose that (S, V) is a semiring-semimodule pair and consider $T = S \times V$. Define on T the operations

$$(s, u) \cdot (s', v) = (ss', u + sv), \quad (s, u) + (s', v) = (s + s', u + v)$$

and constants $0 = (0, 0)$ and $1 = (1, 0)$. Equipped with these operations and constants, T satisfies the equations

$$(x + y) + z = x + (y + z), \quad x + y = y + x, \quad x + 0 = x, \tag{1}$$
$$(x \cdot y) \cdot z = x \cdot (y \cdot z), \quad x \cdot 1 = x, \quad 1 \cdot x = x, \tag{2}$$
$$(x + y) \cdot z = (x \cdot z) + (y \cdot z), \tag{3}$$
$$0 \cdot x = 0. \tag{4}$$

Elgot[4] also defined the unary operation $\P$ on T: $(s, u)\P = (s, 0)$. Thus, $\P$ selects the "first component" of the pair (s, u), while multiplication with 0 on the right selects the "second component", for $(s, u) \cdot 0 = (0, u)$, for all $u \in V$. The new operation satisfies:

$$x\P \cdot (y + z) = (x\P \cdot y) + (x\P \cdot z), \tag{5}$$

$$x = x\P + (x \cdot 0), \quad x\P \cdot 0 = 0, \tag{6}$$

$$(x + y)\P = x\P + y\P, \quad (x \cdot y)\P = x\P \cdot y\P. \tag{7}$$

Note that when V is idempotent, also

$$x \cdot (y + z) = x \cdot y + x \cdot z$$

holds.

Elgot[4] defined a *quemiring* to be an algebraic structure T equipped with the above operations $\cdot, +, \P$ and constants $0, 1$ satisfying the equations (1)–(4) and (5)–(7). It follows from the axioms that $x\P\P = x\P$, for all x in a quemiring T. Moreover, $x\P = x$ iff $x \cdot 0 = 0$.

When T is a quemiring, $S = T\P = \{x\P \mid x \in T\}$ is easily seen to be a semiring. Moreover, $V = T0 = \{x \cdot 0 \mid x \in T\}$ contains 0 and is closed under $+$, moreover, $sx \in V$ for all $s \in S$ and $x \in V$. Each $x \in T$ may be written in a unique way as the sum of an element of $T\P$ and a sum of an element of $T0$, viz. $x = x\P + x \cdot 0$. Sometimes, we will identify $S \times \{0\}$ with S and $\{0\} \times V$ with V. It is shown in Elgot [4] that T is isomorphic to the quemiring $S \times V$ determined by the semiring-semimodule pair (S, V).

Suppose now that (S, V) is a starsemiring-omegasemimodule pair. Then we define on $T = S \times V$ a *generalized star operation*:

$$(s, v)^{\otimes} = (s^*, s^{\omega} + s^* v)$$

for all $(s, v) \in T$.

2 ω-Algebraic Systems

In the sequel, T is a quemiring, $Y = \{y_1, \ldots, y_n\}$ is a set of (quemiring) variables, $T\P = S$ and $T0 = V$. A *product term* t has the form $t(y_1, \ldots, y_n) = s_0 y_{i_1} s_1 \ldots s_{k-1} y_{i_k} s_k$, $k \geq 0$, where $s_j \in S - \{0\}$, $0 \leq j < k$, $s_k \in S$, and $y_{i_j} \in Y$. The elements s_j are referred to as *coefficients* of the product term. If $k \geq 1$, we do not write down coefficients that are equal to 1.

A *sum-product term* p is a finite sum of product terms t_j, i.e.,

$$p(y_1, \ldots, y_n) = \sum_{1 \leq j \leq m} t_j(y_1, \ldots, y_n) \cdot$$

The coefficients of all the product terms t_j, $1 \leq j \leq m$, are referred to as the *coefficients* of the sum-product term p. Observe that each sum-product term represents a polynomial of the *polynomial quemiring over the quemiring T in the set of variables Y* in the sense of Lausch, Nöbauer [10], Chapter 1.4. For a subset

$S' \subseteq S$, we denote the collection of all sum-product terms with coefficients in S' by $S'(Y)$. Observe that the sum-product terms in $S(Y)$ represent exactly the polynomials of the subquemiring of the polynomial quemiring that is generated by $S \cup Y$.

We are only interested in the mappings induced by sum-product terms. These mappings are *polynomial functions on* T in the sense of Lausch, Nöbauer [10], Chapter 1.6.

Each product term t (resp. sum-product term p) with variables $y_1, \ldots, y_n$ induces a mapping $\bar{t}$ (resp. $\bar{p}$) from T^n into T. For a product term t represented as above, the mapping $\bar{t}$ is defined by

$$\bar{t}(\tau_1, \ldots, \tau_n) = s_0 \tau_{i_1} s_1 \ldots s_{k-1} \tau_{i_k} s_k \,,$$

and for a sum-product term p, represented by a finite sum of product terms t_j as above, the mapping $\bar{p}$ is defined by

$$\bar{p}(\tau_1, \ldots, \tau_n) = \sum_{1 \leq j \leq m} \bar{t}_j(\tau_1, \ldots, \tau_n)$$

for all $(\tau_1, \ldots, \tau_n) \in T^n$.

Let (S, V) be a semiring-semimodule pair and let $S \times V$ be the quemiring determined by it. Let $S' \subseteq S$. An S'-*algebraic system* (*with variables* $y_1, \ldots, y_n$) *over the quemiring* $S \times V$ is a system of equations

$$y_i = p_i, \ 1 \leq i \leq n \,,$$

where each p_i is a sum-product term in $S'(Y)$. A *solution* to this S'-algebraic system is given by $(\tau_1, \ldots, \tau_n) \in T^n$ such that $\tau_i = \bar{p}_i(\tau_1, \ldots, \tau_n)$, $1 \leq i \leq n$.

Often it is convenient to write the S'-algebraic system $y_i = p_i$, $1 \leq i \leq n$, in matrix notation. Defining the two column vectors

$$y = \begin{pmatrix} y_1 \\ \vdots \\ y_n \end{pmatrix} \quad \text{and} \quad p = \begin{pmatrix} p_1 \\ \vdots \\ p_n \end{pmatrix}$$

we can write $y_i = p_i$, $1 \leq i \leq n$, in the matrix notation

$$y = p(y) \quad \text{or} \quad y = p \,.$$

A *solution* to $y = p(y)$ is now given by $\tau \in T^n$ such that $\tau = \bar{p}(\tau)$ with $\bar{p} = (\bar{p}_i)_{1 \leq i \leq n}$.

Consider now a product term $t(y_1, \ldots, y_n) = s_0 y_{i_1} s_1 \ldots s_{k-1} y_{i_k} s_k$ and let $\tau_i = (\sigma_i, \omega_i) \in S \times V$, $1 \leq i \leq n$. Then

$$\bar{t}(\tau_1, \ldots, \tau_n) = s_0(\sigma_{i_1}, \omega_{i_1})s_1 \ldots s_{k-1}(\sigma_{i_k}, \omega_{i_k})s_k =$$
$$(s_0 \sigma_{i_1} s_1 \ldots s_{k-1}\sigma_{i_k} s_k, \ s_0\omega_{i_1} + s_0\sigma_{i_1} s_1\omega_{i_2} + \cdots + s_0\sigma_{i_1} s_1 \ldots s_{k-2}\sigma_{i_{k-1}} s_{k-1}\omega_{i_k}) \,.$$

By definition, for $\sigma = (\sigma_1, \ldots, \sigma_n) \in S^n$,

$$t_\sigma(z_1, \ldots, z_n) = s_0 z_{i_1} + s_0\sigma_{i_1} s_1 z_{i_2} + \cdots + s_0\sigma_{i_1} s_1 \ldots s_{k-2}\sigma_{i_{k-1}} s_{k-1} z_{i_k}$$

and, if $p(y_1, \ldots, y_n) = \sum_{1 \leq j \leq m} t_j(y_1, \ldots, y_n)$,

$$p_\sigma(z_1, \ldots, z_n) = \sum_{1 \leq j \leq m} (t_j)_\sigma(z_1, \ldots, z_n).$$

Here $z_1, \ldots, z_n$ are variables over the semimodule V. We now obtain

$$\bar{t}(\tau_1, \ldots, \tau_n) = \bar{t}(\sigma_1, \ldots, \sigma_n) + \bar{t}_\sigma(\omega_1, \ldots, \omega_n)$$

and

$$\bar{p}(\tau_1, \ldots, \tau_n) = \bar{p}(\sigma_1, \ldots, \sigma_n) + \bar{p}_\sigma(\omega_1, \ldots, \omega_n).$$

Moreover,

$$\bar{p}(\tau_1, \ldots, \tau_n)\P = \bar{p}(\sigma_1, \ldots, \sigma_n) \quad \text{and} \quad \bar{p}(\tau_1, \ldots, \tau_n).0 = \bar{p}_\sigma(\omega_1, \ldots, \omega_n).$$

In the next theorem, y (resp. x and z) denotes a column vector $\begin{pmatrix} y_1 \\ \vdots \\ y_n \end{pmatrix}$ (resp.

$\begin{pmatrix} x_1 \\ \vdots \\ x_n \end{pmatrix}$ and $\begin{pmatrix} z_1 \\ \vdots \\ z_n \end{pmatrix}$), where the y_i (resp. x_i and z_i) are variables over $S \times V$
(resp. S and V).

In the sequel, S' will always denote a subset of S containing 0 and 1. The
S'-linear systems (over V) occuring in the next theorem are defined in Ésik,
Kuich [7] before Theorem 4.1. The S'-algebraic systems (over S) occuring in the
next theorem are defined in Kuich [9].

Theorem 2.1 *Let $S \times V$ be a quemiring and let $y = p(y)$ be an S'-algebraic
system over $S \times V$. Then $(\sigma, \omega) \in (S \times V)^n$ is a solution of $y = p(y)$ iff σ is a
solution of the S'-algebraic system $x = p(x)$ over S and ω is a solution of the
$\mathfrak{Alg}(S')$-linear system $z = p_\sigma(z)$ over V.*

Proof. $\tau = (\sigma, \omega)$ is a solution $\Leftrightarrow \tau = \bar{p}(\tau) = \bar{p}(\sigma) + \bar{p}_\sigma(\omega) \Leftrightarrow \sigma = \bar{p}(\sigma)$ and
$\omega = \bar{p}_\sigma(\omega)$. □

The following definition is given just for the purpose of the present paper.
A semiring-semimodule pair (S, V) is called *continuous* if (S, V) is a complete
semiring-semimodule pair and S is a continuous semiring. A quemiring is called
continuous if it is determined by a continuous semiring-semimodule pair.

Consider an S'-algebraic system $y = p(y)$ over a continuous quemiring $S \times V$.
Then the least solution of the S'-algebraic system $x = p(x)$ over S, say σ, exists.
Moreover, write the $\mathfrak{Alg}(S')$-linear system $z = p_\sigma(z)$ over V in the form $z = Mz$,
where M is an $n \times n$-matrix. Then, by Theorem 4.1 of Ésik, Kuich [7], M^{ω_k}
for $0 \leq k \leq n$ is a solution of $z = p_\sigma(z)$. Hence, by Theorem 2.1, (σ, M^{ω_k}),
$0 \leq k \leq n$, is a solution of $y = p(y)$. Given a $k \in \{0, 1, \ldots, n\}$, we call this
solution the *solution of order k of $y = p(y)$*. By ω-$\mathfrak{Alg}(S')$ we denote the collection
of all components of solutions of order k of S'-algebraic systems over $S \times V$.

We now consider a continuous semiring-semimodule pair $(S\langle\!\langle A^*\rangle\!\rangle, S\langle\!\langle A^\omega\rangle\!\rangle)$, where S is a commutative (continuous) semiring and A is an alphabet, and the continuous quemiring $S\langle\!\langle A^*\rangle\!\rangle \times S\langle\!\langle A^\omega\rangle\!\rangle$.

Let $SA^* = \{sw \mid s \in S, \ w \in A^*\}$. Then $\omega\text{-}\mathfrak{Alg}(SA^*)$ is equal to the collection of the components of the solutions of order k of SA^*-algebraic systems over $S\langle\!\langle A^*\rangle\!\rangle \times S\langle\!\langle A^\omega\rangle\!\rangle$ $y_i = p_i$, $1 \le i \le n$, where p_i is a polynomial in $S\langle\!\langle (A \cup Y)^*\rangle\!\rangle$. This is due to the commutativity of S: any polynomial function that is induced by a sum-product term of $SA^*(Y)$ is also induced by a polynomial of $S\langle\!\langle (A \cup Y)^*\rangle\!\rangle$ and vice versa. We denote $\omega\text{-}\mathfrak{Alg}(SA^*)$ by $S^{\omega\text{-alg}}\langle\!\langle A^*, A^\omega\rangle\!\rangle$. The SA^*-algebraic systems are called ω-algebraic systems (over S and A) and the power series in $S^{\omega\text{-alg}}\langle\!\langle A^*, A^\omega\rangle\!\rangle$ are called ω-algebraic power series (over S and A).

Consider now a product term in $S\langle\!\langle (A \cup Y)^*\rangle\!\rangle$

$$t(y_1, \ldots, y_n) = sw_0 y_{i_1} w_1 \ldots w_{k-1} y_{i_k} w_k\,,$$

where $s \in S$ and $w_i \in A^*$, $1 \le i \le k$. By definition, for $x = (x_i)_{1 \le i \le n}$,
$$t_x(x_1, \ldots, x_n, z_1, \ldots, z_n) = sw_0 z_{i_1} + sw_0 x_{i_1} w_1 z_{i_2} + \cdots + sw_0 x_{i_1} w_1 \ldots w_{k-2} x_{i_{k-1}} w_{k-1} z_{i_k}\,,$$ and, if $p(y_1, \ldots, y_n) = \sum_{1 \le j \le m} t_j(y_1, \ldots, y_n)$, then

$$p_x(x_1, \ldots, x_n, z_1, \ldots, z_n) = \sum_{1 \le j \le m} (t_j)_x(x_1, \ldots, x_n, z_1, \ldots, z_n)\,.$$

Here $x_1, \ldots, x_n$ (resp. $z_1, \ldots, z_n$) are variables over S (resp. V). Observe that, for $\sigma \in (S\langle\!\langle A^*\rangle\!\rangle)^n$, we obtain $p_x(\sigma_1, \ldots, \sigma_n, z_1, \ldots, z_n) = p_\sigma(z_1, \ldots, z_n)$.

Given an ω-algebraic system $y = p(y)$ over $S\langle\!\langle A^*\rangle\!\rangle \times S\langle\!\langle A^\omega\rangle\!\rangle$, we call $x = p(x)$, $z = p_x(x, z)$ the mixed ω-algebraic system over $(S\langle\!\langle A^*\rangle\!\rangle, S\langle\!\langle A^\omega\rangle\!\rangle)$ induced by $y = p(y)$.

Write $z = p_x(x, z)$ in the form $z = M(x)z$, where $M(x)$ is an $n \times n$-matrix. Then $(\sigma, M(\sigma)^{\omega_k})$ for $0 \le k \le n$ is a solution of $x = p(x)$, $z = p_x(x, z)$. Moreover, it is the solution of order k of $y = p(y)$.

3 ω-Context-Free Grammars

A mixed ω-context-free grammar

$$G = (n, A, P, j, k)$$

is given by

(i) an alphabet $X = \{x_1, \ldots, x_n\}$ of variables for finite derivations and an alphabet $Z = \{z_1, \ldots, z_n\}$ of variables for infinite derivations, $n \ge 1$, $X \cap Z = \emptyset$;

(ii) an alphabet A of terminal symbols, $A \cap (X \cup Z) = \emptyset$;

(iii) a finite set of productions of the form $x \to \alpha$, $x \in X$, $\alpha \in (X \cup A)^*$, or $z \to \alpha z'$, $z, z' \in Z$, $\alpha \in (X \cup A)^*$;

(iv) the startvariable x_j (resp. z_j) for finite (resp. infinite) derivations, $1 \le i \le n$;

(v) the set of repeated variables for infinite derivations $\{z_1, \ldots, z_k\}$, $0 \le k \le n$.

A *finite leftmost derivation* (with respect to G) $\alpha \Rightarrow_L^* w$, $\alpha \in (X \cup A)^*$, $w \in A^*$, is defined as usual. An *infinite leftmost derivation* (with respect to G) $\pi : z \Rightarrow_L^\omega w$, $z \in Z$, $w \in A^\omega$, is defined as follows:

$$\pi : z \Rightarrow_L \alpha_1 z_{i_1} \Rightarrow_L^* w_1 z_{i_1} \Rightarrow_L w_1 \alpha_2 z_{i_2} \Rightarrow_L^* w_1 w_2 z_{i_2} \Rightarrow_L \cdots \Rightarrow_L^*$$
$$w_1 w_2 \ldots w_m z_{i_m} \Rightarrow_L w_1 w_2 \ldots w_m \alpha_{m+1} z_{i_{m+1}} \Rightarrow_L^* \cdots ,$$

where $z \to \alpha_1 z_{i_1}, z_{i_1} \to \alpha_2 z_{i_2}, \ldots, z_{i_m} \to \alpha_{m+1} z_{i_{m+1}}, \ldots \in P$, $w_1, w_2, \ldots, w_m, \ldots$ $\in A^*$ and $w = w_1 w_2 \ldots w_m \ldots$. Let $\mathrm{INV}(\pi) = \{z \in Z \mid z$ is infinitely often rewritten in $\pi\}$. Then $L(G) = \{w \in A^* \mid x_j \Rightarrow_L^* w\} \cup \{w \in A^\omega \mid \pi : z_j \Rightarrow_L^\omega w, \mathrm{INV}(\pi) \cap \{z_1, \ldots, z_k\} \neq \emptyset\}$.

We now discuss the connection between mixed ω-algebraic systems over $(S\langle\!\langle A^* \rangle\!\rangle, S\langle\!\langle A^\omega \rangle\!\rangle)$, where S is $\mathbb{B}$ or $\mathbb{N}^\infty$, and mixed ω-context-free grammars. Define, for a given mixed ω-context-free grammar $G_{j,k} = (n, A, P, j, k)$, $1 \leq j \leq n$, $0 \leq k \leq n$, the mixed ω-algebraic system $x_i = p_i(x_1, \ldots, x_n)$, $z_i = q_i(x_1, \ldots, x_n, z_1, \ldots, z_n)$, $1 \leq i \leq n$, over $(S\langle\!\langle A^* \rangle\!\rangle, S\langle\!\langle A^\omega \rangle\!\rangle)$ by

$$(p_i, \alpha) = 1 \text{ if } x_i \to \alpha \in P, \quad (p_i, \alpha) = 0 \text{ otherwise},$$
$$(q_i, \alpha) = 1 \text{ if } z_i \to \alpha \in P, \quad (q_i, \alpha) = 0 \text{ otherwise}.$$

Conversely, given a mixed ω-algebraic system $x_i = p_i(x_1, \ldots, x_n)$, $z_i = q_i(x_1, \ldots, x_n, z_1, \ldots, z_n)$, $1 \leq i \leq n$, define the mixed ω-context-free grammars $G_{j,k} = (n, A, P, j, k)$, $1 \leq j \leq n$, $0 \leq k \leq n$, by $x_i \to \alpha \in P$ iff $(p_i, \alpha) \neq 0$ and $z_i \to \alpha \in P$ iff $(z_i, \alpha) \neq 0$. Whenever we speak of a mixed ω-context-free grammar corresponding to a mixed ω-algebraic system or vice versa, then we mean the correspondence in the sense of the above definition.

In the next theorem we use the isomorphism between $\mathbb{B}\langle\!\langle A^* \rangle\!\rangle \times \mathbb{B}\langle\!\langle A^\omega \rangle\!\rangle$ and $\mathfrak{P}(A^*) \times \mathfrak{P}(A^\omega)$.

Theorem 3.1 *Let $G_{j,k} = (n, A, P, j, k)$, $1 \leq j \leq n$, $0 \leq k \leq n$, be a mixed ω-context-free grammar and $x_i = p_i(x_1, \ldots, x_n)$, $z_i = q_i(x_1, \ldots, x_n, z_1, \ldots, z_n)$, $1 \leq i \leq n$, be the mixed ω-algebraic system over $(\mathbb{B}\langle\!\langle A^* \rangle\!\rangle, \mathbb{B}\langle\!\langle A^\omega \rangle\!\rangle)$ corresponding to it. Let (σ, τ) be the solution of order k, $0 \leq k \leq n$, of $x_i = p_i$, $z_i = q_i$, $1 \leq i \leq n$. Then $L(G_{j,k}) = \sigma_j + \tau_j$, $1 \leq j \leq n$, $0 \leq i \leq k$.*

Proof. By Theorem 2 of Ginsburg, Rice [8], we obtain $\sigma_j = \{w \in A^* \mid x_j \Rightarrow_L^* w\}$, $1 \leq j \leq n$, and by Ésik, Kuich [7] we obtain $\tau_j = \{w \in A^\omega \mid \pi : z_j \Rightarrow_L^* w, \mathrm{INV}(\pi) \cap \{z_1, \ldots, z_k\} \neq \emptyset\}$, $1 \leq j \leq n$, $0 \leq k \leq n$. $\square$

If our basic quemiring is $\mathbb{N}^\infty \langle\!\langle A^* \rangle\!\rangle \times \mathbb{N}^\infty \langle\!\langle A^\omega \rangle\!\rangle$ we can draw some stronger conclusions.

Theorem 3.2 *Let $G_{j,k} = (n, A, P, j, k)$, $1 \leq j \leq n$, $0 \leq k \leq n$, be a mixed ω-context-free grammar and $x_i = p_i(x_1, \ldots, x_n)$, $z_i = q_i(x_1, \ldots, x_n, z_1, \ldots, z_n)$, $1 \leq i \leq n$ be the mixed ω-algebraic system over $(\mathbb{N}^\infty \langle\!\langle A^* \rangle\!\rangle, \mathbb{N}^\infty \langle\!\langle A^\omega \rangle\!\rangle)$ corresponding to it. Let (σ, τ) be the solution of order k, $0 \leq k \leq n$, of $x_i = p_i$, $z_i = q_i$, $1 \leq i \leq n$. Denote by $d_j(w)$, $w \in A^*$ (resp. $w \in A^\omega$) the number (possibly ∞) of distinct finite leftmost derivations (resp. infinite leftmost derivations*

π with $INV(\pi) \cap \{z_1, \ldots, z_k\} \neq \emptyset$) from the variable x_j (resp. z_j), $1 \leq j \leq n$. Then

$$\sigma_j = \sum_{w \in A^*} d_j(w)w \quad and \quad \tau_j = \sum_{w \in A^\omega} d_j(w)w, \qquad 1 \leq j \leq n.$$

Proof. By Theorem IV.1.5 of Salomaa, Soittola [11] and Ésik, Kuich [7].   □

An ω-*context-free grammar* (*with repeated variables*) $G = (\Phi, A, P, S, F)$ is a usual context-free grammar (Φ, A, P, S) augmented by a *set* $F \subseteq \Phi$ *of repeated variables.* (See also Cohen, Gold [3].)

An *infinite leftmost derivation* π *with respect to* G, starting from some string α is given by

$$\pi : \alpha \Rightarrow_L \alpha_1 \Rightarrow_L \alpha_2 \Rightarrow_L \cdots,$$

where $\alpha, \alpha_i \in (\Phi \cup A)^*$ and $\Rightarrow_L$ is defined as usual. This infinite leftmost derivation π can be uniquely written as

$$\alpha = \beta_0 B_0 \gamma_0 \Rightarrow_L^* v_0 B_0 \gamma_0 \Rightarrow_L v_0 \beta_1 B_1 \gamma_1 \gamma_0 \Rightarrow_L^*$$
$$v_0 v_1 B_1 \gamma_1 \gamma_0 \Rightarrow_L v_0 v_1 \beta_2 B_2 \gamma_2 \gamma_1 \gamma_0 \Rightarrow_L^* \cdots,$$

where $v_i \in A^*$, $\beta_i, \gamma_i \in (\Phi \cup A)^*$, $B_i \to \beta_{i+1} B_{i+1} \gamma_{i+1} \in P$, $\beta_i \Rightarrow_L^* v_i$, the specific occurence of the variable B_i is not rewritten in the subderivation $\beta_i B_i \gamma_i \Rightarrow_L^* v_i B_i \gamma_i$ and the variables of γ_i are never rewritten in the infinite leftmost derivation π. This occurence of the variable B_i is called the *i-th significant variable* of π. (Observe that the infinite derivation tree of π has a unique infinite path determining the B_i's.) We write also, for this infinite leftmost derivation, $\pi : \alpha \Rightarrow_L^\omega w$ for $w = w_0 w_1 \ldots w_n \ldots$. By definition, $INV(\pi) = \{A \in \Phi \mid A$ is rewritten infinitely often in $\pi\}$. The ω-*language* $L(G)$ *generated by the* ω-*context-free grammar* G is defined by

$$L(G) = \{w \in A^* \mid S \Rightarrow_L^* w\} \cup \{w \in A^\omega \mid \pi : S \Rightarrow_L^\omega w, \ INV(\pi) \cap F \neq \emptyset\}.$$

An ω-*language* L is called ω-*context-free* if it is generated by an ω-context-free grammar. (Usually, an ω-language is a subset of A^ω. In our paper, it is a subset of $A^* \cup A^\omega$.)

The connection between an ω-algebraic system over $S\langle\langle A^* \rangle\rangle \times S\langle\langle A^\omega \rangle\rangle$ and an ω-context-free grammar is as usual. Define, for a given ω-context-free grammar $G_j = (\{y_1, \ldots, y_n\}, A, P, y_j, \{y_1, \ldots, y_k\})$ the ω-algebraic system $y_i = p_i(y_1, \ldots, y_n)$, $1 \leq i \leq n$, over $S\langle\langle A^* \rangle\rangle \times S\langle\langle A^\omega \rangle\rangle$ by $(p_i, \alpha) = 1$ if $y_i \to \alpha \in P$, $(p_i, \alpha) = 0$ otherwise. Conversely, given an ω-algebraic system $y_i = p_i(y_1, \ldots, y_n)$, $1 \leq i \leq n$, define the ω-context-free grammars $G_{j,k} = (\{y_1, \ldots, y_n\}, A, P, y_j, \{y_1, \ldots, y_k\})$, $1 \leq j \leq n$, $0 \leq k \leq n$, by $y_i \to \alpha$ iff $(p_i, \alpha) \neq 0$.

Each ω-context-free grammar G *induces* a mixed ω-context-free grammar G' as follows. Let $G = (\Phi, A, P, S, F)$, where without loss of generality, $\Phi = \{y_1, \ldots, y_n\}$, $S = y_j$, and $F = \{y_1, \ldots, y_k\}$. Then $G' = (n, A, P', j, k)$, where P' is defined as follows. Let $y_i \to \alpha = w_0 y_{i_1} w_1 \ldots w_{t-1} y_{i_t} w_t \in P$, where

$y_i, y_{i_1}, \ldots, y_{i_t} \in \Phi$ and $w_0, w_1, \ldots, w_t \in A^*$. Then we define the following set of productions

$$U_{y_i \to \alpha} = \{x_i \to w_0 x_{i_1} w_1 \ldots w_{t-1} x_{i_t} w_t\} \cup$$
$$\{z_i \to w_0 z_{i_1}, z_i \to w_0 x_{i_1} w_1 z_{i_2}, \ldots, z_i \to w_0 x_{i_1} w_1 x_{i_2} \ldots w_{t-1} z_{i_t}\},$$

and, moreover,

$$P' = \bigcup_{y_i \to \alpha \in P} U_{y_i \to \alpha}.$$

It is clear that, for a finite leftmost derivation $y_i \Rightarrow_L^* w$, $w \in A^*$ in G, there exists a finite leftmost derivation $x_i \Rightarrow_L^* w$ in G' using only the x-productions. Moreover, for each infinite leftmost derivation in G

$$y_i \Rightarrow_L \beta_1 y_{i_1} \gamma_1 \Rightarrow_L^* w_1 y_{i_1} \gamma_1 \Rightarrow_L w_1 \beta_2 y_{i_2} \gamma_2 \gamma_1 \Rightarrow_L^*$$
$$w_1 w_2 y_{i_2} \gamma_2 \gamma_1 \Rightarrow_L w_1 w_2 \beta_3 y_{i_3} \gamma_3 \gamma_2 \gamma_1 \Rightarrow_L^* \cdots$$

where y_i is the 0-th, and y_{i_j} is the j-th significant variable, there exists the following infinite leftmost derivation in G':

$$z_i \Rightarrow_L \bar{\beta}_1 z_{i_1} \Rightarrow_L^* w_1 z_{i_1} \Rightarrow_L w_1 \bar{\beta}_2 z_{i_2} \Rightarrow_L^* w_1 w_2 z_{i_2} \Rightarrow_L w_1 w_2 \bar{\beta}_3 z_{i_3} \Rightarrow_L^* \cdots,$$

where, if in β_i the y's are replaced by x's, we get $\bar{\beta}_i$. Here $z_i \to \bar{\beta}_1 z_{i_1} \in U_{y_i \to \beta_1 y_{i_1} \gamma_1}$ and $z_{i_j} \to \bar{\beta}_{j+1} z_{i_{j+1}} \in U_{y_{i_j} \to \beta_{j+1} y_{i_{j+1}} \gamma_{j+1}}$. Both infinite leftmost derivations generate $w_1 w_2 w_3 \cdots \in A^\omega$.

Vice versa, to each infinite leftmost derivation $z_i \Rightarrow_L^\omega w$ in G' there exists, in the same manner, an infinite leftmost derivation in G $y_i \Rightarrow_L^\omega w$, $w \in A^\omega$. Moreover, if P' is the disjoint union of the $U_{y_i \to \alpha}$ for all $y_i \to \alpha \in P$, then the correspondence between infinite leftmost derivations in G and in G' is one-to-one.

For an infinite leftmost derivation π in an ω-context-free grammar G, define $\text{INSV}(\pi) = \{y_i \in \Phi \mid y_i$ appears infinitely often as a significant variable in $\pi\}$. Clearly, if for all infinite leftmost derivations π of the ω-context-free grammar $G = (\Phi, A, P, S, F)$, $\text{INV}(\pi) \cap F \neq \emptyset$ iff $\text{INSV}(\pi) \cap F \neq \emptyset$, then $L(G') = L(G)$, where G' is the mixed ω-context-free grammar induced by G.

Theorem 3.3 *Let $G_{j,k} = (\{y_1, \ldots, y_n\}, A, P, y_j, \{y_1, \ldots, y_k\})$, $1 \leq j \leq n$, $0 \leq k \leq n$, be an ω-context-free grammar and $y_i = p_i(y_1, \ldots, y_n)$, $1 \leq i \leq n$, be the ω-algebraic system over $\mathbb{B}\langle\langle A^* \rangle\rangle \times \mathbb{B}\langle\langle A^\omega \rangle\rangle$ corresponding to it. Assume that, for each infinite leftmost derivation π, $\text{INV}(\pi) \cap \{y_1, \ldots, y_k\} \neq \emptyset$ iff $\text{INSV}(\pi) \cap \{y_1, \ldots, y_k\} \neq \emptyset$. Let (σ, τ) be the solution of order k, $0 \leq k \leq n$, of the ω-algebraic system over $(\mathbb{B}\langle\langle A^* \rangle\rangle, \mathbb{B}\langle\langle A^\omega \rangle\rangle)$ induced by $y_i = p_i$, $1 \leq i \leq n$. Then $L(G_{j,k}) = \sigma_j + \tau_j$, $1 \leq j \leq n$, $0 \leq i \leq k$.*

Theorem 3.4 *Let $G_{j,k} = (\{y_1, \ldots, y_n\}, A, P, y_j, \{y_1, \ldots, y_k\})$, $1 \leq j \leq n$, $0 \leq k \leq n$, be an ω-context-free grammar and $y_i = p_i(y_1, \ldots, y_n)$, $1 \leq i \leq n$, be the ω-algebraic system over $\mathbb{N}^\infty\langle\langle A^* \rangle\rangle \times \mathbb{N}^\infty\langle\langle A^\omega \rangle\rangle$ corresponding to it. Assume that, for each infinite leftmost derivation π, $\text{INV}(\pi) \cap \{y_1, \ldots, y_k\} \neq \emptyset$ iff $\text{INSV}(\pi) \cap \{y_1, \ldots, y_k\} \neq \emptyset$. Denote by $d_j(w)$, $w \in A^*$ (resp. $w \in A^\omega$) the* number (possibly ∞) of distinct finite leftmost derivations (resp. infinite leftmost derivations π with $INSV(\pi) \cap \{y_1, \ldots, y_k\} \neq \emptyset$) from the variable y_j, $1 \leq j \leq n$. Then

$$\sigma_j = \sum_{w \in A^*} d_j(w)w \quad and \quad \tau_j = \sum_{w \in A^\omega} d_j(w)w, \qquad 1 \leq j \leq n.$$

Observe, that if $k = n$ or $n = 1$, then the assumption $INV(\pi) \cap \{y_1, \ldots, y_k\} \neq \emptyset$ iff $INSV(\pi) \cap \{y_1, \ldots, y_k\} \neq \emptyset$ for all π is satisfied.

Example 3.1 (see also Cohen, Gold [3], Example 3.1.6). Consider the ω-algebraic system over $\mathbb{B}\langle\langle A^* \rangle\rangle \times \mathbb{B}\langle\langle A^\omega \rangle\rangle$ where $A = \{a, b\}$: $y_1 = ay_1b + ab$, $y_2 = y_1y_2$. It induces the mixed ω-algebraic system over $(\mathbb{B}\langle\langle A^* \rangle\rangle, \mathbb{B}\langle\langle A^\omega \rangle\rangle)$ $x_1 = ax_1b + ab$, $x_2 = x_1x_2$, $z_1 = az_1$, $z_2 = z_1 + x_1z_2$. The least solution of $x_1 = ax_1b + ab$, $x_2 = x_1x_2$ is given by $\sigma = \left(\sum_{n \geq 1} a^nb^n, 0 \right)^T$. The z-equations can be written in the form $z = Mz$, where $M = \begin{pmatrix} a & 0 \\ \varepsilon & x_1 \end{pmatrix}$. We obtain $M^{\omega_1} = \begin{pmatrix} a^\omega \\ x_1^*a^\omega \end{pmatrix}$ and

$M^{\omega_2} = \begin{pmatrix} a^\omega \\ x_1^\omega + x_1^*a^\omega \end{pmatrix}$.

The ω-context-free grammar G corresponding to the ω-algebraic system has productions $y_1 \rightarrow ay_1b$, $y_1 \rightarrow ab$, $y_2 \rightarrow y_1y_2$. The infinite leftmost derivations are

(i) $y_1 \Rightarrow_L ay_1b \Rightarrow_L aay_1bb \Rightarrow_L \cdots \Rightarrow_L a^ny_1b^n \Rightarrow_L \ldots$, i.e., $y_1 \Rightarrow_L^\omega a^\omega$, with repeated variable y_1;

(ii) $y_2 \Rightarrow_L y_1y_2 \Rightarrow_L^* a^{n_1}b^{n_1}y_2 \Rightarrow_L a^{n_1}b^{n_1}y_1y_2 \Rightarrow_L^* a^{n_1}b^{n_1} \ldots a^{n_t}b^{n_t}y_2 \Rightarrow_L \ldots$, i.e., $y_1 \Rightarrow_L^\omega a^{n_1}b^{n_1} \ldots a^{n_t}b^{n_t} \ldots$, with repeated variables y_1, y_2;

(iii) $y_2 \Rightarrow_L^* a^{n_1}b^{n_1} \ldots a^{n_t}b^{n_t}y_2 \Rightarrow_L a^{n_1}b^{n_1} \ldots a^{n_t}b^{n_t}y_1y_2 \Rightarrow_L^\omega a^{n_1}b^{n_1} \ldots a^{n_t}b^{n_t}a^\omega$, i.e., $y_2 \Rightarrow_L^\omega a^{n_1}b^{n_1} \ldots a^{n_t}b^{n_t}a^\omega$, $t \geq 0$, with repeated variable y_1.

If y_1 is the only repeated variable, and y_1 or y_2 is the start variable, then $L(G_{1,1}) = \sum_{n \geq 1} a^nb^n + a^\omega$ or $L(G_{2,1}) = \left(\sum_{n \geq 1} a^nb^n \right)^\omega \cup \left(\sum_{n \geq 1} a^nb^n \right)^* a^\omega$, respectively. If the repeated variables are y_1 and y_2, and y_1 or y_2 is the start variable then we obtain again $L(G_{1,2}) = \sum_{n \geq 1} a^nb^n + a^\omega$ or $L(G_{2,2}) = \left(\sum_{n \geq 1} a^nb^n \right)^\omega \cup \left(\sum_{n \geq 1} a^nb^n \right)^* a^\omega$, respectively. Compare this with the solutions of order 1 or 2 of the ω-algebraic system $y_1 = ay_1b + ab$, $y_2 = y_1y_2$: $\left(\sum_{n \geq 1} a^nb^n, 0 \right)^T + \left(a^\omega, \left(\sum_{n \geq 1} a^nb^n \right)^\omega a^\omega \right)^T$ or $\left(\sum_{n \geq 1} a^nb^n, 0 \right)^T + \left(a^\omega, \left(\sum_{n \geq 1} a^nb^n \right)^\omega + \left(\sum_{n \geq 1} a^nb^n \right)^* a^\omega \right)^T$, respectively. If y_1 is the only repeated variable and y_2 is the start variable then $\left(\sum_{n \geq 1} a^nb^n \right)^\omega$ is missing. That is due to the fact that in the derivations (ii) each y_1 derives a finite word $a^{n_j}b^{n_j}$ by a finite leftmost subderivation $y_1 \Rightarrow_L^* a^{n_j}b^{n_j}$ and never is a significant variable.

If all variables are repeated variables that does not matter: each infinite leftmost derivation contributes to the generated language. Hence, if the repeated variables are y_1, y_2 and the start variable is y_1 or y_2, the infinite parts of the solutions of order 1 or 2 correspond to the generated languages by Theorem 3.3.

□

In the next example there is only one variable. Hence, we can apply Theorems 3.3 and 3.4.

Example 3.2. Consider the ω-algebraic system $y_1 = ay_1y_1 + b$ over $\mathbb{N}^\infty\langle\!\langle A^*\rangle\!\rangle \times \mathbb{N}^\infty\langle\!\langle A^\omega\rangle\!\rangle$, where $A = \{a, b\}$. The least solution of the algebraic system $x_1 = ax_1x_1 + b$ over $\mathbb{N}^\infty\langle\!\langle A^*\rangle\!\rangle$ is given by $\sigma = D^*b$, where D is the characteristic series of the restricted Dyck language (see Berstel [1]). The mixed ω-algebraic system over $(\mathbb{N}^\infty\langle\!\langle A^*\rangle\!\rangle, \mathbb{N}^\infty\langle\!\langle A^\omega\rangle\!\rangle)$ $x_1 = ax_1x_1 + b$, $z_1 = az_1 + ax_1z_1$ has the solution of order 1 $(D^*b, (a + ax_1)^\omega(D^*b)) = (D^*b, (a + aD^*b)^\omega) = (D^*b, (a + D)^\omega)$, since $aD^*b = D$.

The ω-context-free grammar corresponding to $y_1 = ay_1y_1 + b$ has productions $y_1 \to ay_1y_1$, $y_1 \to b$ and generates the language $D^*b + (a + D)^\omega = D^*b + (a^*D)^\omega + (a^*D)^*a^\omega$.

Since each word in $(a^*D)^*$ and in $(a^*D)^\omega$ has a unique factorization into words of a^*D, all coefficients of $D^*b + (a + D)^\omega$ are 0 or 1, i. e., the ω-context-free grammar with productions $y_1 \to ay_1y_1$, $y_1 \to b$ is an "unambiguous" ω-context-free grammar.

□

Let (S, V) be a continuous starsemiring-omegasemimodule pair and inspect the solutions of order k: If (σ, ω) is a solution of order k of an S'-algebraic system over $S \times V$ then $\sigma \in \mathfrak{Alg}(S')$ and ω is the k-th automata theoretic solution of a finite $\mathfrak{Alg}(S')$-linear system. Hence, by Theorems 3.9, 3.10, 3.2 of Ésik, Kuich [6] and by Theorem 4.4 of Ésik, Kuich [7], ω is of the form $\omega = \sum_{1 \le j \le m} s_j t_j^\omega$ with $s_j, t_j \in \mathfrak{Rat}(\mathfrak{Alg}(S')) = \mathfrak{Alg}(S')$. Hence, again by Theorem 3.9 of Ésik, Kuich [6] and by Theorem 3.10 of Ésik, Kuich [7] we obtain the following result.

Theorem 3.5 *Let (S, V) be a continuous starsemiring-omegasemimodule pair. Then the following statements are equivalent for $(s, v) \in S \times V$:*

(i) $(s, v) = ||\mathfrak{A}||$, where $\mathfrak{A}$ is a finite $\mathfrak{Alg}(S')$-automaton,
(ii) $(s, v) = ||\mathfrak{A}||_1$, where $\mathfrak{A}$ is a finite $\mathfrak{Alg}(S')$-automaton,
(iii) $(s, v) \in \omega$-$\mathfrak{Alg}(S')$,
(iv) $s \in \mathfrak{Alg}(S')$ and $v = \sum_{1 \le k \le m} s_k t_k^\omega$, where $s_k, t_k \in \mathfrak{Alg}(S')$.

Theorem 3.6 *Let (S, V) be a continuous starsemiring-omegasemimodule pair. Then ω-$\mathfrak{Alg}(S')$ is an ω-rationally closed quemiring.*

Proof. Since, by assumption, $0, 1 \in S'$ we infer that $0, 1 \in \omega$-$\mathfrak{Alg}(S')$. Assume now that (σ_1, ω_1) and (σ_2, ω_2) are in ω-$\mathfrak{Alg}(S')$. Then, by Theorem 3.5, $\sigma_1, \sigma_2 \in \mathfrak{Alg}(S')$ and $\omega_1 = \sum_{1 \le k \le m_1} s_k^1 t_k^{1^\omega}$, $\omega_2 = \sum_{1 \le k \le m_2} s_k^2 t_k^{2^\omega}$ for some $s_k^1, s_k^2, t_k^1, t_k^2 \in \mathfrak{Alg}(S')$. We obtain

$$(\sigma_1, \omega_1) + (\sigma_2, \omega_2) = (\sigma_1 + \sigma_2, \sum_{1 \le k \le m_1} s_k^1 t_k^{1^\omega} + \sum_{1 \le k \le m_2} s_k^2 t_k^{2^\omega})$$

and

$$(\sigma_1, \omega_1) \cdot (\sigma_2, \omega_2) = (\sigma_1 \sigma_2, \sum_{1 \le k \le m_1} s_k^1 {t_k^1}^\omega + \sigma_1 \cdot \sum_{1 \le k \le m_2} s_k^2 {t_k^2}^\omega).$$

Hence, $(\sigma_1, \omega_1) + (\sigma_2, \omega_2)$ and $(\sigma_1, \omega_1) \cdot (\sigma_2, \omega_2)$ are again in $\omega\text{-}\mathfrak{Alg}(S')$.
Moreover, we obtain

$$(\sigma_1, \omega_1)\P = (\sigma_1, 0)$$

and

$$(\sigma_1, \omega_1)^\otimes = (\sigma_1^*, \sigma_1^\omega + \sigma_1^* \cdot \sum_{1 \le k \le m_1} s_k^1 {t_k^1}^\omega).$$

Hence, $(\sigma_1, \omega_1)\P$ and $(\sigma_1, \omega_1)^\otimes$ are again in $\omega\text{-}\mathfrak{Alg}(S')$ and $\omega\text{-}\mathfrak{Alg}(S')$ is rationally closed. □

Notation 3.1.5, Definition 2.2.1 and Theorem 4.1.8(a) of Cohen, Gold [3] and Theorem 3.5(iv) yield the next result.

Theorem 3.7 $CFL_\omega = \{L0 \subseteq A^\omega \mid L0 \in \mathbb{B}^{\omega\text{-}alg}\langle\!\langle A^*, A^\omega \rangle\!\rangle, \ A \text{ an alphabet}\}.$

Let $t \in \mathbb{B}^{alg}\langle\!\langle A^* \rangle\!\rangle$. Then t is the x_2-component of the least solution of an algebraic system $x_i = p_i(x_2, \ldots, x_n)$, $2 \le i \le n$, over $\mathbb{B}^{alg}\langle\!\langle A^* \rangle\!\rangle$. Consider the ω-algebraic system over $\mathbb{B}\langle\!\langle A^* \rangle\!\rangle \times \mathbb{B}\langle\!\langle A^\omega \rangle\!\rangle$:

$$y_1 = y_2 y_1, \qquad y_i = p_i(y_2, \ldots, y_n), \ 2 \le i \le n,$$

and consider the induced mixed ω-algebraic system over $(\mathbb{B}\langle\!\langle A^* \rangle\!\rangle, \mathbb{B}\langle\!\langle A^\omega \rangle\!\rangle)$:

$$z_1 = z_2 + x_2 z_1, \ z_i = (p_i)_x(x_1, \ldots, x_n, z_1, \ldots, z_n), \ 2 \le i \le n,$$
$$x_1 = x_2 x_1, \qquad x_i = p_i(x_2, \ldots, x_n), \ 2 \le i \le n.$$

The first component of the least solution of $x_1 = x_2 x_1$, $x_i = p_i(x_2, \ldots, x_n)$, $2 \le i \le n$, is 0. We now compute the solution of order 1 of $z_1 = z_2 + x_2 z_1$, $z_i = (p_i)_x(x_1, \ldots, x_n, z_1, \ldots, z_n)$, $2 \le i \le n$. We write the system in the form $z = Mz$ and obtain

$$M = \begin{pmatrix} x_2 & \varepsilon & 0 & \ldots & 0 \\ \hline 0 & & & & \\ \vdots & & & M' & \\ 0 & & & & \end{pmatrix}.$$

Hence, the first component of M^{ω_1} is x_2^ω and the first component of the solution of order 1 is given by $(0, t^\omega)$.

Consider now the ω-context-free grammar G corresponding to $y_1 = y_2 y_1$, $y_i = p_i$, $2 \le i \le n$, with the set of repeated variables $\{y_1\}$ and start variable y_1. The only infinite leftmost derivations π, where y_1 appears infinitely often, are of the form

$$\pi : y_1 \Rightarrow_L y_2 y_1 \Rightarrow_L^* w_1 y_1 \Rightarrow_L w_1 y_2 y_1 \Rightarrow_L^* w_1 w_2 y_1 \Rightarrow_L \ldots.$$

The only significant variable of such a derivation π is y_1, i.e., $\mathrm{INSV}(\pi) = \{y_1\}$, and $\mathrm{INSV}(\pi) \cap \{y_1\} \ne \emptyset$ iff $\mathrm{INV}(\pi) \cap \{y_1\} \ne \emptyset$. Hence, $L(G_{1,1}) = t^\omega$ by Theorem 3.3.

The usual constructions yield then, for $s + v$, where $v = \sum_{1 \leq k \leq n} s_k t_k^\omega$, $s, s_k, t_k \in \mathbb{B}^{\mathrm{alg}} \langle\!\langle A^* \rangle\!\rangle$, an ω-context-free grammar G' such that $L(G') = s + v$.

Hence, we have given a construction proving again Theorem 3.7. But additionally, G' has the nice property that for each infinite leftmost derivation π, we obtain $\mathrm{INSV}(\pi) \cap F \neq \emptyset$ iff $\mathrm{INV}(\pi) \cap F \neq \emptyset$, where F is the set of repeated variables of G'.

References

1. Berstel, J.: Transductions and Context-Free Languages. Teubner, 1979.
2. Bloom, S. L., Ésik, Z.: Iteration Theories. EATCS Monographs on Theoretical Computer Science. Springer, 1993.
3. Cohen, R. S., Gold, A. Y.: Theory of ω-languages I: Characterizations of ω-context-free languages. JCSS 15(1977) 169–184.
4. Elgot, C.: Matricial theories. J. Algebra 42(1976) 391–422.
5. Ésik, Z., Kuich, W.: On iteration semiring-semimodule pairs. To appear.
6. Ésik, Z., Kuich, W.: A semiring-semimodule generalization of ω-regular languages I. Technical Report, Technische Universität Wien, 2003.
7. Ésik, Z., Kuich, W.: A semiring-semimodule generalization of ω-regular languages II. Technical Report, Technische Universität Wien, 2003.
8. Ginsburg, S., Rice, H. G.: Two families of languages related to ALGOL. J. Assoc. Comput. Mach. 9(1962) 350–371.
9. Kuich, W.: Semirings and formal power series: Their relevance to formal languages and automata theory. In: Handbook of Formal Languages (Eds.: G. Rozenberg and A. Salomaa), Springer, 1997, Vol. 1, Chapter 9, 609–677.
10. Lausch, H., Nöbauer, W.: Algebra of Polynomials. North-Holland, 1973.
11. Salomaa, A., Soittola, M.: Automata-Theoretic Aspects of Formal Power Series. Springer, 1978.

Integer Weighted Finite Automata, Matrices, and Formal Power Series over Laurent Polynomials

Vesa Halava

Department of Mathematics and TUCS - Turku Centre for Computer Science,
University of Turku, FIN-20014, Turku, Finland
vehalava@utu.fi

Abstract. It is well known that the family of regular languages (over alphabet A), accepted by finite automata, coincides with the set of supports of the rational and recognizable formal power series over $\mathbb{N}$ with the set of variables A. Here we prove that there is a corresponding presentation for languages accepted by integer weighted finite automata, where the weights are from the additive group of integers, via the matrices over Laurent polynomials with integer coefficients.

1 Introduction

It is well known that the family of languages accepted by a finite automata (over alphabet A), can be defined also with the set of recognizable formal power series over $\mathbb{N}$, which on the other hand is equal with the set of rational formal power series over $\mathbb{N}$, where A is considered as a noncommutative set of variables. This connection is proved by using the matrix representation of the finite automata.

Here we give a similar representation for the family of languages accepted with the integer weighted finite automata, see [4, 5]. In these automata the weights are from the additive group of integers and a word is accepted, if it has a successful path in the underlying automaton and the weight of the path adds up to zero. We show that there is a connection between these languages and the recognizable and rational formal power series with coefficients from the ring of the Laurent polynomials with integer coefficients. The proof uses the representation of the integer weighted finite automata with matrices over the Laurent polynomials. The difference between these two constructions is in the definition of the language defined with the series.

Next we give the basic definitions on words and languages. Let A be a finite set of symbols, called an *alphabet*. A *word* over A is a finite sequence of symbols in A. We denote by A^* the set of all words over A. Note that also the *empty word*, denoted by ε, is in A^*.

Let $u = u_1 \ldots u_n$ and $v = v_1 \ldots v_m$ be two words in A^*, where each u_i and v_j are in A for $1 \leq i \leq n$ and $1 \leq j \leq m$. The *concatenation* of u and v is the word $u \cdot v = uv = u_1 \ldots u_n v_1 \ldots v_m$. The operation of concatenation is associative on A^*, and thus A^* is a semigroup (containing an identity element

J. Karhumäki et al. (Eds.): Theory Is Forever (Salomaa Festschrift), LNCS 3113, pp. 81–88, 2004.

ε). Let $A^+ = A^* \setminus \{\varepsilon\}$ be the semigroup of all nonempty words over A. A subset L of A^* is called a *language*.

2 Formal Power Series

Here we give the needed definitions and notations on formal power series. As a general reference and for the details, we give [2, 9, 11] .

Let K be a semiring and A an alphabet. A *formal power series* S is a function

$$A^* \to K.$$

Note that here A is considered as a (noncommutative) set of variables. The image of a word w under S is denoted by (S, w) and it is called the *coefficient* of w in S. The *support* of S is the language

$$\mathrm{supp}(S) = \{w \in A^* \mid (S, w) \neq 0\}.$$

The set of formal series over A with coefficients in K is denoted by $K\langle\langle A \rangle\rangle$. A formal series with a finite support is called a *polynomial*. The set of polynomials is denoted by $K\langle A \rangle$.

Let S and T be two formal series in $K\langle\langle A \rangle\rangle$. Then their *sum* is given by

$$(S + T, w) = (S, w) + (T, w)$$

and their product by

$$(ST, w) = \sum_{uv = w} (S, u)(T, v).$$

We also define two external operations of K in $K\langle\langle A \rangle\rangle$. Assume that a is in K and S in $K\langle\langle A \rangle\rangle$, then the series aS and Sa are defined by

$$(aS, w) = a(S, w) \quad \text{and} \quad (Sa, w) = (S, w)a.$$

A formal series S can also be written in the sum form $S = \sum a_w w$ over all $w \in A^*$ such that a_w is the coefficient of w in K, i.e. $(S, w) = a_w$.

A formal series S in $K\langle\langle A \rangle\rangle$ is called *proper* if the coefficient of the empty word vanishes, that is $(S, \epsilon) = 0$. Let S be proper formal series. Then the family $(S^n)_{n \geq 0}$ is *locally finite* (see [2]), and we can define the sum of this family, denoted by

$$S^* = \sum_{n \geq 0} S^n$$

and it is called the *star* of S. Note that $S^0 = 1$, $S^1 = S$ and $S^n = SS^{n-1}$, where 1 is the identity of K under product.

The *rational operations* in $K\langle\langle A \rangle\rangle$ are the sum, the product and the star. A formal series is called *rational* if it is an element of the *rational closure* of $K\langle A \rangle$, i.e. it can be defined using the polynomials $K\langle A \rangle$ and the rational operations. The family of rational series is denoted by $K^{\mathrm{rat}}\langle\langle A \rangle\rangle$.

As usual, we denote by $K^{m \times n}$ the set of the $m \times n$ matrices over K.

A formal series $S \in K\langle\langle A \rangle\rangle$ is called *recognizable* if there exists an integer $n \geq 1$, and a monoid morphism $\mu : A^* \to K^{n \times n}$, into the multiplicative structure of $K^{n \times n}$, and two vectors $\imath, \rho \in K^n$ such that for all words w,

$$(S, w) = \imath \mu(w) \rho^T.$$

The triple $(\imath, \mu, \rho)$ is called a *linear representation* of S with *dimension* n. The set of recognizable series over K is denoted by $K^{\mathrm{rec}}\langle\langle A \rangle\rangle$

The next theorem is fundamental in the theory of rational series. It was first proved by Kleene in 1956 for languages that are those series with coefficients in the Boolean semiring. It was later extented by Schützenberger to arbitrary semirings. For details, see [2, 9, 11].

Theorem 1. *A formal series is recognizable if and only if it is rational.*

3   Finite Automaton

A (nondeterministic) *finite automaton* is a quintuple $\mathcal{A} = (Q, A, \delta, q_A, F)$, where Q is a *finite* set of *states*, A is a finite *input alphabet*, $\delta \colon Q \times A \to 2^Q$ is a *transition function*, $q_A \in Q$ is an *initial state* and F is the set of *final states*. A transition $p \in \delta(q, a)$, where $p, q \in Q$ and $a \in A$, will also be written as (q, a, p), in which case $\delta \subseteq Q \times A \times Q$ is regarded as a relation (and sometimes also as an alphabet). Without loss of generality, we can assume that

$$Q = \{1, 2, \ldots, n\} \text{ for some } n \geq 1, \text{ and } q_A = 1.$$

Indeed, renaming of the states will not change the accepted language.

A *path* π of $\mathcal{A}$ (from q_1 to q_{n+1}) is a sequence

$$\pi = t_1 t_2 \ldots t_k \text{ where } t_i = (q_i, a_i, q_{i+1}) \in \delta \tag{1}$$

for $i = 1, 2, \ldots, k$. If we consider δ as an alphabet, then we can write $\pi \in \delta^*$. The *label* of the path π in (1) is the word $\|\pi\| = a_1 a_2 \ldots a_k$. Let

$$\mathcal{A}(w : p \to q) = \{\pi \mid \pi \text{ a path from } p \text{ to } q \text{ with } \|\pi\| = w\}.$$

Moreover, a path $\pi \in \mathcal{A}(w : p \to q)$ is *successful* (for w), if $p = 1$ and $q \in F$. The language *accepted* by $\mathcal{A}$ is the subset $L(\mathcal{A}) \subseteq A^*$ consisting of the labels of the successful paths of $\mathcal{A}$:

$$L(\mathcal{A}) = \{w \in A^* \mid \pi \in \mathcal{A}(w : 1 \to q) \text{ for some } q \in F\}.$$

It is well-known that each finite automata has a matrix representation obtained as in the following. Let $\mathcal{A} = (Q, A, \delta, 1, F)$ be a finite automaton with n states, i.e., $Q = \{1, 2, \ldots, n\}$. Define for all $a \in A$, the matrix $M_a \in \mathbb{N}^{n \times n}$ by

$$(M_a)_{ij} = \begin{cases} 1, & \text{if } j \in \delta(i, a), \\ 0, & \text{otherwise.} \end{cases} \tag{2}$$

We define a monoid morphism $\mu : A^* \to \mathbb{N}^{n \times n}$ by setting $\mu(a) = M_a$, where the operation in $\mathbb{N}^{n \times n}$ is the usual matrix multiplication.

Let $\iota = (1, 0, \ldots, 0)$, where only the first term is nonzero, and let $\rho = (\rho_1, \rho_2, \ldots, \rho_n)$ in $\mathbb{N}^n$ where

$$\rho_i = \begin{cases} 1, & \text{if } q_i \in F, \\ 0, & \text{otherwise.} \end{cases} \tag{3}$$

The triple (ι, μ, ρ) is then called the *linear representation* of A.

For the proof for the following theorem, see [2, 9, 11].

Theorem 2. *A language L is accepted by a finite automaton if and only if there exists a linear representation (ι, μ, ρ) such that*

$$w \in L \iff \iota\mu(w)\rho^T \neq 0.$$

Note that we could have defined the matrices over the boolean semiring $\mathbb{B}$ instead of the semiring $\mathbb{N}$, and then replacing $\iota\mu(w)\rho^T \neq 0$ by $\iota\mu(w)\rho^T = 1$. But using the ring $\mathbb{N}$, we achieve the following advantage.

Theorem 3. *For a finite automaton A having a linear representation (ι, μ, ρ), the value $\iota\mu(w)\rho^T$ equals the number of different successful paths in A for w.*

By Theorem 2 and the fundamental theorem, Theorem 1, we get the following corollary.

Corollary 1. *$L \subseteq A^*$ is a regular language if and only if there exists a formal series $S_L \in \mathbb{N}^{rat}\langle\langle A \rangle\rangle = \mathbb{N}^{rec}\langle\langle A \rangle\rangle$ such that $L = \mathrm{supp}(S_L)$.*

Note that it follows that the regular languages are closed under the rational operations, since $\mathbb{N}^{rat}\langle\langle A \rangle\rangle$ is.

4   Laurent Polynomials and Weighted Automata

In this section we give a corresponding representation for the languages accepted by the integer weighted finite automata. We begin with some definitions.

A *Laurent polynomial* $p \in \mathbb{Z}[x, x^{-1}]$ with coefficients in $\mathbb{Z}$ is a series

$$p(x) = \ldots a_{-2}x^{-2} + a_{-1}x^{-1} + a_0 + a_1 x + a_2 x^2 + \ldots,$$

where there are only finitely many nonzero coefficients $a_i \in \mathbb{Z}$. The *constant term* of the Laurent polynomial $p \in \mathbb{Z}[x, x^{-1}]$ is a_0. The family of Laurent polynomials with coefficients in $\mathbb{Z}$ forms a ring with respect to the operations of sum and multiplication, that are defined in the usual way. Indeed, the sum is defined componentwise and the multiplication is the Cauchy product of the polynomials:

$$\left(\sum_{i=-\infty}^{\infty} a_i x^i \right) \left(\sum_{i=-\infty}^{\infty} b_i x^i \right) = \sum_{i=-\infty}^{\infty} \left(\sum_{j+k=i} a_j b_k \right) x^i.$$

Note that in the definition of Laurent polynomials we could have used also arbitrary ring instead of $\mathbb{Z}$, but here we need only the integer case. Actually, we concentrate on matrices over Laurent polynomials with integer coefficients, that is, the elements of $\mathbb{Z}[x, x^{-1}]^{n \times n}$ for $n \geq 1$. A Laurent polynomial matrix

$$M = (c_{ij})_{n \times n} \in \mathbb{Z}[x, x^{-1}]^{n \times n}$$

is a $n \times n$-square matrix the entries of which are Laurent polynomials from $\mathbb{Z}[x, x^{-1}]$. For these matrices, multiplication is defined in the usual way using the multiplication of the ring $\mathbb{Z}[x, x^{-1}]$. Indeed, if $M_1 = (c_{ij})_{n \times n}$ and $M_2 = (d_{ij})_{n \times n}$, then

$$M_1 \cdot M_2 = (e_{ij})_{n \times n},$$

where

$$e_{ij} = \sum_{k=1}^{n} c_{ik} d_{kj} \in \mathbb{Z}[x, x^{-1}].$$

Also the sum for these matrices can be defined, but we are interested in the semigroups generated by a finite number of Laurent polynomials under multiplication.

Next we consider a generalization of finite automata where the transitions have integer weights. The type of automata we consider is closely related to the 1-turn counter automata as considered by Baker and Book [1], Greibach [3], and especially by Ibarra [8]. Also, regular valence grammars are related to these automata, see [7]. Moreover, the extended finite automata of Mitrana and Stiebe [10] are generalizations of these automata.

Consider the additive group of $\mathbb{Z}$ of integers. A $(\mathbb{Z}\text{-})weighted finite automaton$ $\mathcal{A}^\gamma$ consists of a finite automaton $\mathcal{A} = (Q, A, \delta, 1, F)$ as above, except that here δ may be a *finite multiset* of transitions in $Q \times A \times Q$, and a *weight function* $\gamma \colon \delta \to \mathbb{Z}$. We let δ be a multiset in order to be able to define (finitely) many different weights for each transition of $\mathcal{A}$. For example, it is possible that for $t_1, t_2 \in \delta$, $t_1 = (i, a, j) = t_2$ and $\gamma(t_1) \neq \gamma(t_2)$.

Let $\pi = t_1 t_2 \ldots t_k$ be a path of $\mathcal{A}$, where $t_i = (q_i, a_i, q_{i+1})$ for $i = 1, 2, \ldots, k$. The *weight* of π is the element

$$\gamma(\pi) = \gamma(t_1) + \gamma(t_2) + \cdots + \gamma(t_k).$$

Furthermore, we let

$$L(\mathcal{A}^\gamma) = \{w \in A^* \mid \gamma(\pi) = 0, \ \pi \in \mathcal{A}(w : 1 \to q) \text{ for some } q \in F\},$$

be the *language of* $\mathcal{A}^\gamma$. In other words, a word is accepted by $\mathcal{A}^\gamma$ if and only if there is a successful path of weight 0 in $\mathcal{A}^\gamma$.

Next we shall introduce a matrix representation of integer weighted finite automata with the matrices over the Laurent polynomials $\mathbb{Z}[x, x^{-1}]$.

Let $\mathcal{A}^\gamma$ be a weighted finite automaton, where $\mathcal{A} = (Q, A, \delta, 1, F)$ and $\gamma \colon \delta \to \mathbb{Z}$. Let again $Q = \{1, 2, \ldots, n\}$. Define for each element $a \in A$ and a pair of states $i, j \in Q$ the Laurent polynomial

$$p_{ij}^a = \sum_{t=(i,a,j)\in\delta} x^{\gamma(t)} \, .$$

Moreover, define the Laurent polynomial matrix $M_a \in \mathbb{Z}[x, x^{-1}]^{n\times n}$ for all $a \in A$ by

$$(M_a)_{ij} = p_{ij}^a. \tag{4}$$

Let $\mu\colon A^* \to \mathbb{Z}[x, x^{-1}]^{n\times n}$ be the morphism defined by $\mu(a) = M_a$. Let $\imath$ and ρ be the vectors as in (3). The triple $(\imath, \mu, \rho)$ is called a *Laurent representation* of A^γ. For completeness sake, we give here the proof of the following result of [6].

Lemma 1. *Let $(\imath, \mu, \rho)$ be a Laurent representation of A^γ, and let $w \in A^*$. Then the coefficient of x^z in $\mu(w)_{ij}$ is equal to the number of paths $\pi \in A(w : i \to j)$ of weight z.*

Proof. We write $M_u = \mu(w)$ for each word w. We prove the claim by induction on the length of the words. The claim is trivial, if $w \in A$. Assume then that the claim holds for the words $u, v \in A^+$, and let $(M_u)_{ij} = p_{ij}^u = \sum_z \alpha_{ij}^z x^z$, where α_{rs}^z is the number of paths from $A(u : i \to j)$ of weight z. Similarly, let $(M_v)_{ij} = p_{ij}^v = \sum_z \beta_{ij}^z x^z$, where β_{rs}^z is the number of paths from $A(v : i \to j)$ of weight z. Now,

$$(M_u M_v)_{ij} = \sum_{k=1}^n p_{ik}^u p_{kj}^v = \sum_{k=1}^n \left(\sum_{z_1} \alpha_{ik}^{z_1} x^{z_1} \sum_{z_2} \beta_{kj}^{z_2} x^{z_2} \right)$$

$$= \sum_{k=1}^n \sum_{z_1, z_2} \alpha_{ik}^{z_1} \beta_{kj}^{z_2} x^{z_1+z_2} = \sum_{z_1, z_2} \sum_{k=1}^n \alpha_{ik}^{z_1} \beta_{kj}^{z_2} x^{z_1+z_2} \, .$$

In other words, the coefficient of x^z is equal to $\sum_{z_1+z_2=z} \sum_{k=1}^n \alpha_{ik}^{z_1} \beta_{kj}^{z_2}$, wherefrom the claim easily follows.

The following result is an immediate corollary to Lemma 1.

Theorem 4. *Let $(\imath, \mu, \rho)$ be a Laurent representation of A^γ, and let $w \in A^*$. Then the constant term c of $\imath\mu(w)\rho^T$ equals the number of different successful paths of w in A^γ. In particular, $w \in L(A^\gamma)$ if and only if $c > 0$.*

Since $\mathbb{Z}[x, x^{-1}]$ is a ring we can also study the formal power series $\mathbb{Z}[x, x^{-1}]\langle\langle A \rangle\rangle$. Note that the zero element of $\mathbb{Z}[x, x^{-1}]$ is the zero polynomial, where all the coefficients are 0. By Theorem 1, we get the following corollary.

Corollary 2. *A language $L \subseteq A^*$ accepted with A^γ if and only if there exists a formal power series $S_L \in \mathbb{Z}[x, x^{-1}]^{rat}\langle\langle A \rangle\rangle = \mathbb{Z}[x, x^{-1}]^{rec}\langle\langle A \rangle\rangle$ such that*

$$w \in L \iff (S_L, w) = \sum_{i=m}^n a_i x^i \quad and \quad a_0 \neq 0.$$

Note that the this corollary does not give any closure properties on the family of languages accepted with integer weighted finite automata. The closure properties of these languages were studied in [4]. For example, the family is not closed under star.

Note also that the undecidability result in [5] gives undecidability result for matrices over Laurent polynomials, see [6].

Actually, for the power series $S_L \in \mathbb{Z}[x, x^{-1}]\langle\langle A \rangle\rangle$ in Corollary 2, supp$(S_L) = L(\mathcal{A})$, i.e., the support of S_L is the regular language accepted by the underlying automaton of $\mathcal{A}^\gamma$. By reordering the terms according to powers of the variable x, we get

$$S_L = \sum_{z=m}^{n} L_z x^z, \tag{5}$$

where L_z is the sum of words of the language

$$\{w \in A^* \mid \pi \in \mathcal{A}(w : 1 \to q) \text{ for some } q \in F, \ \gamma(\pi) = z\} \subseteq A^*,$$

with multiplicities from $\mathbb{Z}$. We denote these languages simply by L_z. Now $L_0 = L(\mathcal{A}^\gamma) = L$ and $L(\mathcal{A}) = \cup_{z \in \mathbb{Z}} L_z$. Note that this union can be infinite, since the sum (5) can be infinite, even for both directions. It follows also by the rationality of S_L that the languages L_z are in the family of languages accepted with integer weighted finite automata, since $x^{-z} S_L \in \mathbb{Z}[x, x^{-1}]^{\mathrm{rat}}\langle\langle A \rangle\rangle$.

Acknowledgements

I want to thank Dr. Tero Harju for the several comments and suggestion for this work during our coffee breaks.

References

1. B. Baker and R. Book, *Reversal-bounded multipushdown machines*, J. Comput. System Sci. **8** (1974), 315–332.
2. J. Berstel and C. Reutenauer, *Rational series and their languages*, Springer-Verlag, 1988.
3. S. A. Greibach, *An infinite hierarchy of context-free languages*, J. Assoc. Comput. Mach. **16** (1969), 91–106.
4. V. Halava and T. Harju, *Languages accepted by integer weighted finite automata*, Jewels are forever, Springer, Berlin, 1999, pp. 123–134.
5. V. Halava and T. Harju, *Undecidability in integer weighted finite automata*, Fund. Inform. **38** (1999), no. 1-2, 189–200.
6. V. Halava and T. Harju, *Undecidability in matrices over Laurent polynomials*, Tech. Report 600, Turku Centre for Computer Science, March 2004, submitted.
7. V. Halava, T. Harju, H. J. Hoogeboom, and M. Latteux, *Valence languages generated by generalized equality sets*, Tech. Report 502, TUCS, 2002, to appear in JALC.
8. O. H. Ibarra, *Restricted one-counter machines with undecidable universe problems*, Math. Systems Theory **13** (1979), 181–186.

9. W. Kuich and A. Salomaa, *Semirings, automata, languages*, Springer-Verlag, 1986.
10. V. Mitrana and R. Stiebe, *The accepting power of finite automata over groups*, New Trends in Formal Language (G. Păun and A. Salomaa, eds.), Lecture Notes in Comput. Sci., vol. 1218, Springer-Verlag, 1997, pp. 39–48.
11. A. Salomaa and M. Soittola, *Automata–theoretic aspects of formal power series*, Springer–Verlag, 1978.

Two Models for Gene Assembly in Ciliates

Tero Harju[1,3], Ion Petre[2,3], and Grzegorz Rozenberg[4]

[1] Department of Mathematics, University of Turku
Turku 20014 Finland
`harju@utu.fi`
[2] Department of Computer Science, Åbo Akademi University
Turku 20520 Finland
`ipetre@abo.fi`
[3] Turku Centre for Computer Science
Turku 20520 Finland
[4] Leiden Institute for Advanced Computer Science, Leiden University
Niels Bohrweg 1, 2333 CA Leiden, the Netherlands, and
Department of Computer Science, University of Colorado at Boulder
Boulder, Co 80309-0347, USA
`rozenber@liacs.nl`

Abstract. Two models for gene assembly in ciliates have been proposed and investigated in the last few years. The DNA manipulations postulated in the two models are very different: one model is *intramolecular* – a single DNA molecule is involved here, folding on itself according to various patterns, while the other is *intermolecular* – two DNA molecules may be involved here, hybridizing with each other. Consequently, the assembly strategies predicted by the two models are completely different. Interestingly however, the final result of the assembly (including the assembled gene) is always the same. We compare in this paper the two models for gene assembly, formalizing both in terms of pointer reductions. We also discuss invariants and universality results for both models.

1 Introduction

Ciliates are unicellular eukaryotic organisms, see, e.g. [23]. This is an ancient group of organisms, estimated to have originated around two billion years ago. It is also a very diverse group – some 8000 species are currently known and many others are likely to exist. Their diversity can be appreciated by comparing their genomic sequences: some ciliate types differ genetically more than humans differ from fruit flies! Two characteristics unify ciliates as a single group: the possession of hairlike cilia used for motility and food capture, and the presence of two kinds of functionally different nuclei in the same cell, a micronucleus and a macronucleus, see [15], [24], [25]; the latter feature is unique to ciliates. The macronucleus is the "household" nucleus – all RNA transcripts are produced in the macronucleus. The micronucleus is a germline nucleus and has no known function in the growth or in the division of the cell. The micronucleus is activated only in the process of sexual reproduction, where at some stage the micronuclear

J. Karhumäki et al. (Eds.): Theory Is Forever (Salomaa Festschrift), LNCS 3113, pp. 89–101, 2004.

genome gets transformed into the macronuclear genome, while the old macronuclear genome is destroyed. This process is called *gene assembly*, it is the most involved DNA processing known in living organisms, and it is most spectacular in the *Stichotrichs* species of ciliates (which we consider in this paper). What makes this process so complex is the unusual rearrangements that ciliates have engineered in the structure of their micronuclear genome. While genes in the macronucleus are contiguous sequences of DNA placed (mostly) on their own molecules (and some of them are the shortest DNA molecules known in Nature), the genes in the micronucleus are placed on long chromosomes and they are broken into pieces called *MDSs*, separated by noncoding blocks called *IESs*, see [15, 21–26]. Adding to the complexity, the order of the MDSs is permuted and MDSs may be inverted. One of the amazing features of this process is that ciliates appear to use "linked lists" in gene assembly, see [29, 30], similarly as in software engineering!

Two different models have been proposed for gene assembly. The first one, proposed by Landweber and Kari, see [19, 20], is intermolecular: the DNA manipulations here may involve two molecules exchanging parts of their sequences through recombination. The other one, proposed by Ehrenfeucht, Prescott, and Rozenberg, see [11, 27], is intramolecular: here, all manipulations involve one single DNA molecule folding on itself and swapping parts of its sequence through recombination. In the intermolecular model one traditionally attempts to capture both the process of identifying pointers and the process of using pointers by operations that accomplish gene assembly. In the intramolecular model one assumes that the pointer structure of a molecule is known, i.e., the pointers have been already identified. This implies some important differences between the models: e.g., the intramolecular representations of genes contain only pointers, with two occurrences for each pointer, and moreover, processing a pointer implies its removal from the processed string; these properties do not hold in the intermolecular model. Finally, the bulk of the work on the intermolecular model [1–3, 16–18] is concerned with the computational power of the operations in the sense of computability theory; e.g., it is proved in [18–20] that the model has the computational power of the Turing machine. On the other hand, research on the intramolecular model, see [4–6, 8–10, 12–14] and especially [7], deals with representations and properties of the gene assembly process (represented by various kinds of reduction systems). We believe that the two approaches together shed light on the computational nature of gene assembly in ciliates.

In this paper, we take a novel approach on the intermolecular model aiming to compare the assembly strategies predicted by each model. Therefore, we formalize both models in terms of MDS-IES descriptors and describe the gene assembly in terms of pointer reductions. We prove a universality result showing that the assembly power of the two models is the same: any gene that can be assembled in one model can also be assembled in the other. Nevertheless, the assembly strategies and the gene patterns throughout the process are completely different in the two models. Somewhat surprisingly, we show that the two models agree on the final results of the assembly process.

2   The Structure of Micronuclear Genes

We shall now take a formal approach to gene assembly. The central role in this process is played by *pointers*. These are short sequences at the ends of MDSs (i.e., at the border of an MDS and an IES) – the pointer in the end of an MDS M coincides as a nucleotide sequence with the pointer in the beginning of the MDS following M in the macronuclear gene, see [22, 25]. For the purpose of an adequate formal representation, the first (last, resp.) MDS begins (ends, resp.) with a specific marker b (e, resp.). It is enough for our purposes to describe any MDS by the pair of pointers or markers flanking it at its ends. The gene will then be described as a sequence of such pairs interspersed with strings describing the sequence of IES – we thus obtain *MDS-IES descriptors* formally defined in the following. For more details we refer to [7].

For an alphabet Σ and a string u over Σ, we will denote by $[u]$ the circular version of string u – we refer to [7] for a formal definition. Let $\overline{\Sigma} = \{\overline{a} \mid a \in \Sigma\}$ and $u = a_1 a_2 \ldots a_n$, $a_i \in \Sigma \cup \overline{\Sigma}$. The *inverse* of u is the string $\overline{u} = \overline{a}_n \ldots \overline{a}_2 \overline{a}_1$, where $\overline{\overline{a}} = a$, for all $a \in \Sigma$. The empty string will be denoted by Λ.

Let $\mathcal{M} = \{b, e, \overline{b}, \overline{e}\}$ denote the set of the *markers* and their inverses. For each index $\kappa \geq 2$, let

$$\Delta_\kappa = \{2, 3, \ldots, \kappa\} \quad \text{and} \quad \Pi_\kappa = \Delta_\kappa \cup \overline{\Delta}_\kappa.$$

An element $p \in \Pi_\kappa$ is called a *pointer*. Also let

$$\Gamma_\kappa = \{(b, e),\} \cup \{(b, i), (i, e) \mid 2 \leq i \leq \kappa\} \cup \{(i, j) \mid 2 \leq i < j \leq \kappa\}$$

and $\overline{\Gamma}_\kappa = \{(\overline{\beta}, \overline{\alpha}) \mid (\alpha, \beta) \in \Gamma_\kappa\}$. A string δ over $\Gamma_\kappa \cup \overline{\Gamma}_\kappa$ is called an *MDS descriptor* if

(a) δ has exactly one occurrence from the set $\{b, \overline{b}\}$ and exactly one occurrence from the set $\{e, \overline{e}\}$;

(b) δ has either zero, or two occurrences from $\{p, \overline{p}\}$, for any pointer $p \in \Pi_\kappa$.

Let $\Omega_\kappa = \{I_1, I_2, \ldots, I_{\kappa-1}\}$ and $\overline{\Omega}_\kappa = \{\overline{I} \mid I \in \Omega_\kappa\}$. Any string ι over $\Omega_\kappa \cup \overline{\Omega}_\kappa$ is called an *IES-descriptor* if for any $I \in \Omega_\kappa$, ι contains at most one occurrence from $\{I, \overline{I}\}$.

A string γ over $\Gamma_\kappa \cup \overline{\Gamma}_\kappa \cup \Omega_\kappa \cup \overline{\Omega}_\kappa$ is called an *MDS-IES descriptor* if

$$\gamma = \iota_1(p_1, q_1)\iota_2(p_2, q_2)\ldots\iota_n(p_n, q_n)\iota_{n+1},$$

where $\iota_1 \iota_2 \ldots \iota_{n+1}$ is an IES-descriptor, and $(p_1, q_1) \ldots (p_n, q_n)$ is an MDS-descriptor. We say that γ is *assembled* if $\gamma = \iota_1(m, m')\iota_2$ for some IES-descriptors ι_1, ι_2 and $m, m' \in \mathcal{M}$. If $(m, m') = (b, e)$, then we say that γ is assembled in the *orthodox order* and if $(m, m') = (\overline{e}, \overline{b})$, then we say that γ is assembled in the *inverted order*.

A circular string $[\gamma]$ is an *(assembled) MDS-IES descriptor* if γ is so.

Example 1. The MDS-IES descriptor associated to the micronuclear *actin I* gene in *S.nova*, shown in Fig. 1, is $M_3 I_1 M_4 I_2 M_6 I_3 M_5 I_4 M_7 I_5 M_9 I_6 \overline{M}_2 I_7 M_1 I_8 M_8$. □

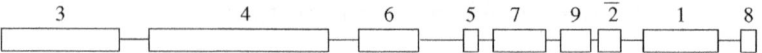

Fig. 1. Structure of the micronuclear gene encoding actin protein in the stichotrich *Sterkiella nova*. The nine MDSs are in a scrambled disorder

3   Two Models for Gene Assembly

We briefly present in this section the intramolecular and the intermolecular models for gene assembly in ciliates. We then formalize both models in terms of pointer reductions and MDS-IES descriptors. For more details we refer to [7, 11, 19, 20, 27].

3.1   The Intramolecular Model

Three intramolecular operations were postulated in [11] and [27] for gene assembly: ld, hi, and dlad. In each of these operations, a linear DNA molecule containing a specific pattern is folded on itself in such a way that recombination can take place. Operations hi and dlad yield as a result a linear DNA molecule, while ld yields one linear and one circular DNA molecule, see Figs. 2-4. The specific patterns required by each operation are described below:

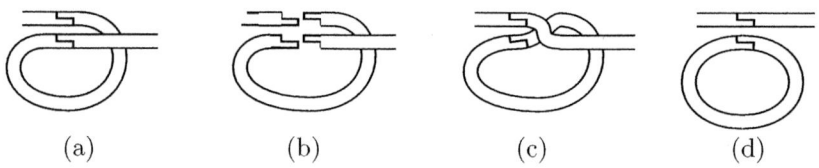

(a)   (b)   (c)   (d)

Fig. 2. Illustration of the ld molecular operation

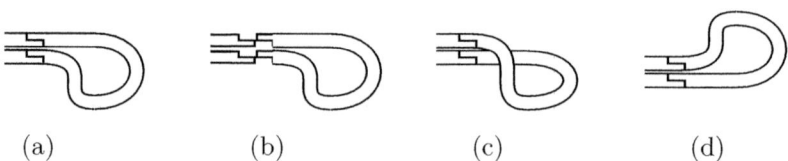

(a)   (b)   (c)   (d)

Fig. 3. Illustration of the hi molecular operation

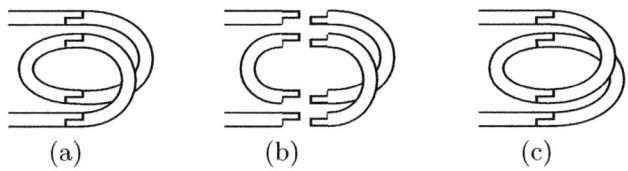

$$(a) \qquad\qquad (b) \qquad\qquad (c)$$

Fig. 4. Illustration of the dlad molecular operation

(i) The ld operation is applicable to molecules in which two occurrences (on the same strand) of the same pointer p flank one IES. The molecule is folded so that the two occurrences of p are aligned to guide the recombination, see Fig. 2. As a result, one circular molecule is excised.

(ii) The hi operation is applicable to molecules in which a pointer p has two occurrences, of which exactly one is inverted. The folding is done as in Fig. 3 so that the two occurrences of p are aligned to guide the recombination.

(iii) The dlad operation is applicable to molecules in which two pointers p and q have interspersed occurrences (on the same strand): $p - q - p - q$. The folding is done as in Fig. 4 so that the two occurrences of p and q are aligned to guide the double recombination.

Operations ld, hi, and dlad can be formalized in terms of reduction rules $\underline{\mathrm{ld}}$, $\underline{\mathrm{hi}}$, and $\underline{\mathrm{dlad}}$ for MDS-IES descriptors as follows:

(1) For each $p \in \Pi_\kappa$, the $\underline{\mathrm{ld}}$-rule for p is defined by:

$$\underline{\mathrm{ld}}_p(\delta_1(q,p)\iota_1(p,r)\delta_2) = \delta_1(q,r)\delta_2 + [\iota_1],$$
$$\underline{\mathrm{ld}}_p(\iota_1(p,m)\iota_2(m',p)\iota_3) = \iota_1\iota_3 + [(m',m)\iota_2],$$

where $q,r \in \Pi_\kappa \cup \mathcal{M}$, δ_1, δ_2 are MDS-IES descriptors, $\iota_1, \iota_2, \iota_3$ are IES descriptors, and $m, m' \in \mathcal{M}$.

(2) For each $p \in \Pi_\kappa$, the $\underline{\mathrm{hi}}$-rule for p is defined by:

$$\underline{\mathrm{hi}}_p(\delta_1(p,q)\delta_2(\overline{p},\overline{r})\delta_3) = \delta_1\overline{\delta_2}(\overline{q},\overline{r})\delta_3,$$
$$\underline{\mathrm{hi}}_p(\delta_1(q,p)\delta_2(\overline{r},\overline{p})\delta_3) = \delta_1(q,r)\overline{\delta_2}\delta_3,$$

where $q,r \in \Pi_\kappa$ and $\delta_1, \delta_2 \in (\Gamma_\kappa \cup \Omega)^{\maltese}$.

(3) For each $p,q \in \Pi_\kappa$, $p \neq q$, the $\underline{\mathrm{dlad}}$ rule for p and q is defined by:

$$\underline{\mathrm{dlad}}_{p,q}(\delta_1(p,r_1)\delta_2(q,r_2)\delta_3(r_3,p)\delta_4(r_4,q)\delta_5) = \delta_1\delta_4(r_4,r_2)\delta_3(r_3,r_1)\delta_2\delta_5,$$
$$\underline{\mathrm{dlad}}_{p,q}(\delta_1(p,r_1)\delta_2(r_2,q)\delta_3(r_3,p)\delta_4(q,r_4)\delta_5) = \delta_1\delta_4\delta_3(r_3,r_1)\delta_2(r_2,r_4)\delta_5,$$
$$\underline{\mathrm{dlad}}_{p,q}(\delta_1(r_1,p)\delta_2(q,r_2)\delta_3(p,r_3)\delta_4(r_4,q)\delta_5) = \delta_1(r_1,r_3)\delta_4(r_4,r_2)\delta_3\delta_2\delta_5,$$
$$\underline{\mathrm{dlad}}_{p,q}(\delta_1(r_1,p)\delta_2(r_2,q)\delta_3(p,r_3)\delta_4(q,r_4)\delta_5) = \delta_1(r_1,r_3)\delta_4\delta_3\delta_2(r_2,r_4)\delta_5,$$
$$\underline{\mathrm{dlad}}_{p,q}(\delta_1(p,r_1)\delta_2(q,p)\delta_4(r_4,q)\delta_5) = \delta_1\delta_4(r_4,r_1)\delta_2\delta_5,$$
$$\underline{\mathrm{dlad}}_{p,q}(\delta_1(p,q)\delta_3(r_3,p)\delta_4(q,r_4)\delta_5) = \delta_1\delta_4\delta_3(r_3,r_4)\delta_5,$$
$$\underline{\mathrm{dlad}}_{p,q}(\delta_1(r_1,p)\delta_2(q,r_2)\delta_3(p,q)\delta_5) = \delta_1(r_1,r_2)\delta_3\delta_2\delta_5,$$

where $r_1, r_2, r_3, r_4, r_5 \in \Pi_\kappa$, and $\delta_1, \delta_2, \delta_3, \delta_4, \delta_5 \in (\Gamma_\kappa \cup \Omega)^{\maltese}$.

Note that each operation removes one or two pointers from the MDS-IES descriptor. When assembled (on a linear or on a circular string), the descriptor has no pointers anymore. Thus, the whole process of gene assembly may be viewed as a process of pointer removals.

If a composition φ of ld, hi, and dlad operations is applicable to an MDS-IES descriptor γ, then $\varphi(\gamma)$ is a set of linear and circular MDS-IES descriptors. We say that φ is a *successful reduction* for γ if no pointers occur in any of the descriptors in $\varphi(\gamma)$.

Example 2. Consider the MDS-IES descriptor $\delta = (b, 2)I_1(2, 3)I_2(4, e)I_3(3, 4)$. An assembly strategy for this descriptor in the intramolecular model is the following:

$$\underline{\text{dlad}}_{3,4}(\delta) = (b, 2)I_1(2, e)I_3I_2,$$
$$\underline{\text{ld}}_2(\underline{\text{dlad}}_{3,4}(\delta)) = (b, e)I_3I_2 + [I_1].$$

$\square$

3.2   The Intermolecular Model

Three operations were postulated in [19] and [20] for gene assembly. One of these operations is intramolecular: it is a sort of a generalized version of the ld operation, while the other two are intermolecular: they involve recombination between two different DNA molecules, linear or circular, see Figs. 5-6. We describe these operations below in terms of pointers, similarly as for the intramolecular model.

(i) In the first operation a DNA molecule containing two occurrences of the same pointer x (on the same strand) is folded so that they get aligned to guide the recombination, see Fig. 5. Note that unlike in ld, the two occurrences of x may have more than just one IES between them.

(ii) The second operation is the inverse of the first one: two DNA molecules, one linear and one circular, each containing one occurrence of a pointer x get aligned so that the two occurrences of x guide the recombination, yielding one linear molecule – see Fig. 5.

(iii) The third operation is somewhat similar to the second one: two *linear* DNA molecules, each containing one occurrence of a pointer x get aligned so that the two occurrences of x guide the recombination, yielding two linear molecules, see Fig. 6.

Note that the three molecular operations in this model are reversible, unlike the operations in the intramolecular model – this is one of the main differences between the two models.

We formalize now this intermolecular model in terms of reduction rules for MDS-IES descriptors. The three operations defined above are modelled by the following reduction rules for MDS-IES descriptors:

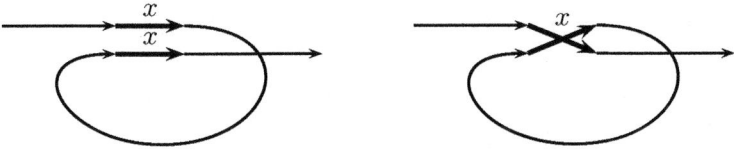

Fig. 5. Illustration of the intramolecular operation of the Landweber-Kari model

Fig. 6. Illustration of the intermolecular operation of the Landweber-Kari model

$$\delta_1(q,p)\delta_2(p,r)\delta_3 \xrightarrow{p} \delta_1(q,r)\delta_3 + [\delta_2], \tag{1}$$

$$\delta_1(p,q)\delta_2(r,p)\delta_3 \xrightarrow{p} \delta_1\delta_3 + [\delta_2(r,q)], \tag{2}$$

$$\delta_1(p,q)\delta_2 + [(r,p)\delta_3] \xrightarrow{p} \delta_1\delta_3(r,q)\delta_2, \tag{3}$$

$$\delta_1(q,p)\delta_2 + [(p,r)\delta_3] \xrightarrow{p} \delta_1(q,r)\delta_3\delta_2, \tag{4}$$

$$\delta_1(p,q)\delta_2 + \delta_3(r,p)\delta_4 \xrightarrow{p} \delta_1\delta_4 + \delta_3(r,q)\delta_2, \tag{5}$$

where $\delta_1, \delta_2, \delta_3 \in (\Gamma_\kappa \cup \overline{\Gamma}_\kappa \cup \Omega_\kappa \cup \overline{\Omega}_\kappa)^*$.

Note that each reduction rule above *removes* one pointer, thus making the whole process irreversible. Although the intermolecular model was specifically intended to be reversible, this restriction helps in unifying the notation for (and the reasoning about) the two models and it suffices for the results presented in this paper.

If a composition φ of the reduction rules (1)-(5) is applicable to an MDS-IES descriptor γ, then $\varphi(\gamma)$ is a set of linear and circular MDS-IES descriptors. We say that φ is a *successful reduction* for γ if no pointers occur in any of the descriptors in $\varphi(\gamma)$.

Example 3. Consider the MDS-IES descriptor $\delta = (b,2)I_1(2,3)I_2(4,e)I_3(3,4)$ of Example 2. An assembly strategy for this descriptor in the intermolecular model is the following:

$$\delta \xrightarrow{3} (b,2)I_1(2,4) \ + \ [I_2(4,e)I_3] \xrightarrow{4} (b,2)I_1(2,e)I_3I_2 \xrightarrow{2} (b,e)I_3I_2 \ + \ [I_1].$$

Note that although the assembly strategy is very different from the one in Example 2, the final result of the assembly, $\{(b,e)I_3I_2, [I_1]\}$ is the same in the two models.   □

4   Reduction Strategies in the Two Models

The obvious difficulty with the intermolecular model is that it cannot deal with DNA molecules in which a pointer is inverted – this is the case, e.g., for the *actin I* gene in *S.nova*. Nevertheless, we can show that inverted pointers can be handled in this model, provided the input molecule (or its MDS-IES descriptor) is available in two copies. Moreover, *we consider all linear descriptors modulo inversion*. The first assumption is essentially used in research on the intermolecular model, see [16–18, 20]. The second assumption is quite natural whenever we model double-stranded DNA molecules. As a matter of fact, we use the two assumptions to conclude that for each input descriptor, both the descriptor and its inversion are available. Then the hi-rule can be simulated using the intermolecular rules as follows.

Let $\delta = \delta_1(p,q)\delta_2(\overline{p},\overline{r})\delta_3$ (the other case is treated similarly) be an MDS-IES descriptor to which $\underline{\text{hi}}_p$ is applicable. Therefore, we assume that also $\overline{\delta} = \overline{\delta}_3(r,p)\overline{\delta}_2(\overline{q},\overline{p})\overline{\delta}_1$ is available. Then we obtain

$$\delta + \overline{\delta} \xrightarrow{p} \delta_1\,\overline{\delta}_2\,(\overline{q},\overline{p})\,\overline{\delta}_1 + \overline{\delta}_3\,(r,q)\delta_2\,(\overline{p},\overline{r})\,\delta_3$$
$$\xrightarrow{\overline{p}} \delta_1\overline{\delta}_2(\overline{q},\overline{r})\delta_3 + \overline{\delta}_3(r,q)\delta_2\overline{\delta}_1 = \underline{\text{hi}}_p(\delta) + \overline{\underline{\text{hi}}_p(\delta)}\,.$$

Note that, having two copies of the initial string available, this rule yields two copies of $\underline{\text{hi}}_p(w)$.

We also observe that the ld-rule is a particular case of intermolecular rules (1) and (2), obtained by setting $\delta_2 = \Lambda$. Moreover, the dlad-rule can be simulated using intermolecular rules as follows.

Let $\delta = \delta_1(p,r_1)\delta_2(q,r_2)\delta_3(r_3,p)\delta_4(r_4,q)\delta_5)$ be an MDS-IES descriptor to which $\underline{\text{dlad}}_{p,q}$ is applicable – all other cases can be treated similarly. Then

$$\delta \xrightarrow{p} \delta_1\delta_4(r_4,q)\delta_5 + [\delta_2(q,r_2)\delta_3(r_3,r_1)] = \delta_1\delta_4(r_4,q)\delta_5 + [\delta_3(r_3,r_1)\delta_2(q,r_2)]$$
$$\xrightarrow{q} \delta_1\delta_4(r_4,r_2)\delta_3(r_3,r_1)\delta_2\delta_5 = \underline{\text{dlad}}_{p,q}(w)\,.$$

The following results is thus proved.

Theorem 1. *Let δ be an MDS-IES descriptor having a successful reduction in the intramolecular model. If two copies of δ are available, then δ has a successful reduction in the intermolecular model.*

The following universality result has been proved in [8], see [7] for more details.

Theorem 2. *Any MDS-IES descriptor has a successful reduction in the intramolecular model.*

Theorems 1 and 2 give the following universality result for the intermolecular model.

Corollary 1. *Any MDS-IES descriptor available in two copies has a successful reduction in the intermolecular model.*

5   Invariants

In the following two examples we consider the *actin I* gene in *S.nova* and investigate assembly strategies for this gene in the intra- and inter-molecular models.

Example 4. Consider the actin gene in *Sterkiella nova*, see Fig. 1, having the MDS–IES descriptor

$$\delta = (3,4)I_1(4,5)I_2(6,7)I_3(5,6)I_4(7,8)I_5(9,e)I_6(\overline{3},\overline{2})I_7(b,2)I_8(8,9).$$

Consider then an assembly strategy for δ, e.g., $\underline{\mathsf{ld}}_4\ \underline{\mathsf{dlad}}_{5,6}\ \underline{\mathsf{ld}}_7\ \underline{\mathsf{dlad}}_{8,9}\ \underline{\mathsf{hi}}_{\overline{2}}\ \underline{\mathsf{hi}}_3$:

$$\underline{\mathsf{ld}}_4(\delta) = (3,5)I_2(6,7)I_3(5,6)I_4(7,8)I_5(9,e)I_6(\overline{3},\overline{2})I_7(b,2)I_8(8,9)\ +\ [I_1],$$
$$\underline{\mathsf{dlad}}_{5,6}(\underline{\mathsf{ld}}_4(\delta)) = I_0(3,7)I_3I_2I_4(7,8)I_5(9,e)I_6(\overline{3},\overline{2})I_7(b,2)I_8(8,9)\ +\ [I_1],$$
$$\underline{\mathsf{ld}}_7(\underline{\mathsf{dlad}}_{5,6}(\underline{\mathsf{ld}}_4(\delta))) = (3,8)I_5(9,e)I_6(\overline{3},\overline{2})I_7(b,2)I_8(8,9)\ +\ [I_1]\ +\ [I_3I_2I_4],$$
$$\underline{\mathsf{dlad}}_{8,9}(\underline{\mathsf{ld}}_7(\underline{\mathsf{dlad}}_{5,6}(\underline{\mathsf{ld}}_4(\delta)))) = (3,e)I_6(\overline{3},\overline{2})I_7(b,2)I_8I_5\ +\ [I_1]\ +\ [I_3I_2I_4],$$
$$\underline{\mathsf{hi}}_{\overline{2}}(\underline{\mathsf{dlad}}_{8,9}(\underline{\mathsf{ld}}_7(\underline{\mathsf{dlad}}_{5,6}(\underline{\mathsf{ld}}_4(\delta))))) = (3,e)I_6(\overline{3},\overline{b})\overline{I}_7I_8I_5\ +\ [I_1]\ +\ [I_3I_2I_4],$$
$$\underline{\mathsf{hi}}_3(\underline{\mathsf{hi}}_{\overline{2}}(\underline{\mathsf{dlad}}_{8,9}(\underline{\mathsf{ld}}_7(\underline{\mathsf{dlad}}_{5,6}(\underline{\mathsf{ld}}_4(\delta)))))) = \overline{I}_6(\overline{e},\overline{b})\overline{I}_7I_8I_5\ +\ [I_1]\ +\ [I_3I_2I_4].$$

Thus, the gene is assembled in the inverted order, placed in a linear DNA molecule, with the IES $\overline{I}_6$ preceding it and the sequence of IESs $\overline{I}_7\,I_8\,I_5$ succeeding it. Two circular molecules are also produced: $[I_1]$ and $[I_3I_2I_4]$. □

Example 5. Consider the same actin gene in *Sterkiella nova* with the MDS–IES descriptor

$$\delta = (3,4)I_1(4,5)I_2(6,7)I_3(5,6)I_4(7,8)I_5(9,e)I_6(\overline{3},\overline{2})I_7(b,2)I_8(8,9)\,.$$

Then δ can be assembled in the intermolecular model as follows:

$$\delta \xrightarrow{5} (3,4)I_1(4,6)I_4(7,8)I_5(9,e)I_6(\overline{3},\overline{2})I_7(b,2)I_8(8,9) + [I_2(6,7)I_3]$$
$$\xrightarrow{8} (3,4)I_1(4,6)I_4(7,9) + [I_5(9,e)I_6(\overline{3},\overline{2})I_7(b,2)I_8] + [I_2(6,7)I_3]$$
$$\xrightarrow{4} (3,6)I_4(7,9) + [I_1] + [I_5(9,e)I_6(\overline{3},\overline{2})I_7(b,2)I_8] + [I_2(6,7)I_3]$$
$$\xrightarrow{7} (3,6)I_4I_3I_2(6,9) + [I_1] + [I_5(9,e)I_6(\overline{3},\overline{2})I_7(b,2)I_8]$$
$$\xrightarrow{6} (3,9) + [I_4I_3I_2] + [I_1] + [I_5(9,e)I_6(\overline{3},\overline{2})I_7(b,2)I_8]$$
$$\xrightarrow{9} (3,e)I_6(\overline{3},\overline{2})I_7(b,2)I_8I_5 + [I_4I_3I_2] + [I_1]\,.$$

Since $\overline{\delta}$ is also available, the assembly continues as follows. Here, for a (circular) string τ, we use $2\cdot\tau$ to denote $\tau+\tau$:

$$\delta + \bar{\delta} \;\; \rightarrow \ldots \rightarrow (3,e)I_6(\bar{3},\bar{2})I_7(b,2)I_8I_5 + \bar{I}_5\bar{I}_8(\bar{2},\bar{b})\bar{I}_7(2,3)\bar{I}_6(\bar{e},\bar{3})$$
$$+ 2 \cdot [I_4I_3I_2] + 2 \cdot [I_1]$$

$$\xrightarrow{\;\bar{2}\;} \;\; (3,e)I_6(\bar{3},\bar{b})\bar{I}_7(2,3)\bar{I}_6(\bar{e},\bar{3}) + \bar{I}_5\bar{I}_8\bar{I}_7(b,2)I_8I_5$$
$$+ 2 \cdot [I_4I_3I_2] + 2 \cdot [I_1]$$

$$\xrightarrow{\;2\;} \;\; (3,e)I_6(\bar{3},\bar{b})\bar{I}_7I_8I_5 + \bar{I}_5\bar{I}_8I_7(b,3)\bar{I}_6(\bar{e},\bar{3}) + 2 \cdot [I_4I_3I_2] + 2 \cdot [I_1]$$

$$\xrightarrow{\;3\;} \;\; (3,e)I_6 + \bar{I}_5\bar{I}_8I_7(b,3)\bar{I}_6(\bar{e},\bar{b})\bar{I}_7I_8I_5 + 2 \cdot [I_4I_3I_2] + 2 \cdot [I_1]$$

$$\xrightarrow{\;3\;} \;\; \bar{I}_6(\bar{e},\bar{b})\bar{I}_7I_8I_5 + \bar{I}_5\bar{I}_8I_7(b,e) + 2 \cdot [I_4I_3I_2] + 2 \cdot [I_1]$$

$$= \;\; 2 \cdot (\bar{I}_6(\bar{e},\bar{b})\bar{I}_7I_8I_5 + +[I_1] + [I_3I_2I_4]) \,.$$

Note that this intermolecular assembly predicts the same context for the assembled string, the same set of residual strings, and the same linearity of the assembled string as the intramolecular assembly considered in Example 4.   □

It is clear from the above two examples, see also Examples 2 and 3, that the two models for gene assembly predict completely different assembly strategies for the same micronuclear gene. As it turns out however, the predicted final result of the assembly, i.e., the linearity of the assembled gene and the exact nucleotide sequences of all excised molecules, is the same in the two models, see [7] for details. The following is a result from [7], see also [10].

Theorem 3. *Let δ be an MDS–IES descriptor. If φ_1 and φ_2 are any two successful assembly strategies for δ, intra- or inter-molecular, then*

(1) if $\varphi_1(\delta)$ is assembled in a linear descriptor, then so is $\varphi_2(\delta)$;
(2) if $\varphi_1(\delta)$ is assembled in a linear descriptor in orthodox order, then so is $\varphi_2(\delta)$;
(3) The sequence of IESs flanking the assembled gene is the same in $\varphi_1(\delta)$ and $\varphi_2(\delta)$;
(4) The sequence of IESs in all excised descriptors is the same in $\varphi_1(\delta)$ and $\varphi_2(\delta)$;
(5) There si an equal number of circular descriptors in $\varphi_1(\delta)$ and $\varphi_2(\delta)$.

Example 6. Consider the MDS–IES descriptor

$$\delta = (\overline{10},\bar{8})I_1(\bar{3},\bar{b})I_2(\bar{5},\bar{3})I_3(10,11)I_4(5,8)I_5(11,e) \,.$$

A successful assembly strategy for δ in the intramolecular model is the following:

$$\underline{\mathsf{hi}}_{\overline{10}}(\delta) = \bar{I}_3(3,5)\bar{I}_2(b,3)\bar{I}_1(8,11)I_4(5,8)I_5(11,e),$$

$$\underline{\mathsf{dlad}}_{8,11}(\underline{\mathsf{hi}}_{\overline{10}}(\delta)) = \bar{I}_3(3,5)\bar{I}_2(b,3)\bar{I}_1I_5I_4(5,e),$$

$$\underline{\mathsf{dlad}}_{3,5}(\underline{\mathsf{dlad}}_{8,11}(\underline{\mathsf{hi}}_{\overline{10}}(\delta))) = \bar{I}_3\bar{I}_1I_5I_4\bar{I}_2(b,e).$$

Thus, δ is always assembled in a linear molecule, and no IES is excised during the assembly process, i.e., no $\underline{\mathrm{ld}}$ is ever applied in a process of assembling δ. Moreover, the assembled descriptor will always be preceded by the IES sequence $\overline{I}_3\overline{I}_1 I_5 I_4 \overline{I}_2$ and followed by the empty IES sequence.   □

Example 7. Consider the MDS–IES descriptor

$$\delta = (\overline{10}, \overline{8})I_1(\overline{3}, \overline{b})I_2(\overline{5}, \overline{3})I_3(10, 11)I_4(5, 8)I_5(11, e)$$

from Example 6. Then δ can be assembled in the intermolecular model as follows:

$$\delta \xrightarrow{\overline{3}} \quad (\overline{10}, \overline{8})I_1 I_3(10, 11)I_4(5, 8)I_5(11, e) + [I_2(\overline{5}, \overline{b})]$$
$$\xrightarrow{11} \quad (\overline{10}, \overline{8})I_1 I_3(10, e) + [I_4(5, 8)I_5] + [I_2(\overline{5}, \overline{b})].$$

Since also $\overline{\delta}$ is available, the assembly continues as follows:

$$\delta + \overline{\delta} \to \ldots \to \quad (\overline{10}, \overline{8})I_1 I_3(10, e) + (\overline{e}, \overline{10})\overline{I}_3\overline{I}_1(8, 10)$$
$$+ 2 \cdot [I_4(5, 8)I_5] + 2 \cdot [I_2(\overline{5}, \overline{b})]$$
$$\xrightarrow{\overline{10}} \quad \overline{I}_3\overline{I}_1(8, 10) + (\overline{e}, \overline{8})I_1 I_3(10, e) + 2 \cdot [I_4(5, 8)I_5] + 2 \cdot [I_2(\overline{5}, \overline{b})]$$
$$\xrightarrow{10} \quad \overline{I}_3\overline{I}_1(8, e) + (\overline{e}, \overline{8})I_1 I_3 + [I_4(5, 8)I_5] + [\overline{I}_5(\overline{8}, \overline{5})\overline{I}_4] + 2 \cdot [I_2(\overline{5}, \overline{b})]$$
$$\xrightarrow{8, \overline{8}} \quad \overline{I}_3\overline{I}_1 I_5 I_4(5, e) + (\overline{e}, \overline{5})\overline{I}_4\overline{I}_5 I_1 I_3 + [(\overline{5}, \overline{b})I_2] + [\overline{I}_2(b, 5)]$$
$$\xrightarrow{5, \overline{5}} \quad \overline{I}_3\overline{I}_1 I_5 I_4\overline{I}_2(b, e) + (\overline{e}, \overline{b})I_2\overline{I}_4\overline{I}_5 I_1 I_3$$
$$= \quad 2 \cdot \overline{I}_3\overline{I}_1 I_5 I_4\overline{I}_2(b, e).$$

Note that, again, we obtain the same context for the assembled string, the same set of residual strings, and the same linearity of the assembled string as the intramolecular assembly considered in Example 6.   □

References

1. Daley, M., *Computational Modeling of Genetic Processes in Stichotrichous Ciliates.* PhD thesis, University of London, Ontario, Canada (2003)
2. Daley, M., and Kari, L., Some properties of ciliate bio-operations. *Lecture Notes in Comput. Sci.* **2450** (2003) 116–127
3. Daley, M., Ibarra, O. H., and Kari, L., Closure propeties and decision questions of some language classes under ciliate bio-operations. *Theoret. Comput. Sci.,* to appear
4. Ehrenfeucht, A., Harju, T., Petre, I., Prescott, D. M., and Rozenberg, G., Formal systems for gene assembly in ciliates. *Theoret. Comput. Sci.* **292** (2003) 199–219
5. Ehrenfeucht, A., Harju, T., Petre, I., and Rozenberg, G., Patterns of micronuclear genes in cliates. *Lecture Notes in Comput. Sci.* **2340** (2002) 279–289

6. Ehrenfeucht, A., Harju, T., Petre, I., and Rozenberg, G., Characterizing the micronuclear gene patterns in ciliates. *Theory of Comput. Syst.* **35** (2002) 501–519

7. Ehrenfeucht, A., Harju, T., Petre, I., Prescott, D. M., and Rozenberg, G., *Computation in Living Cells: Gene Assembly in Ciliates*, Springer (2003).

8. Ehrenfeucht, A., Petre, I., Prescott, D. M., and Rozenberg, G., Universal and simple operations for gene assembly in ciliates. In: V. Mitrana and C. Martin-Vide (eds.) *Words, Sequences, Languages: Where Computer Science, Biology and Linguistics Meet*, Kluwer Academic, Dortrecht, (2001) pp. 329–342

9. Ehrenfeucht, A., Petre, I., Prescott, D. M., and Rozenberg, G., String and graph reduction systems for gene assembly in ciliates. *Math. Structures Comput. Sci.* **12** (2001) 113–134

10. Ehrenfeucht, A., Petre, I., Prescott, D. M., and Rozenberg, G., Circularity and other invariants of gene assembly in cliates. In: M. Ito, Gh. Paun and S. Yu (eds.) *Words, semigroups, and transductions*, World Scientific, Singapore, (2001) pp. 81–97

11. Ehrenfeucht, A., Prescott, D. M., and Rozenberg, G., Computational aspects of gene (un)scrambling in ciliates. In: L. F. Landweber, E. Winfree (eds.) *Evolution as Computation*, Springer, Berlin, Heidelberg, New York (2001) pp. 216–256

12. Harju, T., Petre, I., and Rozenberg, G., Gene assembly in ciliates: molecular operations. In: G.Paun, G. Rozenberg, A.Salomaa (Eds.) *Current Trends in Theoretical Computer Science*, (2004).

13. Harju, T., Petre, I., and Rozenberg, G., Gene assembly in ciliates: formal frameworks. In: G.Paun, G. Rozenberg, A.Salomaa (Eds.) *Current Trends in Theoretical Computer Science*, (2004).

14. Harju, T., and Rozenberg, G., Computational processes in living cells: gene assembly in ciliates. *Lecure Notes in Comput. Sci.* **2450** (2003) 1–20

15. Jahn, C. L., and Klobutcher, L. A., Genome remodeilng in ciliated protozoa. *Ann. Rev. Microbiol.* **56** (2000), 489–520.

16. Kari, J., and Kari, L. Context free recombinations. In: C. Martin-Vide and V. Mitrana (eds.) *Where Mathematics, Computer Science, Linguistics, and Biology Meet*, Kluwer Academic, Dordrecht, (2000) 361–375

17. Kari, L., Kari, J., and Landweber, L. F., Reversible molecular computation in ciliates. In: J. Karhumäki, H. Maurer, G. Păun and G. Rozenberg (eds.) *Jewels are Forever*, Springer, Berlin HeidelbergNew York (1999) pp. 353–363

18. Kari, L., and Landweber, L. F., Computational power of gene rearrangement. In: E. Winfree and D. K. Gifford (eds.) *Proceedings of DNA Bases Computers, V* American Mathematical Society (1999) pp. 207–216

19. Landweber, L. F., and Kari, L., The evolution of cellular computing: Nature's solution to a computational problem. In: *Proceedings of the 4th DIMACS Meeting on DNA-Based Computers*, Philadelphia, PA (1998) pp. 3–15

20. Landweber, L. F., and Kari, L., Universal molecular computation in ciliates. In: L. F. Landweber and E. Winfree (eds.) *Evolution as Computation*, Springer, Berlin Heidelberg New York (2002)

21. Prescott, D. M., Cutting, splicing, reordering, and elimination of DNA sequences in hypotrichous ciliates. *BioEssays* **14** (1992) 317–324

22. Prescott, D. M., The unusual organization and processing of genomic DNA in hypotrichous ciliates. *Trends in Genet.* **8** (1992) 439–445

23. Prescott, D. M., The DNA of ciliated protozoa. *Microbiol. Rev.* **58**(2) (1994) 233–267

24. Prescott, D. M., The evolutionary scrambling and developmental unscabling of germlike genes in hypotrichous ciliates. *Nucl. Acids Res.* **27** (1999), 1243 – 1250.

25. Prescott, D. M., Genome gymnastics: unique modes of DNA evolution and processing in ciliates. *Nat. Rev. Genet.* 1(3) (2000) 191–198

26. Prescott, D. M., and DuBois, M., Internal eliminated segments (IESs) of Oxytrichidae. *J. Eukariot. Microbiol.* **43** (1996) 432–441

27. Prescott, D. M., Ehrenfeucht, A., and Rozenberg, G., Molecular operations for DNA processing in hypotrichous ciliates. *Europ. J. Protistology* **37** (2001) 241–260

28. Prescott, D. M., Ehrenfeucht, A., and Rozenberg, G., Template-guided recombination for IES elimination and unscrambling of genes in stichotrichous ciliates. Technical Report 2002-01, LIACS, Leiden University (2002)

29. Prescott, D. M., and Rozenberg, G., How ciliates manipulate their own DNA – A splendid example of natural computing. *Natural Computing* **1** (2002) 165–183

30. Prescott, D. M., and Rozenberg, G., Encrypted genes and their reassembly in ciliates. In: M. Amos (ed.) *Cellular Computing*, Oxford University Press, Oxford (2003)

On Self-Dual Bases of the Extensions of the Binary Field

Mika Hirvensalo* and Jyrki Lahtonen

Department of Mathematics, University of Turku, FIN-20014, Turku, Finland.
TUCS – Turku Centre for Computer Science.
{mikhirve,lahtonen}@utu.fi

Abstract. There are at least two points of view when representing elements of $\mathbb{F}_{2^n}$, the field of 2^n elements. We could represent the (nonzero) elements as powers of a generating element, the exponent ranging from 0 to $2^n - 2$. On the other hand, we could represent the elements as strings of n bits. In the former representation, multiplication becomes a very easy task, whereas in the latter one, addition is obvious. In this note, we focus on representing $\mathbb{F}_{2^n}$ as strings of n bits in such a way that the natural basis $(1, 0, \ldots, 0)$, $(0, 1, \ldots, 0)$, $\ldots$, $(0, 0, \ldots, 1)$ becomes self-dual. We also outline an idea which leads to a very simple algorithm for finding a self-dual basis. Finally we study multiplication tables for the natural basis and present necessary and sufficient conditions for a multiplication table to give $\mathbb{F}_2^n$ a field structure in such a way that the natural basis is self-dual.

1 Introduction and Preliminaries

The bit strings, and operations on them have an essential role in theoretical computer science. The reason for this is very evident: to compute is to operate finite sets and all finite sets can be encoded into binary strings. Therefore, all properties (subsets) of finite sets can be represented by Boolean functions, that is, by functions from bit strings to bits.

The most difficult problems in theoretical computer science can be represented in terms of Boolean functions. For instance, it is known that all Boolean functions can be represented by *Boolean circuits* [7], but far too little is known about the number of *gates* to implement those functions. Quite a simple counting argument shows that on n variables, *most* Boolean functions need $\frac{2^n}{2n}$ gates to be implemented [7], and yet the best known lower bound is only linear in n.

If the scalar multiplication and addition are defined bitwise, the bit strings of length n form a vector space of dimension n over the binary field. However, for many applications it would be helpful if the bit strings could have even a stronger algebraic structure. Numerous very good examples of such applications can be found in the theory of error-correcting codes [5].

---

* Supported by the Academy of Finland under grant 44087.

J. Karhumäki et al. (Eds.): Theory Is Forever (Salomaa Festschrift), LNCS 3113, pp. 102–111, 2004.
© Springer-Verlag Berlin Heidelberg 2004

We will use notation $\mathbb{F}_q$ for the finite field of q elements, but we are going to concentrate only on the case $q = 2^n$. As a *binary field* we understand the two-element field $\mathbb{F}_2$. An n-dimensional vector space over $\mathbb{F}_2$ is denoted, as usual, by $\mathbb{F}_2^n$. The *characters* of the additive group of $\mathbb{F}_2^n$ are well-known and easy to define (see [2], for example): for each element $\boldsymbol{y} = (y_1, \ldots, y_n) \in \mathbb{F}_2^n$, there exists a character $\chi_{\boldsymbol{y}}$ defined as

$$\chi_{\boldsymbol{y}}(\boldsymbol{x}) = (-1)^{\boldsymbol{x} \cdot \boldsymbol{y}},$$

where $\boldsymbol{x} \cdot \boldsymbol{y} = x_1 y_1 + \ldots + x_n y_n$. Notice that $\boldsymbol{x} \cdot \boldsymbol{y}$ belongs to $\mathbb{F}_2$, but $(-1)^{\boldsymbol{x} \cdot \boldsymbol{y}}$ is interpreted in the most obvious way.

It turns out that all functions $\mathbb{F}_2^n \to \mathbb{C}$ can be represented as linear combinations of the characters [2]. The characters, in fact, have even a more important role: there is an obvious way to introduce a Euclidean vector space structure for the set of functions $\mathbb{F}_2^n \to \mathbb{C}$, and the characters form an orthonormal basis of that Euclidean space. This role of the characters allows us to use discrete analogues of Fourier analysis, see [1] and [3] for instance.

To introduce more algebraic structure in $\mathbb{F}_2^n$, it is always possible to define (usually in many ways) the multiplication in $\mathbb{F}_2^n$ in such a way that $\mathbb{F}_2^n$ becomes the field $\mathbb{F}_{2^n}$, extension of $\mathbb{F}_2$ of degree n [5].

The *trace* of an element $\alpha \in \mathbb{F}_{2^n}$ over the prime field $\mathbb{F}_2$ is defined as

$$\mathrm{Tr}(\alpha) = \alpha + \alpha^2 + \alpha^{2^2} + \ldots + \alpha^{2^{n-1}},$$

and it is a well-known fact that always $\mathrm{Tr}(\alpha) \in \mathbb{F}_2$, and that $\mathrm{Tr} : \mathbb{F}_{2^n} \to \mathbb{F}_2$ is a linear mapping satisfying $|\mathrm{Ker}(\mathrm{Tr})| = 2^{n-1}$. As an easy consequence of the definition, we have also that $\mathrm{Tr}(\alpha^2) = \mathrm{Tr}(\alpha)$ for each $\alpha \in \mathbb{F}_{2^n}$.

As a vector space over $\mathbb{F}_2$, field $\mathbb{F}_{2^n}$ has an n-element basis $B = \{\alpha_1, \ldots, \alpha_n\}$ over $\mathbb{F}_2$. We say that the basis B is *self-dual*, if

$$\mathrm{Tr}(\alpha_i \alpha_j) = \begin{cases} 1, \text{ if } i = j, \text{ and} \\ 0, \text{ if } i \neq j. \end{cases}$$

For some practical applications, bases of special types are valuable. For instance, multiplication in a bit-string representation is in some sense "computationally cheap" if the chosen basis is so-called *normal basis* [6]. In some other situations, a self-dual basis is extremely welcome. Consider, for example, the characters of the additive group of $\mathbb{F}_{2^n}$. As it is well-known, they all are of form

$$\psi_y(x) = (-1)^{\mathrm{Tr}(xy)},$$

where $y \in \mathbb{F}_{2^n}$ [5]. Assuming that the basis $\{\alpha_1, \ldots, \alpha_n\}$ is self-dual, we can represent

$$x = x_1 \alpha_1 + \ldots + x_n \alpha_n$$

and

$$y = y_1 \alpha_1 + \ldots + y_n \alpha_n,$$

where x_i, $y_i \in \mathbb{F}_2$. Using this representation we can find out that

$$\operatorname{Tr}(xy) = \operatorname{Tr}(\sum_{i=1}^{n} x_i \alpha_i \sum_{j=1}^{n} y_j \alpha_j) = \sum_{i=1}^{n} \sum_{j=1}^{n} x_i y_j \operatorname{Tr}(\alpha_i \alpha_j) = \sum_{i=1}^{n} x_i y_i.$$

But then

$$\psi_y(x) = (-1)^{x_1 y_1 + \ldots + x_n y_n},$$

which is to say that the character value (character determined by the element $y = y_1 \alpha_1 + \ldots + y_n \alpha_n$) on element $x = x_1 \alpha_1 + \ldots + x_n \alpha_n$ is exactly the same as the corresponding character value in the additive group $\mathbb{F}_2^n$.

2   Finding a Self-Dual Basis

In this section, we outline the idea of a very simple algorithm to find a self-dual basis of $\mathbb{F}_{2^n}$. It seems that similar ideas were already present in [4].

The most crucial observation for the algorithm is the following.

Lemma 1. *Matrix*

$$\begin{pmatrix} 1 & 0 & 0 \\ 0 & 0 & 1 \\ 0 & 1 & 0 \end{pmatrix} \in \mathbb{F}_2^{3 \times 3}$$

can be diagonalized.

Proof. A straightforward calculation shows that

$$\begin{pmatrix} 1 & 0 & 1 \\ 1 & 1 & 0 \\ 1 & 1 & 1 \end{pmatrix} \begin{pmatrix} 1 & 0 & 0 \\ 0 & 0 & 1 \\ 0 & 1 & 0 \end{pmatrix} \begin{pmatrix} 1 & 1 & 1 \\ 0 & 1 & 1 \\ 1 & 0 & 1 \end{pmatrix} = \begin{pmatrix} 1 & 0 & 0 \\ 0 & 1 & 0 \\ 0 & 0 & 1 \end{pmatrix}$$

and that the diagonalizing matrix is indeed invertible.   □

Let now $B = \{\alpha_1, \ldots, \alpha_n\}$ be any basis of $\mathbb{F}_{2^n}/\mathbb{F}_2$, and define a matrix $M \in \mathbb{F}_2^{n \times n}$ as

$$M_{ij} = \operatorname{Tr}(\alpha_i \alpha_j).$$

Clearly M is symmetric. Matrix M is also invertible, for otherwise its rows would be linearly dependent, and therefore we could find elements $c_1, \ldots, c_n \in \mathbb{F}_2$ not all zero such that

$$\sum_{i=1}^{n} c_i \operatorname{Tr}(\alpha_i \alpha_j) = 0$$

for each j. But this would mean that there would be an element $\gamma = c_1 \alpha_1 + \ldots + c_n \alpha_n \in \mathbb{F}_2^n$ such that $\operatorname{Tr}(\gamma \alpha_j) = 0$ for each basis element α_j. It would follow that mapping $\alpha \mapsto \operatorname{Tr}(\gamma \alpha)$ is identically zero, which implies that $\gamma = 0$. But this would contradict the fact that there is a nonzero element c_i.

If M is a diagonal matrix, then B is already a self-dual basis (since M is invertible, there cannot be zeros in the diagonal). Assume then that M is not diagonal. Since B is a basis, not all the diagonal elements are 0, for otherwise $\mathrm{Tr}(\alpha_i^2) = \mathrm{Tr}(\alpha_i) = 0$ for each basis element α_i, which would contradict the property $|\mathrm{Ker}(\mathrm{Tr})| = 2^{n-1}$.

Because there is at least one nonzero element in the diagonal, there exists an invertible matrix $E_1 \in \mathbb{F}_2^{n \times n}$ such that $E_1 M E_1^T$ can be written as

$$E_1 M E_1^T = \begin{pmatrix} 1 & 0 \\ 0 & M_1 \end{pmatrix},$$

where $M_1 \in \mathbb{F}_2^{(n-1) \times (n-1)}$ is symmetric.

Now matrix M_1 must have nonzero elements, for otherwise M would not be invertible. If M_1 has a nonzero diagonal element, we can again find a matrix $E_2 \in \mathbb{F}_2^{n \times n}$ such that

$$E_2 E_1 M E_1^T E_2^T = \begin{pmatrix} 1 & 0 & 0 \\ 0 & 1 & \\ 0 & & M_2 \end{pmatrix},$$

where $M_2 \in \mathbb{F}_2^{(n-2) \times (n-2)}$ is a symmetric matrix.

If M_1 does not contain any nonzero element in its diagonal, we can find, because M_1 is symmetric, an invertible matrix $E_2' \in \mathbb{F}_2^{(n-1) \times (n-1)}$ such that

$$E_2' M_1 E_2'^T = \begin{pmatrix} 0 & 1 & 0 \\ 1 & 0 & \\ 0 & & M_2 \end{pmatrix},$$

where M_2 is an $(n-3) \times (n-3)$-matrix. Matrix E_2' can be straightforwardly extended to an invertible $n \times n$-matrix E_2 which satisfies

$$E_2 E_1 M E_1^T E_2^T = \begin{pmatrix} 1 & 0 & 0 & \\ 0 & 0 & 1 & 0 \\ 0 & 1 & 0 & \\ & 0 & & M_2 \end{pmatrix}$$

On the other hand, by Lemma 1, the 3×3-matrix in the left upper corner is diagonalizable, which implies that there exists an invertible matrix $E_3 \in \mathbb{F}_2^{n \times n}$ such that

$$E_3 M E_3^T = \begin{pmatrix} 1 & 0 & 0 & \\ 0 & 1 & 0 & 0 \\ 0 & 0 & 1 & \\ & 0 & & M_2 \end{pmatrix}.$$

Continuing the same reasoning, we can eventually find an invertible matrix $E \in \mathbb{F}_2^{n \times n}$ such that EME^T is diagonal. Since EME^T is invertible, necessarily $EME^T = I$ is the identity matrix.

Now that E is invertible, set $\{\beta_1, \ldots, \beta_n\}$ defined by

$$\beta_i = \sum_{k=1}^n E_{ik} \alpha_k$$

is also a basis of $\mathbb{F}_{2^n}/\mathbb{F}_2$, and

$$\beta_i\beta_j = \sum_{k=1}^{n} E_{ik}\alpha_k \sum_{l=1}^{n} E_{jl}\alpha_l = \sum_{k=1}^{n}\sum_{l=1}^{n} E_{ik}E_{jl}\alpha_k\alpha_l.$$

It follows that

$$\mathrm{Tr}(\beta_i\beta_j) = \sum_{k=1}^{n}\sum_{l=1}^{n} E_{ik}E_{jl}\,\mathrm{Tr}(\alpha_k\alpha_l)$$

$$= \sum_{k=1}^{n}\sum_{l=1}^{n} E_{ik}M_{kl}E_{jl}$$

$$= (EME^T)_{ij} = I_{ij},$$

which proves that basis $\{\beta_1,\ldots,\beta_n\}$ is self-dual.

3   Multiplication Tables

Assume now that $\{\alpha_1,\ldots,\alpha_n\}$ is a self-dual basis of $\mathbb{F}_{2^n}/\mathbb{F}_2$, and choose a *coordinate representation* $e_1 = (1,0,\ldots,0)$, $e_2 = (0,1,\ldots,0)$, ..., $e_n = (0,0,\ldots,1)$ for the basis elements $\alpha_1, \alpha_2 \ldots, \alpha_n$. Moreover, define matrices $C^{(1)}, \ldots, C^{(n)} \in \mathbb{F}_2^{n\times n}$ by condition

$$\alpha_i \cdot \alpha_j = \sum_{k=1}^{n} C_{ij}^{(k)}\alpha_k. \tag{1}$$

Now that e_i is merely a renaming of α_i, we can build up a multiplication operation in $\mathbb{F}_2^n$ by first defining

$$e_i \cdot e_j = \sum_{k=1}^{n} C_{ij}^{(k)}e_k,$$

and then extending this to be a bilinear mapping $\mathbb{F}_2^n \times \mathbb{F}_2^n \to \mathbb{F}_2^n$. This clearly defines a field structure for $\mathbb{F}_2^n$ in such a way, that the vectors $e_1, \ldots, e_n$ of the natural basis form a self-dual basis of $\mathbb{F}_2^n$.

It is clear that the matrices $C^{(k)}$ are symmetric. Examples of multiplication tables of $\mathbb{F}_2^3$ and $\mathbb{F}_2^4$ are shown in Figures 1 and 3. The corresponding matrices are shown in Figures 2 and 4, respectively.

As mentioned above, the matrices found this way are symmetric, but even more interesting symmetries can be found. Consider, for instance, Figure 1 and notice that, by the definition, matrix $C^{(1)}$ is formed by taking the first *columns* which are under vectors $(1,0,0)$, $(0,1,0)$, and $(0,0,1)$ (which are above the horizontal line), respectively. Again by definition, matrix $C^{(2)}$ is formed by taking the second columns under those vectors, and $C^{(3)}$ by picking up the last columns. Interestingly, the same matrices $C^{(1)}$, $C^{(2)}$, and $C^{(3)}$ can be obtained by directly forming 3×3-matrices of the row vectors which lie under vectors $(1,0,0)$, $(0,1,0)$, and $(0,0,1)$ (which are above the horizontal line).

$$
\begin{array}{c|ccc}
\cdot & (1,0,0) & (0,1,0) & (0,0,1) \\
\hline
(1,0,0) & (0,1,0) & (1,0,1) & (0,1,1) \\
(0,1,0) & (1,0,1) & (0,0,1) & (1,1,0) \\
(0,0,1) & (0,1,1) & (1,1,0) & (1,0,0)
\end{array}
$$

Fig. 1. Multiplication table of $\mathbb{F}_2^3$

$$
C^{(1)} = \begin{pmatrix} 0 & 1 & 0 \\ 1 & 0 & 1 \\ 0 & 1 & 1 \end{pmatrix}, \quad
C^{(2)} = \begin{pmatrix} 1 & 0 & 1 \\ 0 & 0 & 1 \\ 1 & 1 & 0 \end{pmatrix}, \quad
C^{(3)} = \begin{pmatrix} 0 & 1 & 1 \\ 1 & 1 & 0 \\ 1 & 0 & 0 \end{pmatrix}
$$

Fig. 2. Matrices corresponding to the multiplication table of $\mathbb{F}_2^3$

$$
\begin{array}{c|cccc}
\cdot & (1,0,0,0) & (0,1,0,0) & (0,0,1,0) & (0,0,0,1) \\
\hline
(1,0,0,0) & (1,1,0,1) & (1,0,0,1) & (0,0,1,1) & (1,1,1,1) \\
(0,1,0,0) & (1,0,0,1) & (0,1,1,1) & (0,1,1,0) & (1,1,0,0) \\
(0,0,1,0) & (0,0,1,1) & (0,1,1,0) & (1,1,1,0) & (1,0,0,1) \\
(0,0,0,1) & (1,1,1,1) & (1,1,0,0) & (1,0,0,1) & (1,0,1,1)
\end{array}
$$

Fig. 3. Multiplication table of $\mathbb{F}_2^4$

$$
C^{(1)} = \begin{pmatrix} 1 & 1 & 0 & 1 \\ 1 & 0 & 0 & 1 \\ 0 & 0 & 1 & 1 \\ 1 & 1 & 1 & 1 \end{pmatrix}, \quad
C^{(2)} = \begin{pmatrix} 1 & 0 & 0 & 1 \\ 0 & 1 & 1 & 1 \\ 0 & 1 & 1 & 0 \\ 1 & 1 & 0 & 0 \end{pmatrix},
$$

$$
C^{(3)} = \begin{pmatrix} 0 & 0 & 1 & 1 \\ 0 & 1 & 1 & 0 \\ 1 & 1 & 1 & 0 \\ 1 & 0 & 0 & 1 \end{pmatrix}, \quad
C^{(4)} = \begin{pmatrix} 1 & 1 & 1 & 1 \\ 1 & 1 & 0 & 0 \\ 1 & 0 & 0 & 1 \\ 1 & 0 & 1 & 1 \end{pmatrix}
$$

Fig. 4. Matrices corresponding to the multiplication table of $\mathbb{F}_2^4$

Another interesting property is that the matrices $C^{(1)}$, $C^{(2)}$, and $C^{(3)}$ sum up to the identity matrix. A more peculiar feature is that one of the matrices (in this example $C^{(1)}$) generate a multiplicative group of order seven, and that group augmented with the zero matrix forms the field of eight elements. The following theorem and its proof clarify the phenomena mentioned above.

Theorem 1. *Let* $\{\alpha_1, \ldots, \alpha_n\}$ *be a self-dual basis of* $\mathbb{F}_{2^n}/\mathbb{F}_2$, *and matrices* $C_{ij}^{(k)}$ *defined as in (1). Then the following conditions hold:*

1. $C_{ij}^{(k)} = C_{kj}^{(i)}$ *for each* i, j, k.
2. *If* $(a_1, \ldots, a_n) \neq (0, \ldots, 0)$, *then* $\det(a_1 C^{(1)} + \ldots + a_n C^{(n)}) \neq 0$.

3. $C^{(i)}C^{(j)} = C^{(j)}C^{(i)}$ for each i, j.

4. $C^{(1)} + \ldots + C^{(n)} = I$.

5. $C^{(i)^T} = C^{(i)}$ for each $i \in \{1, \ldots, n\}$.

Proof. Condition 5 is clear by the definition of the matrices $C^{(i)}$. Taking the traces of the both sides of the equation

$$\alpha_i \alpha_j = \sum_{l=1}^{n} C_{ij}^{(l)} \alpha_l \tag{2}$$

gives

$$\mathrm{Tr}(\alpha_i \alpha_j) = \sum_{l=1}^{n} C_{ij}^{(l)} Tr(\alpha_l) = \sum_{l=1}^{n} C_{ij}^{(l)},$$

which proves 4. Multiplication of (2) by α_k and taking the traces gives

$$\mathrm{Tr}(\alpha_i \alpha_j \alpha_k) = \sum_{l=1}^{n} C_{ij}^{(l)} \, \mathrm{Tr}(\alpha_l \alpha_k) = C_{ij}^{(k)},$$

which shows that the condition 1 is satisfied. In fact, the above equation shows directly that when referring to the matrix element $C_{ij}^{(k)}$, we can permute i, j, and k in any way.

We can now show that the matrices $C^{(i)}$ in fact generate a *matrix representation* of the field $\mathbb{F}_{2^n}$ (the zero matrix must, of course, be included, too): Any mapping $\mathbb{F}_{2^n} \to \mathbb{F}_{2^n}$ defined by rule $x \mapsto \alpha x$ is linear, and can therefore be also considered as a linear mapping $\mathbb{F}_2^n \to \mathbb{F}_2^n$. The matrix of the linear mapping $\mathbb{F}_2^n \to \mathbb{F}_2^n$ corresponding to element α is known as a *matrix representation* of $\alpha \in \mathbb{F}_{2^n}$. It is a well-known fact that the matrices found in this way also form field $\mathbb{F}_{2^n}$ with respect to the ordinary matrix sum and multiplication. The set of matrices found in this way is called a *matrix representation* of field $\mathbb{F}_{2^n}$. In what follows, we use basis $\{\alpha_1, \ldots, \alpha_n\}$ to present those matrices.

Let us fix an element $\alpha = \sum_{i=1}^{n} a_i \alpha_i \in \mathbb{F}_{2^n}$. Then for each basis element α_j we have

$$\alpha \alpha_j = \sum_{i=1}^{n} a_i \alpha_i \alpha_j = \sum_{i=1}^{n} a_i \sum_{k=1}^{n} C_{ij}^{(k)} \alpha_k = \sum_{k=1}^{n} \sum_{i=1}^{n} a_i C_{jk}^{(i)} \alpha_k,$$

which shows that matrix $\sum_{i=1}^{n} a_i C^{(i)}$ is in fact a matrix representation of element $\sum_{i=1}^{n} a_i \alpha_i$. Conditions 2 and 3 follow now immediately. $\square$

Also, a "converse" of the above theorem can be shown true.

Theorem 2. *Let $C^{(1)}, \ldots, C^{(n)} \in \mathbb{F}_2^{n \times n}$ be matrices which satisfy the conditions 1-5 of the previous theorem. If the multiplication in $\mathbb{F}_2^n$ is defined as*

$$e_i \cdot e_j = \sum_{k=1}^{n} C_{ij}^{(k)} e_k$$

and extended to be a bilinear operation, then $\mathbb{F}_2^n$ is becomes a field, and the natural basis is self-dual.

Proof. The additive structure of $\mathbb{F}_2^n$ is trivial, so we only have to consider the multiplicative structure. That the distributive law holds, is straightforward. By the condition 5 we have that

$$e_i \cdot e_j = \sum_{k=1}^{n} C_{ij}^{(k)} e_k = \sum_{k=1}^{n} C_{ji}^{(k)} e_k = e_j \cdot e_i,$$

meaning that the basis elements form a commuting set. It follows directly that all elements of $\mathbb{F}_2^n$ commute in multiplication.

It follows from the properties 1, 5, and 4 that

$$\left(\sum_{i=1}^{n} e_i \right) \cdot e_j = \sum_{i=1}^{n} e_i \cdot e_j = \sum_{i=1}^{n} \sum_{k=1}^{n} C_{ij}^{(k)} e_k = \sum_{k=1}^{n} \sum_{i=1}^{m} C_{jk}^{(i)} e_k = e_j,$$

which shows that $e_1 + \ldots + e_n$ is the unit element with respect to the multiplication.

A direct calculation shows that

$$(e_i \cdot e_j) \cdot e_k = \sum_{r=1}^{n} C_{ij}^{(r)} e_r \cdot e_k = \sum_{s=1}^{n} \sum_{r=1}^{n} C_{ij}^{(r)} C_{rk}^{(s)} e_s.$$

Similarly,

$$e_i \cdot (e_j \cdot e_k) = \sum_{s=1}^{n} \sum_{r=1}^{n} C_{jk}^{(r)} C_{ir}^{(s)} e_s.$$

To prove $(e_i \cdot e_j) \cdot e_k = e_i \cdot (e_j \cdot e_k)$ it remains to show that for each $s \in \{1, \ldots, n\}$, equation

$$\sum_{r=1}^{n} C_{ij}^{(r)} C_{rk}^{(s)} = \sum_{r=1}^{n} C_{jk}^{(r)} C_{ir}^{(s)} \tag{3}$$

holds. But conditions 1 and 5 imply that the left hand side of (3) is equal to

$$\sum_{r=1}^{n} C_{ir}^{(j)} C_{rk}^{(s)} = (C^{(j)} C^{(s)})_{ik},$$

and that the right hand side is equal to

$$\sum_{r=1}^{n} C_{rk}^{(j)} C_{ir}^{(s)} = (C^{(s)} C^{(j)})_{ik} = (C^{(j)} C^{(s)})_{ik},$$

because of the condition 3. It follows that the multiplication is associative.

It remains to demonstrate that each nonzero element has an inverse. For that purpose, assume that an element $a = \sum_{i=1}^{n} a_i e_i$ is nonzero. To find the inverse x of a, we have to solve the equation

$$a \cdot x = 1 = e_1 + \ldots + e_n,$$

which can be written as

$$\sum_{i=1}^{n}\sum_{j=1}^{n} a_i x_j \sum_{k=1}^{n} C_{ij}^{(k)} e_k = \sum_{k=1}^{n} e_k,$$

or, equivalently as

$$\sum_{i=1}^{n}\sum_{j=1}^{n} a_i x_j C_{ij}^{(k)} = 1$$

for each k. Using again 1 and 5, the above can be written as

$$\sum_{j=1}^{n}\sum_{i=1}^{n} a_i C_{kj}^{(i)} x_j = 1. \tag{4}$$

Now that $(a_1, \ldots, a_n) \neq (0, \ldots, 0)$, matrix $a_1 C^{(1)} + \ldots + a_n C^{(n)}$ is invertible by condition 2, and hence the system of equations (4) is solvable. □

Remark 1. Condition 2 of Theorem 1 seems to be the most complicated one, but unfortunately it cannot be derived from the other conditions, even though the matrices were all invertible. To see why this holds, consider the following 4×4-matrices

$$C^{(1)} = \begin{pmatrix} 0\,0\,0\,1 \\ 0\,0\,1\,1 \\ 0\,1\,0\,1 \\ 1\,1\,1\,1 \end{pmatrix}, C^{(2)} = \begin{pmatrix} 0\,0\,1\,1 \\ 0\,0\,1\,0 \\ 1\,1\,1\,1 \\ 1\,0\,1\,0 \end{pmatrix},$$

$$C^{(3)} = \begin{pmatrix} 0\,1\,0\,1 \\ 1\,1\,1\,1 \\ 0\,1\,0\,0 \\ 1\,1\,0\,0 \end{pmatrix}, C^{(4)} = \begin{pmatrix} 1\,1\,1\,1 \\ 1\,0\,1\,0 \\ 1\,1\,0\,0 \\ 1\,0\,0\,0 \end{pmatrix},$$

which are all invertible. A direct calculation shows that they satisfy all conditions but the second one, since $\det(C^{(1)} + C^{(4)}) = 0$. This proves that the above matrices cannot be used to create a field structure to $\mathbb{F}_2^4$. On the other hand, the proof of Theorem 2 shows that the above matrices induce a commutative ring structure in $\mathbb{F}_2^4$, since condition 2 was used only to prove the existence of inverses of non-zero elements.

References

1. Y. Brandman, A. Orlitsky, and J. Hennessy: *A Spectral Lower Bound Technique for the Size of Decision Trees and Two-Level* AND/OR *Circuits*, IEEE Transactions on Computers 39:2, 282–287 (1990).
2. M. Hirvensalo: *Quantum Computing*, Springer (2001).
3. M. Hirvensalo: *Studies on Boolean Functions Related to Quantum Computing*, Ph.D Thesis, University of Turku (2003).
4. A. Lempel: *Matrix factorization over GF(2) and trace-orthogonal bases of GF(2^n)**, SIAM Journal on Computing 4:2, 175–186 (1975).

5. F.J. MacWilliams and N.J.A. Sloane: *The theory of error-correcting codes*, North-Holland (1981).
6. R.C. Mullin, I.M. Onyszchuk, S.A. Vanstone, and R.M. Wilson: *Optimal normal bases in $GF(p^n)^*$*, Discrete Applied Mathematics 22, 149–161 (1989).
7. C. H. Papadimitriou: *Computational Complexity*, Addison-Wesley (1994).

On NFA Reductions

Lucian Ilie[1,*], Gonzalo Navarro[2,**], and Sheng Yu[1,***]

[1] Department of Computer Science, University of Western Ontario
N6A 5B7, London, Ontario, CANADA
ilie|syu@csd.uwo.ca
[2] Department of Computer Science, University of Chile
Blanco Encalada 2120, Santiago, CHILE
gnavarro@dcc.uchile.cl

Abstract. We give faster algorithms for two methods of reducing the number of states in nondeterministic finite automata. The first uses equivalences and the second uses preorders. We develop restricted reduction algorithms that operate on position automata while preserving some of its properties. We show empirically that these reductions are effective in largely reducing the memory requirements of regular expression search algorithms, and compare the effectiveness of different reductions.

1 Introduction

Regular expression handling is at the heart of many applications, such as linguistics, computational biology, pattern recognition, text retrieval, and so on. An elegant theory gives the support to easily and efficiently solve many complex problems by mapping them to regular expressions, then obtaining a nondeterministic finite automaton (NFA) that recognizes it, and finally making it deterministic (a DFA). However, a severe obstacle in any real implementation of the above scheme is the size of the DFA, which can be exponential in the length of the original regular expression.

Although a simple algorithm for minimizing DFAs exists [5], it has the problem of requiring prior construction of the DFA to later minimize it. This can be infeasible because of main memory requirements and construction cost.

A much more promising (and more challenging) alternative is that of directly reducing the NFA before converting it into a DFA. This has the advantage of working over a much smaller structure (of size polynomial in the length of the regular expression) and of building the smaller DFA without the need to go through a larger one first.

However, the NFA state minimization problem is very hard (PSPACE-complete, [10]) and therefore algorithms such as [11, 13, 14] cannot be used in practice. There are also algorithms which build small NFAs from regular expressions,

---

 * Research partially supported by NSERC.
 ** Supported in part by Fondecyt grant 1-020831.
* * * Research partially supported by NSERC.

J. Karhumäki et al. (Eds.): Theory Is Forever (Salomaa Festschrift), LNCS 3113, pp. 112–124, 2004.
© Springer-Verlag Berlin Heidelberg 2004

see [7,4], but they consider the total size, that is, they count both states and transitions, and they increase artificially the number of states to reduce the number of transitions. As the implementation crucially depends on the number of states, such algorithms may not help.

The approach we follow is reducing the size of a given NFA. The idea of reducing the size of NFAs by merging states was first introduced by Ilie and Yu [8] who used equivalence relations. Later, Champarnaud and Coulon [2] modified the idea to work for preorders. In this paper we give fast algorithms to compute these two reductions. We show that the algorithm based on equivalences can be implemented in $O(m \log n)$ time on an NFA with n states and m transitions, while that based on preorders can run in $O(mn)$ time. Both results improve the previous work.

When starting from a regular expression, the initial NFA, which we want to reduce, is the position automaton. Navarro and Raffinot [17, 18] showed that its special properties permit a more compact DFA representation. Our modified reductions are restricted to preserve those properties and hence may produce NFAs with more states than the original reductions.

Finally, we empirically evaluate the impact of the reduction algorithms. We show that the number of NFA states can be reduced by 10%–40%. Those reductions translate into huge reductions in the DFA size, with factors of up to 10^{-6}. We also compare the alternatives of full reduction versus restricted reduction, since the former yields less NFA states but the latter permits a more compact DFA representation. The results show that full reduction is preferable in most cases of interest.

2   Basic Notions

We recall here the basic definitions we need throughout the paper. For further details we refer to [6] or [22].

Let A be an alphabet and A^* the set of all words over A; ε denotes the empty word. A *language* over A is a subset of A^*. A *nondeterministic finite automaton* (*NFA*) is a tuple $M = (Q, A, \delta, I, F)$, where Q is the set of states, $I \subseteq Q$ is the set of initial states, $F \subseteq Q$ is the set of final states, and $\delta : Q \times A \to 2^Q$ is the transition mapping; δ is extended to $\delta : 2^Q \times A^* \to 2^Q$ by $\delta(S, a) = \bigcup_{q \in S} \delta(q, a)$ and $\delta(S, \varepsilon) = S$, $\delta(S, aw) = \delta(\delta(S, a), w)$, for $S \subseteq Q$, $w \in A^*$. The *language* recognized by M is $\mathcal{L}(M) = \{w \in A^* \mid \delta(I, w) \cap F \neq \emptyset\}$. For a state $q \in Q$, we denote

$$\mathcal{L}_L(M, q) = \{w \in A^* \mid q \in \delta(I, w)\},$$
$$\mathcal{L}_R(M, q) = \{w \in A^* \mid \delta(q, w) \cap F \neq \emptyset\};$$

when M is understood, we write simply $\mathcal{L}_L(q)$ and $\mathcal{L}_R(q)$, resp. The *reversed* automaton of M is $M^r = (Q, A, \delta^r, F, I)$, where $q \in \delta^r(p, a)$ iff $p \in \delta(q, a)$.

3   NFA Reduction with Equivalences

The idea of reducing the size of NFAs by merging state was investigated first by Ilie and Yu [8]; see also [9]. We describe it briefly in this section.

Let $M = (Q, A, \delta, I, F)$ be an NFA. We define $\equiv_R$ as the coarsest equivalence relation over Q that satisfies:

(P_1) $\equiv_R \cap (F \times (Q - F)) = \emptyset$,
(P_2) for any $p, q \in Q, a \in A, \left(p \equiv_R q \Rightarrow \forall q' \in \delta(q, a), \exists p' \in \delta(p, a), q' \equiv_R p'\right)$.

The equivalence $\equiv_R$ is the largest equivalence over Q which is right-invariant w.r.t. M; see [8, 9]. Given $\equiv_R$, the algorithm to reduce the automaton M using it is trivial: simply merge all states in the same equivalence class and modify the transitions accordingly. Here is an example.

Example 1. The NFA in Fig. 4 is reduced using $\equiv_R$ as shown in Fig. 1; the equivalence classes are also shown.

classes of $\equiv_R$: $\{0\}$
$\{1, 2, 3, 4, 5, 6\}$

Fig. 1. $\mathbf{A}_R(\tau) = \mathbf{A}_{\mathrm{pos}}(\tau)/_{\equiv_R}$ for $\tau = (a + b)(a^* + ba^* + b^*)^*$

Symmetrically, the relation $\equiv_L$ can be defined using the reversed automaton. The automaton M can be reduced according to either equivalence. As examples in [9] show, M can be reduced more using both equivalences but the problem of finding the best way to do the reduction is open. Fig. 2 gives an example (from [9]) where there is no unique way to reduce optimally using both $\equiv_R$ and $\equiv_L$.

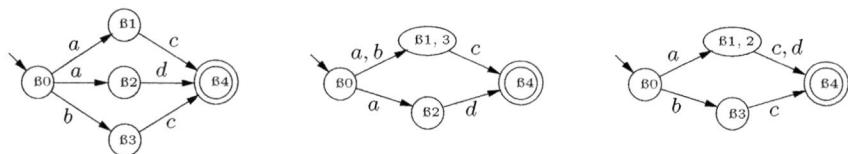

Fig. 2. An NFA and its corresponding quotients modulo $\equiv_R$ and $\equiv_L$

4   Computing Equivalences

The algorithm in [8] for computing $\equiv_R$ runs in low polynomial time but the problem of finding a fast algorithm was left open. We show here that an old very fast algorithm of Paige and Tarjan [19] can be used to solve the problem.

Recall some definitions from [19]. For a binary relation E over a finite set U we denote, for any subset $S \subseteq U$, $E^{-1}(S) = \{x \mid \exists y \in S \text{ such that } xEy\}$. A subset $B \subseteq U$ is called *stable w.r.t.* S if either $B \subseteq E^{-1}(S)$ or $B \cap E^{-1}(S) = \emptyset$. A partition P of U is *stable w.r.t.* S if all the blocks of P are stable w.r.t. S.

P is *stable* if it is stable w.r.t. each of its own blocks. The *relational coarsest partition problem* is that of finding, for a given relation E and a partition P over a set U, the coarsest stable refinement of P. Paige and Tarjan [19] gave an algorithm for this problem which runs in time $\mathcal{O}(m \log n)$ and space $\mathcal{O}(m + n)$, where $n = \mathsf{card}(U)$, $m = \mathsf{card}(E)$. They remarked that the algorithm works also for several relations.

This algorithm applies to our problem of finding $\equiv_R$ as follows. For any $a \in A$, denote $\delta_a = \{(p, q) \in Q \times Q \mid q \in \delta(p, a)\}$. Then $\equiv_R$ is the coarsest stable refinement of the partition $\{F, Q - F\}$ w.r.t. all relations δ_a, $a \in A$.

Therefore, if the number of states in our automaton is n and the number of transitions is m, we have the following theorem.

Theorem 1. *The equivalences $\equiv_R$ and $\equiv_L$ can be computed in time $\mathcal{O}(m \log n)$ and space $\mathcal{O}(m + n)$.*

It is interesting to notice that, to reduce NFAs by equivalences, we employed an idea from deterministic finite automata (DFA) reduction and then, to make it fast, we used an algorithm which was inspired itself from Hopcroft's algorithm [5] to reduce DFAs.

5   NFA Reduction with Preorders

Champarnaud and Coulon [2] noticed that a better reduction can be obtained if the axioms (P_1) and (P_2) above are used to construct a preorder relation instead of an equivalence. Let us denote the largest (w.r.t. inclusion) preorder which verifies (P_1) and (P_2) by $\subseteq_R$. It is then immediate that $p \subseteq_R q$ implies $\mathcal{L}_R(p) \subseteq \mathcal{L}_R(q)$.

As in the case of equivalences, the relation $\subseteq_L$ is symmetrically defined using the reversed automaton. Then, $p \subseteq_L q$ implies $\mathcal{L}_L(p) \subseteq \mathcal{L}_L(q)$.

The reduction with preorders is more complicated than with equivalences. First, we can merge two states p and q as soon as any of the following conditions is met:

(i) $p \subseteq_R q$ and $q \subseteq_R p$,
(ii) $p \subseteq_L q$ and $q \subseteq_L p$,
(iii) $p \subseteq_R q$ and $p \subseteq_L q$.

However, after merging two states, the preorders $\subseteq_R$ and $\subseteq_L$ must be updated such that their relation with the languages $\mathcal{L}_R$ and $\mathcal{L}_L$ (see above) is preserved. For instance, in the case (i), assuming the merged state of p and q is denoted q, the update amounts to removing from $\subseteq_R$ all pairs (q, s) for which $p \not\subseteq_R s$. Case (ii) is handled similarly and (iii) does not need any update.

An open problem here is how to merge the states using the two preorders such that the reduction of the NFA is optimal; see the example in Fig. 2.

Since the preorder requirement is weaker than equivalence, $p \equiv_R q$ implies that $p \subseteq_R q$ and $q \subseteq_R p$. The converse is not true in general (see [2] for an example). Therefore, using preorders we have a chance to obtain a better reduction of the NFA. It remains to investigate how much better.

6   Computing Preorders

We give here an algorithm to compute the preorders $\subseteq_R$ and $\subseteq_L$. Assuming that $Q = \{1, 2, \ldots, n\}$ and the number of transitions is m, our algorithm runs in time $\mathcal{O}(mn)$ and space $\mathcal{O}(n^2)$. The best algorithm given by Champarnaud and Coulon [2] runs in time $\mathcal{O}(mn^2)$.

We shall compute the complement $\not\subseteq_R$ of $\subseteq_R$ by the algorithm PREORDER(M) from Fig. 3; ω is the relation which is $\not\subseteq_R$ at the end. According to the definition of $\subseteq_R$, its complement $\not\subseteq_R$ is the smallest relation over Q such that

(P_1') $(F \times (Q - F)) \subseteq \not\subseteq_R$,
(P_2') for any $i, j \in Q, a \in A$, $\left(\exists i' \in \delta(i, a), \forall j' \in \delta(j, a), i' \not\subseteq_R j' \Rightarrow i \not\subseteq_R j\right)$.

So, we add (i, j) to $\not\subseteq_R$ based on the fact that there is $i' \in \delta(i, a)$ for which the number of those $j' \in \delta(j, a)$ with $i' \not\subseteq_R j'$ is precisely $\mathsf{card}(\delta(j, a))$; that is, all j''s. Therefore, we shall compute some matrices of counters $N(a)$, for any $a \in A$; $N(a)$ is a $n \times n$ matrix such that

$$N(a)_{ij} = \mathsf{card}(\{\ell \in \delta(j, a) \mid i \not\subseteq_R \ell\}),$$

for all $i, j \in Q$. We start with all these counters set to zero and update them anytime there is new information on $\not\subseteq_R$; any new pair added to $\not\subseteq_R$ is enqueued (steps 9 and 19) and later dequeued (step 11) and processed such that all counters involved are adequately updated (step 14). Anytime such a counter $N(a)_{ik}$ reaches maximum value $\mathsf{card}(\delta(k, a))$ (step 15), all pairs (j, k) such that $i \in \delta(j, a)$ are added to $\not\subseteq_R$ if not already there (steps 16–18).

Let us show that the algorithm PREORDER(M) computes correctly the preorder $\not\subseteq_R$. First, it is clear that $\not\subseteq_R$ is obtained by adding all pairs in (P_1') and then using (P_2') as long as pairs can still be added. Assume then

$$\not\subseteq_R = \{(i_1, j_1), \ldots, (i_r, j_r), \ldots, (i_s, j_s)\},$$

where the first r pairs are added because of (P_1') and the remaining ones due to (P_2'). We show that the algorithm PREORDER(M) computes the same relation. Denote by ω the relation computed by the algorithm. Obviously, $\omega \subseteq \not\subseteq_R$. Assume there is (i_t, j_t) in $\not\subseteq_R$ but not in ω; consider such a pair with the lowest index t. It must be that $t > r$ since all pairs in $F \times (Q - F)$ are certainly added to ω. As (i_t, j_t) is in $\not\subseteq_R$, there must be an $i' \in \delta(i_t, a)$ such that (i_t, j_t) was added to $\not\subseteq_R$ because all pairs $(i', j'), j' \in \delta(j_t, a)$, were already in $\not\subseteq_R$. Thus, at least one of those pairs (i', j') is not in ω. Since the index of (i', j') is strictly smaller than t, a contradiction is obtained.

The time complexity of the above algorithm is $\mathcal{O}(m+n^2)$ for the preprocessing and proportional to

$$\sum_{i=1}^{n} \sum_{j=1}^{n} \sum_{a \in A} (\mathsf{card}(\delta^r(i, a)) + \mathsf{card}(\delta^r(j, a))) = \mathcal{O}(mn)$$

for processing. Therefore, the time complexity is $\mathcal{O}(mn)$. The space complexity is $\mathcal{O}(n^2)$. We have proved the following theorem.

PREORDER(M)

- given: an NFA M
- returns: $\not\subseteq_R$

1. **for** $q \in Q, a \in A$ **do**	//1–4: preprocessing
2. compute $\delta^r(q,a)$ as a linked list	
3. compute $\mathrm{card}(\delta(q,a))$	
4. initialize all $N(a)$s with 0s	
5. $\omega \leftarrow \emptyset$, $\mathcal{C} \leftarrow$ NEWQUEUE()	//5–19: processing
6. **for** $i \in F$ **do**	//6–8: initialize ω
7. **for** $j \in Q - F$ **do**	//ω will be $\not\subseteq_R$ at the end
8. $\omega \leftarrow \omega \cup \{(i,j)\}$	
9. ENQUEUE($\mathcal{C}, (i,j)$)	
10. **while** $\mathcal{C} \neq \emptyset$ **do**	
11. $(i,j) \leftarrow$ DEQUEUE($\mathcal{C}$)	//11–19: updates due to
12. **for** $a \in A$ **do**	//(i,j) being added to ω
13. **for** $k \in \delta^r(j,a)$ **do**	
14. $N(a)_{ik} \leftarrow N(a)_{ik} + 1$	//14: update counter
15. **if** $N(a)_{ik} = \mathrm{card}(\delta(k,a))$ **then**	//15–18: update ω
16. **for** $j \in \delta^r(i,a)$ **do**	//when a counter is maximal
17. **if** $(j,k) \notin \omega$ **then**	
18. $\omega \leftarrow \omega \cup \{(j,k)\}$	
19. ENQUEUE($\mathcal{C}, (j,k)$)	
20. **return** ω	

Fig. 3. Algorithm for computing preorders

Theorem 2. *The preorders $\subseteq_R$ and $\subseteq_L$ can be computed in time $\mathcal{O}(mn)$ and space $\mathcal{O}(n^2)$.*

7   Position Automaton

We recall in this section the well-known construction of the position automaton[3], discovered independently by Glushkov [3] and McNaughton and Yamada [12].

Let α be a regular expression. The basic idea of the position automaton is to assume that all occurrences of letters in α are different. For this, all letters are made different by marking each letter with a unique index, called its *position* in α. The set of positions of α is $\mathsf{pos}(\alpha) = \{1, 2, \ldots, |\alpha|_A\}$, where $|\alpha|_A$ is the number of letter occurrences in α. We shall denote also $\mathsf{pos}_0(\alpha) = \mathsf{pos}(\alpha) \cup \{0\}$. The expression obtained from α by marking each letter with its position is denoted $\overline{\alpha} \in \overline{A}^*$, where $\overline{A} = \{a_i \mid a \in A, 1 \leq i \leq |\alpha|_A\}$. For instance, if $\alpha = a(baa + b^*)$, then $\overline{\alpha} = a_1(b_2 a_3 a_4 + b_5^*)$. Notice that $\mathsf{pos}(\alpha) = \mathsf{pos}(\overline{\alpha})$. The same notation is also used for unmarking, that is, $\overline{\overline{a}} = a$.

Three mappings first, last, and follow are then defined as follows (see [3]). For any regular expression α and any $i \in \mathsf{pos}(\alpha)$, we have:

---
[3] This automaton is sometimes called *Glushkov automaton*; e.g., in [18].

$$\text{first}(\alpha) = \{i \mid a_i w \in \mathcal{L}(\overline{\alpha})\},$$
$$\text{last}(\alpha) = \{i \mid w a_i \in \mathcal{L}(\overline{\alpha})\}, \tag{1}$$
$$\text{follow}(\alpha, i) = \{j \mid u a_i a_j v \in \mathcal{L}(\overline{\alpha})\}.$$

We extend follow by $\text{follow}(\alpha, 0) = \text{first}(\alpha)$. Also, let $\text{last}_0(\alpha)$ stand for $\text{last}(\alpha)$ if $\varepsilon \notin \mathcal{L}(\alpha)$ and $\text{last}(\alpha) \cup \{0\}$ otherwise.

The *position automaton* for α is

$$\mathbf{A}_{\text{pos}}(\alpha) = (\text{pos}_0(\alpha), A, \delta_{\text{pos}}, 0, \text{last}_0(\alpha))$$

where

$$\delta_{\text{pos}} = \{(i, a, j) \mid j \in \text{follow}(\alpha, i), a = \overline{a_j}\}.$$

Besides the property of accepting the language expressed by the original regular expression, that is, $\mathcal{L}(\mathbf{A}_{\text{pos}}(\alpha)) = \mathcal{L}(\alpha)$, the position automaton has two very important properties. First, the number of states is always $|\alpha|_A + 1$, which makes it work better then Thompson's automaton [20] in bit-parallel regular expression search algorithms [18]. Second, all transitions incoming to any given state are labelled by the same letter, a property exploited by Navarro and Raffinot [17, 18] in regular expression search algorithms to represent the DFA using $O(2^{|\alpha|_A} + |A|)$ bit-masks of length $|\alpha|_A$, rather than $O(2^{|\alpha|_A} |A|)$.

Example 2. Consider the regular expression $\tau = (a + b)(a^* + ba^* + b^*)^*$. The marked version of τ is $\overline{\tau} = (a_1 + b_2)(a_3^* + b_4 a_5^* + b_6^*)^*$. The values of the mappings first, last, and follow for τ and the corresponding position automaton $\mathbf{A}_{\text{pos}}(\tau)$ are given in Fig. 4.

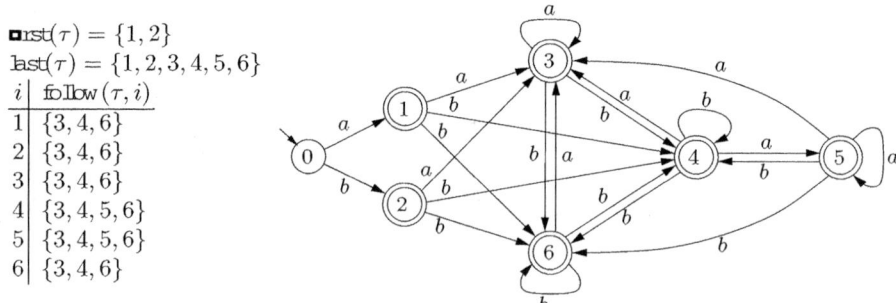

$\text{first}(\tau) = \{1, 2\}$
$\text{last}(\tau) = \{1, 2, 3, 4, 5, 6\}$

i	$\text{follow}(\tau, i)$
1	$\{3, 4, 6\}$
2	$\{3, 4, 6\}$
3	$\{3, 4, 6\}$
4	$\{3, 4, 5, 6\}$
5	$\{3, 4, 5, 6\}$
6	$\{3, 4, 6\}$

Fig. 4. $\mathbf{A}_{\text{pos}}(\tau)$ for $\tau = (a + b)(a^* + ba^* + b^*)^*$

The position automaton can be computed easily in cubic time using the inductive definitions of first, last, and follow, but Brüggemann-Klein [1] showed how to compute it in quadratic time.

8   Reducing the Position Automaton

In this section we show how the position automaton can be reduced using equivalences and/or preorders such that its essential properties are preserved.

Consider a regular expression α and define the equivalence $\sim_\ell$ over $\mathsf{pos}_0(\alpha)$ by $i \sim_\ell j$ iff the letter labelling all transitions incoming to i is the same as the one for j.

The idea is to reduce the position automaton such that the transitions incoming to a given state are still labelled the same. Therefore, any states we merge must be in $\sim_\ell$. Using equivalences, say $\equiv_R$, we merge according to the equivalence $\equiv_R \cap \sim_\ell$. Fig. 5 shows an example.

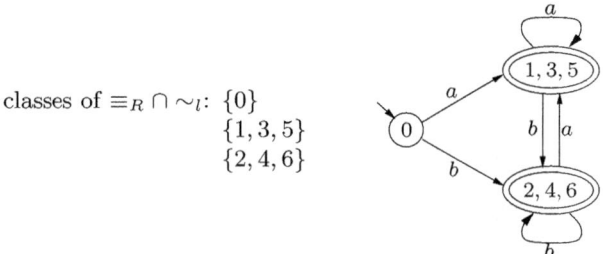

classes of $\equiv_R \cap \sim_l$: $\{0\}$
$\{1,3,5\}$
$\{2,4,6\}$

Fig. 5. $\mathbf{A}_{\mathrm{pos}}(\tau)/_{\equiv_R \cap \sim_l}$ for $\tau = (a+b)(a^* + ba^* + b^*)^*$

Using preorders, we do just as before with the restriction imposed by $\sim_\ell$.

9   Experimental Results

In this section we aim at establishing how significant is the reduction obtained using equivalences, and its relevance to regular expression search algorithms. In particular, we are interested in comparing two choices: *full right-equivalence*, where the properties of the position automaton are not preserved (Section 3), and *restricted right-equivalence*, where those properties are preserved (Section 8). While full right-equivalence can potentially yield larger reductions in number of NFA states, it requires a representation of $O(2^{n_f}|A|)$ cells for the DFA (n_f is the number of reduced NFA states, A is the alphabet). Restricted right-equivalence may yield more states, but permits a more compact representation in $O(2^{n_r}+|A|)$ cells for the DFA (n_r is the number of NFA states after the restricted reduction).

We have tested regular expressions on DNA, having 10 to 100 alphabet symbols, averaging over 10,000 expressions per length, with density of operators from 0.1 to 0.4 (Section 9.1 gives more details on the generation process). For each such expression, we built its position automaton (Section 7), and then applied full and restricted reduction. Figure 6 shows the reductions obtained, as a fraction of the original number of states (which was always $n = 1 + |\alpha|_A$ because of Glushkov's construction). There is a second version of the full reduction, where the NFA is previously modified to include a self-loop at the initial state, for search purposes. This permits no further restricted reduction, but allows a slightly better full reduction, as it can be seen. Both reduction factors tend to stabilize as n grows, being better for higher density of operators in the regular expression. It is also clear that full reduction gives substantially better reductions compared to restricted reduction (10%-20% better).

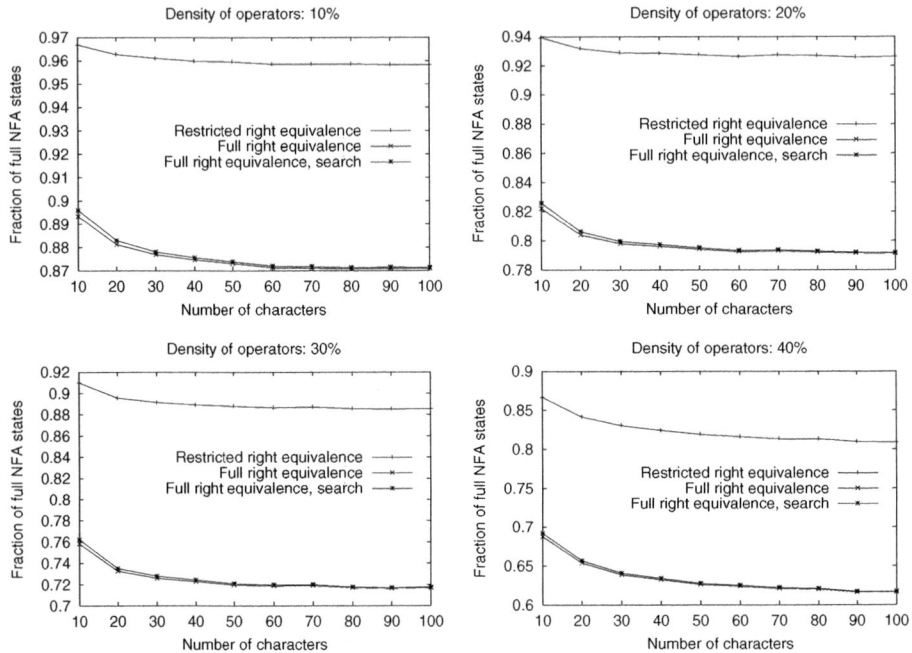

Fig. 6. Reduction factors in number of states obtained over position automata built from regular expressions of lengths 10 to 100 and density of operators from 0.1 to 0.4, built from DNA text

As explained, the above does not immediately mean that full reduction is better, because its DFA representation must have one table per alphabet letter. Figure 7 shows the reduction fraction in the representation of the DFA, compared to that of the position automaton. This time the difference between the search and the original automaton are negligible. As it can be seen, the restricted reduction is convenient only for $n \leq 10$ to $n \leq 30$, depending on the operator density. Note, on the other hand, that those short expression lengths imply that even the original position automaton is not problematic in terms of space or construction cost. That is, full reduction becomes superior precisely when the space problem becomes important.

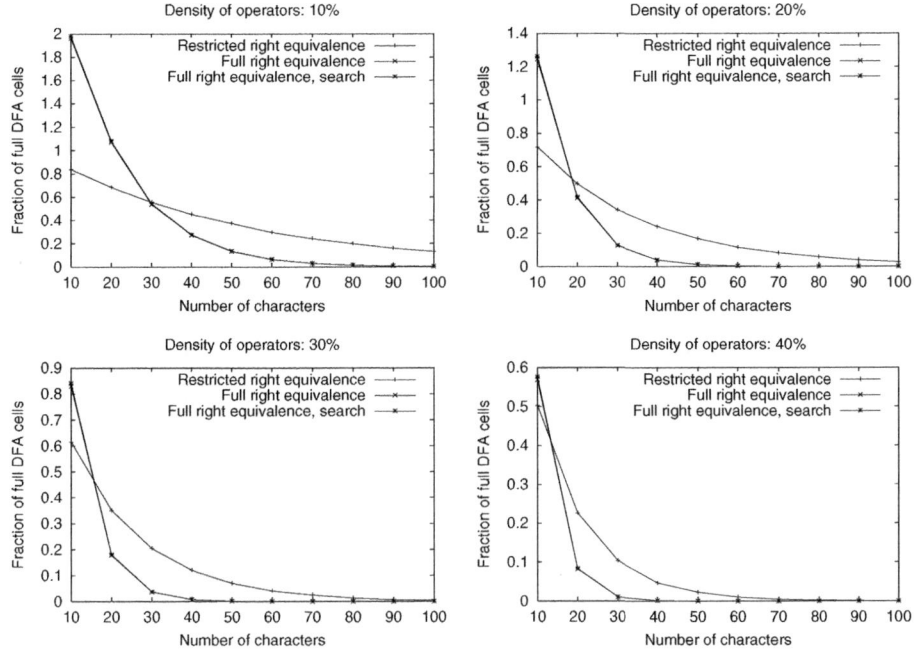

Fig. 7. Reduction factors in DFA sizes obtained over position automata built from regular expressions of lengths 10 to 100 and density of operators from 0.1 to 0.4, built from DNA text

Figure 8 shows the same results in logarithmic scale, to show how large are the savings due to full reductions when n becomes large. The second version of NFA (for searching) is omitted since the difference in DFA size is unnoticeable.

As noted in [17, 18], DFA space and construction cost can be traded for search cost as follows: A single table of $O(2^n)$ entries can be split into k tables of size $O(2^{n/k})$ each, so that each such table has to be accessed for each text character in order to build the original entry. Hence the search cost becomes $O(k)$ per text character. If main memory is limited, a huge DFA actually means larger search time, and a reduction in its size translate into better search times. We have computed the number of tables needed with and without reductions assuming that we dedicate 4 megabytes of RAM to the DFA. For the highest operator density we have obtained speedups of up to 50% in search times, that is, we need 2/3 of the tables needed by the original automaton.

9.1 Generating Regular Expressions

The choice of test patterns is always problematic when dealing with regular expressions, since there is no clear concept of what a random regular expression is and, as far as we know, there is no public repository of regular expressions

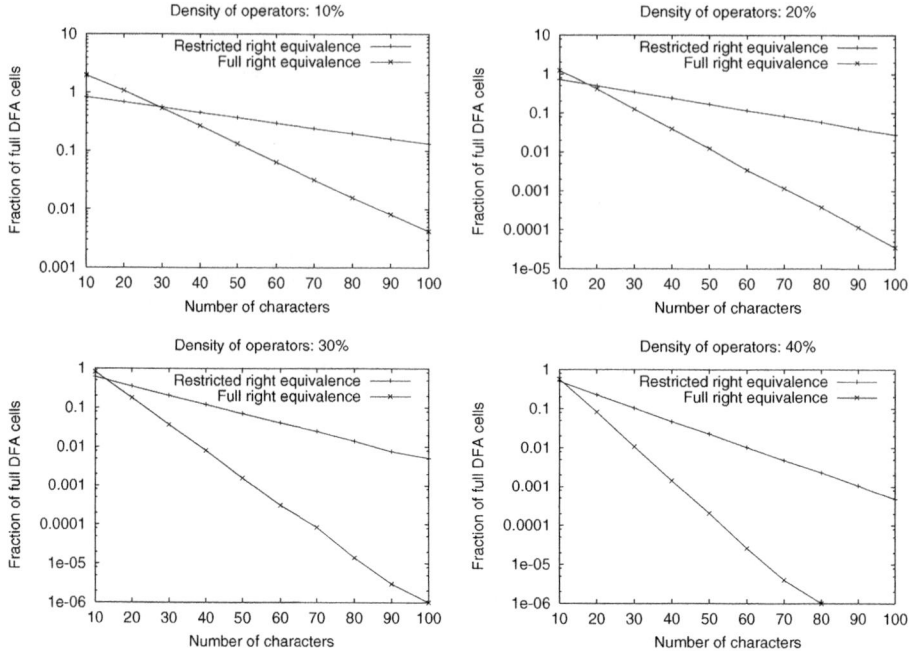

Fig. 8. Reduction factors in DFA sizes obtained over position automata built from regular expressions of lengths 10 to 100 and density of operators from 0.1 to 0.4, built from DNA text. Note the logarithmic scale

available, except for a dozen of trivial examples. We have chosen to generate random regular expressions as follows:

1. We choose a base real-world text, in this case DNA from Homo Sapiens.
2. We choose n and pick a random text substring of length n.
3. We choose an *operator density* $0 \leq \gamma \leq 1$.
4. We apply a recursive procedure to convert a string of length ℓ into a regular expression:
   (a) An empty string is converted into an empty regular expression. In the rest, we assume a nonempty string.
   (b) With probability $1 - \gamma$ we choose that the expression will be the concatenation of two subexpressions: a left part of ℓ' characters and a right part of $\ell - \ell'$ characters, where ℓ' is chosen uniformly in the range $1 \leq \ell' \leq \ell - 1$. We recursively convert both subparts into regular expressions e_1 and e_2. The resulting expression is $e_1 \cdot e_2$. If $\ell = 1$ we simply write down the string character.
   (c) Otherwise, if the parent in the recursion has just generated a Kleene closure operator "$*$", we choose to add a union operator "$|$", if not, we choose with the same probability between a Kleene closure and a union.

(d) If we chose that the expression will have a union operator, we choose a left part of ℓ' characters and a right part of $\ell - \ell'$ characters, where ℓ' is chosen uniformly in the range $0 \leq \ell' \leq \ell$. We recursively convert both subparts into regular expressions e_1 and e_2. The resulting expression is $e_1|e_2$.

(e) If we chose to add a Kleene closure operator "$*$" at the end of the string, we recursively generate a regular expression e_1 for the string. The resulting expression is e_1*.

The above procedure is just one of the many possible alternatives to generate random regular expressions one could argue for, but it has a couple of advantages. First, it permits determining the length n (number of characters of A) in advance. Second, it takes the characters from the text, respecting its distribution. Third, it permits us to choose expressions with more or less operators by varying γ. We show experiments with $\gamma = 0.10$ to $\gamma = 0.40$. Examples obtained from our tests, with $n = 10$, are "$ACAT(T|\varepsilon)TT * AG(T|\varepsilon)$" and "$A(CAT(\varepsilon|T*) * ((\varepsilon|\varepsilon|T) * TA*) * (\varepsilon|\varepsilon|GT))*$", respectively.

10 Conclusion

We have developed faster algorithms to implement two existing NFA reduction techniques. We have also adapted them to work over position automata while preserving their properties that allow a compact DFA representation. Finally, we have empirically assessed the practical impact of the reductions, as well as the convenience of preserving or not the position automata properties.

Future work involves empirically evaluating the impact of using preorders instead of equivalences. The former are more complex and slower to compute, and it is not clear which is the optimal way to apply the different reductions, hence the importance of determining their practical value.

References

1. Brüggemann-Klein, A., Regular expressions into finite automata, *Theoret. Comput. Sci.* **120** (1993) 197 – 213.
2. Champarnaud, J.-M., and F. Coulon, NFA reduction algorithms by means of regular inequalities, in: Z. Ésik, Z. Fülöp, eds., *Proc. of DLT 2003* (Szeged, 2003), Lecture Notes in Comput. Sci. **2710**, Springer-Verlag, Berlin, Heidelberg, 2003, 194 – 205.
3. Glushkov, V.M., The abstract theory of automata, *Russian Math. Surveys* **16** (1961) 1 – 53.
4. Hagenah, C., and Muscholl, A., Computing ϵ-free NFA from regular expressions in $O(n \log^2(n))$ time, *Theor. Inform. Appl.* **34** (4) (2000) 257 – 277.
5. Hopcroft, J., An $n \log n$ algorithm for minimizing states in a finite automaton, *Proc. Internat. Sympos. Theory of machines and computations*, Technion, Haifa, 1971, Academic Press, New York, 1971, 189–196.

6. Hopcroft, J.E., and Ullman, J.D., *Introduction to Automata Theory, Languages, and Computation*, Addison-Wesley, Reading, Mass., 1979.
7. Hromkovic, J., Seibert, S., and Wilke, T., Translating regular expressions into small ε-free nondeterministic finite automata, *J. Comput. System Sci.* **62** (4) (2001) 565 – 588.
8. Ilie, L., and Yu, S., Algorithms for computing small NFAs, in: K. Diks, W. Rytter, eds., *Proc. of the 27th MFCS*, (Warszawa, 2002), Lecture Notes in Comput. Sci., **2420**, Springer-Verlag, Berlin, Heidelberg, 2002, 328 – 340.
9. Ilie, L., and Yu, S., Reducing NFAs by invariant equivalences, *Theoret. Comput. Sci.* **306** (2003) 373 – 390.
10. Jiang, T., and Ravikumar, B., Minimal NFA problems are hard, *SIAM J. Comput.* **22**(6) (1993), 1117 – 1141.
11. Kameda, T., and Weiner, P., On the state minimization of nondeterministic finite automata, *IEEE Trans. Computers* **C-19**(7) (1970) 617 – 627.
12. McNaughton, R., and Yamada, H., Regular expressions and state graphs for automata, *IEEE Trans. on Electronic Computers* **9** (1) (1960) 39 – 47.
13. Melnikov, B. F., A new algorithm of the state-minimization for the nondeterministic finite automata, *Korean J. Comput. Appl. Math.* **6**(2) (1999) 277 – 290.
14. Melnikov, B. F., Once more about the state-minimization of the nondeterministic finite automata, *Korean J. Comput. Appl. Math.* **7**(3) (2000) 655–662.
15. Navarro, G., NR-grep: a Fast and Flexible Pattern Matching Tool, *Software Practice and Experience* **31** (2001) 1265 – 1312.
16. Navarro, G., and Raffinot, M., Fast Regular Expression Search, *Proc. WAE'99*, Lecture Notes Comput. Sci. **1668**, Springer-Verlag, Berlin, Heidelberg, 1999, 198 – 212.
17. Navarro, G., and Raffinot, M., Compact DFA Representation for Fast Regular Expression Search, *Proc. WAE'01*, Lecture Notes Comput. Sci. **2141**, Springer-Verlag, Berlin, Heidelberg, 2001, 1 – 12.
18. Navarro, G., and Raffinot, M., *Flexible Pattern Matching in Strings. Practical On-Line Search Algorithms for Texts and Biological Sequences*, Cambridge University Press, Cambridge, 2002.
19. Paige, R., and Tarjan, R.E, Three Partition Refinement Algorithms, *SIAM J. Comput.* (1987) **16**(6) 973 – 989.
20. Thompson, K., Regular expression search algorithm, *Comm. ACM* **11** (6) (1968) 419 – 422.
21. Wu, S., and Mamber, U., Fast text searching allowing errors, *Comm. ACM* **35**(10) (1992) 83 – 91.
22. Yu, S., Regular Languages, in: G. Rozenberg, A. Salomaa, *Handbook of Formal Languages, Vol. I*, Springer-Verlag, Berlin, 1997, 41 – 110.

Some Results on Directable Automata

Masami Ito[1] and Kayoko Shikishima-Tsuji[2]

[1] Department of Mathematics, Kyoto Sangyo University, Kyoto 603-8555, Japan
[2] Tenri University, Tenri 619-0224, Japan

Abstract. In this paper, we provide some properties of classes of regular languages consisting of directing words of directable automata and some new results on the shortest directing words of nondeterministic directable automata.

1 Introduction

Let X be a nonempty finite set, called an *alphabet*. An element of X is called a *letter*. By X^*, we denote the free monoid generated by X. Let $X^+ = X^* \setminus \{\epsilon\}$ where ϵ denotes the empty word of X^*. For the sake of simplicity, if $X = \{a\}$, then we write a^+ and a^* instead of $\{a\}^+$ and $\{a\}^*$, respectively. Let $L \subseteq X^*$. Then L is called a *language* over X. If $L \subseteq X^*$, then L^+ denotes the set of all concatenations of words in L and $L^* = L^+ \cup \{\epsilon\}$. In particular, if $L = \{w\}$, then we write w^+ and w^* instead of $\{w\}^+$ and $\{w\}^*$, respectively. Let $u \in X^*$. Then u is called a *word* over X. If $u \in X^*$, then $|u|$ denotes the length of u, i.e. the number of letters appearing in u. Notice that we also denote the cardinality of a finite set A by $|A|$.

A *finite automaton* (in short, an *automaton*) $\mathcal{A} = (S, X, \delta)$ consists of the following data: (1) S is a nonempty finite set, called a *state set*. (2) X is a nonempty finite alphabet. (3) δ is a function, called a *state transition function*, of $S \times X$ into S.

The state transition function δ can be extended to the function of $S \times X^*$ into S as follows: (1) $\delta(s, \epsilon) = s$ for any $s \in S$. (2) $\delta(s, au) = \delta(\delta(s, a), u)$ for any $s \in S, a \in X$ and $u \in X^*$.

Let $\mathcal{A} = (S, X, \delta)$ be an automaton, let $s \in S$ and let $u \in X^*$. In what follows, we will write $su^{\mathcal{A}}$ instead of $\delta(s, u)$.

A *finite recognizer* $\mathcal{A} = (S, X, \delta, s_0, F)$ consists of the following data: (1) The triple (S, X, δ) constitutes a finite automaton. (2) $s_0 \in S$ is called the *initial state*. (3) $F \subseteq S$ is called the *set of final states*.

Let $\mathcal{A} = (S, X, \delta, s_0, F)$ be a finite recognizer. Then the language $\mathcal{T}(\mathcal{A}) = \{u \in X^* \mid \delta(s_0, u) \in F\}$ is called the *language accepted by* $\mathcal{A}$.

Let $L \subseteq X^*$. Then L is said to be *regular* if L is accepted by a finite recognizer.

Now we define an directable automaton.

Definition 1. An automaton $\mathcal{A} = (S, X, \delta)$ is said to be *directable* if the following condition is satisfied: There exists $w \in X^*$ such that $sw^{\mathcal{A}} = tw^{\mathcal{A}}$ for any $s, t \in S$.

J. Karhumäki et al. (Eds.): Theory Is Forever (Salomaa Festschrift), LNCS 3113, pp. 125–133, 2004.
© Springer-Verlag Berlin Heidelberg 2004

In the above definition, a word $w \in X^*$ is called a *directing word* of $\mathcal{A}$. Then we have:

Fact *Let $\mathcal{A} = (S, X, \delta)$ be an automaton. Then $\mathcal{A}$ is directable if and only if for any $s, t \in S$, there exists $u \in X^*$ such that $su^{\mathcal{A}} = tu^{\mathcal{A}}$.*

Proposition 1. *Assume that $\mathcal{A} = (S, X, \delta)$ is a directable automata. Then the set of directing words $D(\mathcal{A})$ of $\mathcal{A}$ is a regular language.*

Let $\mathcal{A} = (S, X, \delta)$ be a directable automaton. By $d(\mathcal{A})$, we denote the value $min\{|w| \mid w \in D(\mathcal{A})\}$. Moreover, $d(n)$ denotes the value $max\{d(\mathcal{A}) \mid \mathcal{A} = (S, X, \delta)$ is a directable automaton with n states$\}$. In the definition of $d(n)$, X ranges over all finite nonempty alphabets.

In [2], Černý conjectured the following.

Conjecture 1. For any $n \geq 1, d(n) = (n-1)^2$.

However, the above problem is still open and at present we have only the following result:

Proposition 2. *For any $n \geq 1$, we have $(n-1)^2 \leq d(n) \leq \mathcal{O}(n^3)$.*

The lower bound is due to [2] and the uper bound is due to [7] and [8].

A similar problem for some classes of automata can be disscussed. For instance, an automaton $\mathcal{A} = (S, X, \delta)$ is said to be *commutative* if $s(uv)^{\mathcal{A}} = s(vu)^{\mathcal{A}}$ holds for any $s \in S$ and any $u, v \in X^*$. By $d_{com}(n)$, we denote the value $max\{d(\mathcal{A}) \mid \mathcal{A} = (S, X, \delta)$ is commutative and directable, and $|S| = n\}$. In the definition of $d_{com}(n)$, X ranges over all finite nonempty alphabets. The following result is due to [9] and [10].

Proposition 3. *For any $n \geq 1$, we have $d_{com}(n) = n - 1$.*

2   Nondeterministic Directable Automata

A *nondeterministic automaton* $\mathcal{A} = (S, X, \delta)$ consists of the following data: (1) S, X are the same materials as in the definition of finite automata. (2) δ is a relation such that $\delta(s, a) \subseteq S$ for any $s \in S$ and any $a \in X \cup \{\epsilon\}$.

As in the case of finite automata, δ can be extended to the following relation in a natural way, i.e. $\delta(s, au) = \bigcup_{t \in \delta(s,a)} \delta(t, u)$ for any $s \in S$, any $u \in X^*$ and any $a \in X \cup \{\epsilon\}$. In what follows, we will write $su^{\mathcal{A}}$ instead of $\delta(s, u)$ as in the case of finite automata.

Now we will deal with nondeterministic directable automata and their related languages. For nondeterministic automata, the directability can be defined in several ways. In each case, the directing words constitute a regular language. We will consider six classes of regular languages with respect to the different definitions of directability.

Let $\mathcal{A} = (S, X, \delta)$ be a nondeterministic automaton. In [5], the notion of directing words of $\mathcal{A}$ is given. In the definition, $Sw^{\mathcal{A}}$ denotes $\bigcup_{s \in S} sw^{\mathcal{A}}$ for $w \in X^*$.

Definition 2. (1) A word $w \in X^*$ is D_1-*directing* word@D_i-directing if $sw^{\mathcal{A}} \neq \emptyset$ for any $s \in S$ and $|Sw^{\mathcal{A}}| = 1$. (2) A word $w \in X^*$ is D_2-*directing* if $sw^{\mathcal{A}} = Sw^{\mathcal{A}}$ for any $s \in S$. (3) A word $w \in X^*$ is D_3-*directing* if $\bigcap_{s \in S} sw^{\mathcal{A}} \neq \emptyset$.

Definition 3. Let $i = 1, 2, 3$. Then $\mathcal{A}$ is called a D_i-*directable automaton* if the set of D_i-directing words is not empty.

Let $\mathcal{A} = (S, X, \delta)$ be a nondeterministic automaton. Then, for any $i = 1, 2, 3$, $D_i(\mathcal{A})$ denotes the set of all D_i-directing words. Then we have:

Proposition 4. *For any* $i = 1, 2, 3, D_i(\mathcal{A})$ *is a regular language.*

A nondeterministic automaton $\mathcal{A} = (s, X, \delta)$ is said to be *complete* if $sa^{\mathcal{A}} \neq \emptyset$ for any $s \in S$ and any $a \in X$. As for the D_1-directability of a complete nondeterministic automaton, Burkhard introduced it in [1]. We will investigate the classes of languages consisting of D_1-, D_2- and D_3-directing words of nondeterministic automata and complete nondeterministic automata.

The classes of D_i-directable nondeterministic automata and complete nondeterministic automata are denoted by $\mathbf{Dir}(i)$ and $\mathbf{CDir}(i)$, respectively. Let X be an alphabet. For $i = 1, 2, 3$, we define the following classes of languages:

(1) $\mathcal{L}^X_{\mathrm{ND}(i)} = \{D_i(\mathcal{A}) \mid \mathcal{A} = (S, X, \delta) \in \mathbf{Dir}(i)\}$. (2) $\mathcal{L}^X_{\mathrm{CND}(i)} = \{D_i(\mathcal{A}) \mid \mathcal{A} = (S, X, \delta) \in \mathbf{CDir}(i)\}$.

Let $\mathbf{D}$ be the class of deterministic directable automata. For $\mathcal{A} \in \mathbf{D}$, $D(\mathcal{A})$ denotes the set of all directing words of $\mathcal{A}$. Then we can define the class, i.e. $\mathcal{L}^X_{\mathbf{D}} = \{D(\mathcal{A}) \mid \mathcal{A} = (S, X, \delta) \in \mathbf{D}\}$.

Then, by Propsition 1 and Proposition 4, all the above classes are subclasses of regular languages. Figure 1 depicts the inclusion relations among such 7 classes. In [3], the inclusion relations among more classes are provided.

We will consider the shortest directing words of nondeterministic automata.

Let $i = 1, 2, 3$ and let $\mathcal{A} = (S, X, \delta)$ be a nondeterministic automaton. Then $d_i(\mathcal{A})$ denotes the value $min\{|u| \mid u \in D_i(\mathcal{A})\}$. For any positive integer $n \geq 1$, $d_i(n)$ denotes the value $max\{d_i(\mathcal{A}) \mid \mathcal{A} = (S, X, \delta) : \mathcal{A} \in \mathbf{Dir}(i)$ and $|S| = n\}$. Moreover, $cd_i(n)$ denotes the value $max\{d_i(\mathcal{A}) \mid \mathcal{A} = (S, X, \delta) : \mathcal{A} \in \mathbf{CDir}(i)$ and $|S| = n\}$. Notice that in the definitions of $d_i(n)$ and $cd_i(n)$, X ranges over all finite nonempty alphabets.

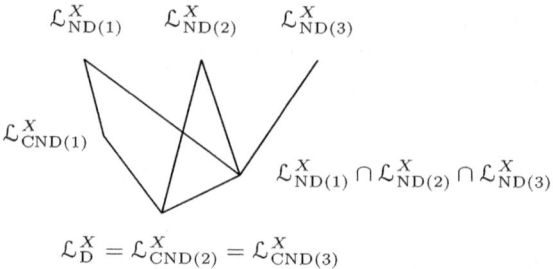

Fig. 1. Inclusion relations

In [1], Burkhard determined the value $cd_1(n)$ as follows:

Proposition 5. *Let $n \geq 1$. Then $cd_1(n) = 2^n - n - 1$.*

For $d_1(n)$, we have the following new result.

Proposition 6. *Let $n \geq 2$. Then $2^n - n \leq d_1(n) \leq \sum_{k=2}^{n} \binom{n}{k}(2^k - 1)$. Notice that $d_1(1) = 0$ and $d_1(2) = 3$.*

Proof Let $n \geq 2$. First, We show that $d_1(n) \leq \sum_{k=2}^{n} \binom{n}{k}(2^k - 1)$. Let $\mathcal{A} = (S, X, \delta)$ be a D_1-directable automaton with n states and let $w = a_1 a_2 \cdots a_r \in D_1(\mathcal{A})$ such that $a_i \in X, i = 1, 2, \ldots, r, r \geq 1$ and $|w| = r = d_1(\mathcal{A})$. Since $w \in D_1(\mathcal{A})$, there exists $s_0 \in S$ such that $sw^{\mathcal{A}} = \{s_0\}$ for any $s \in S$. For any $i = 1, 2, \ldots, r$, we define the set S_i and T_i as follows: (1) $S_i = S(a_1 a_2 \cdots a_i)^{\mathcal{A}}$. (2) $T_i = \{t \in S_i \mid t(a_{i+1}a_{i+2} \cdots a_r)^{\mathcal{A}} = \{s_0\}\}$.

Let $s \in S$ and let $i = 1, 2, \ldots, r$. Since $s(a_1 a_2 \cdots a_i a_{i+1} \cdots a_r)^{\mathcal{A}} = (s(a_1 a_2 \cdots a_i)^{\mathcal{A}})(a_{i+1} \cdots a_r)^{\mathcal{A}} = \{s_0\}$, we have $s(a_1 a_2 \cdots a_i)^{\mathcal{A}} \cap T_i \neq \emptyset$. Let $S = S_0 = T_0$. Consider the set $\{(S_i, T_i) \mid i = 0, 1, 2, \ldots, r - 1\}$. It is obvious that $S_i \neq \emptyset$ for any $i = 0, 1, \ldots, r - 1$. It is also obvious that $|S_0| \neq 1$. Suppose that $|S_i| = 1$ for some $i = 1, 2, \ldots, r - 1$. Then $S_i = T_i = \{t\}$ for some $t \in S$. By the definition of T_i, this means that $a_1 a_2 \cdots a_i \in D_1(\mathcal{A})$, which contradicts the minimality of $|w|$. Therefore, $|S_i| \neq 1$ for any $i = 1, 2, \ldots, n$. Hence the set $\{(S_i, T_i) \mid i = 0, 1, 2, \ldots, r - 1\}$ does not contain any $(\{s\}, \{s\})$ with $s_0 \neq s \in S$.

Now assume that $(S_i, T_i) = (S_j, T_j)$ for some $i, j = 1, 2, \ldots, r - 1, i < j$. Then it can be seen that $a_1 a_2 \cdots a_i a_{j+1} a_{j+2} \cdots a_r \in D_1(\mathcal{A})$, which contradicts the minimality of $|w|$. Hence all $(S_i, T_i), i = 0, 1, 2, \ldots, r - 1$, are distinct. Therefore, $|\{(S_i, T_i) \mid i = 0, 1, 2, \ldots, r - 1\}| \leq \sum_{k=2}^{n} \binom{n}{k}(2^k - 1)$ and hence $r \leq \sum_{k=2}^{n} \binom{n}{k}(2^k - 1)$.

We will show that $2^n - n \leq d_1(n)$. It is obvious that $d_1(2) \geq 2$. Let $n \geq 3$. We will construct a D_1-directable automaton $\mathcal{A} = (S, X, \delta)$ such that $|S| = n$

and $d_1(A) = 2^n - n$. Let S be a finite set with $|S| = n$ and let $\{T_1, T_2, \ldots, T_r\} = \{T \subset S \mid |T| \geq 2\}$. Notice that $r = 2^n - n - 2$. Moreover, we assume that $|T_1| \geq |T_2| \geq \cdots \geq |T_r|$, $\{s_0\} = S \setminus T_1$ and $T_r = \{s_1, s_2\}$. Now we construct the following nondeterministic automaton $A = (S, X, \delta)$: (1) $X = \{a_1, a_2, \ldots, a_r, b\}$. (2) For any $i = 1, 2, \ldots, r - 1$, $sa_i^A = T_{i+1}$ if $s \in T_i$ and $sa_i^A = S$, otherwise. (3) $s_1 a_r^A = s_2 a_r^A = \{s_1\}$ and $sa_r^A = S$ if $s \in S \setminus \{s_1, s_2\}$. (4) $s_0 b^A = \emptyset$ and $sb^A = T_1$ for any $s \in S \setminus \{s_0\}$.

Let $s \in S$ and let $i = 1, 2, \ldots, r$. Notice that $s(a_i b a_1 a_2 \cdots a_r)^A = \{s_1\}$ and hence $a_i b a_1 a_2 \cdots a_r \in D_1(A)$. Moreover, since $s_0 b^A = \emptyset$, we have $bX^* \cap D_1(A) = \emptyset$. Let $i, j = 1, 2, \ldots, r$. Then $S(a_i a_j)^A = S$. On the other hand, $s(a_i b)^A = T_1$ for any $s \in S$. This means that $u \in a_i b X^*$ if u is a shortest D_1-directing word of A. Let $i = 1, 2, \ldots, r - 1$. Then $T_i(a_i a_j)^A = T_{i+1} a_j^A = S$ if $j > i + 1$ and $T_i(a_i a_j)^A = T_{i+1} a_j^A \supseteq T_{j+1}$ if $j \leq i$. Notice that in the latter case $j + 1 \leq i + 1$. This implies that u is not a shortest D_1-directing word of A if $u \in X^* a_i a_j X^*$ where $j \neq i + 1$. Moreover, since $Sb^A = T_1$, u is not a shortest D_1-directing word of A if $u \in XX^+ bX^*$. Consequently, $a_i b a_1 a_2 \cdots a_r$ is a shortest D_1-directing word of A, i.e. $d_1(A) = r + 2 = 2^n - n$. Hence we have $2^n - n \leq d_1(n)$.

Finally, we compute $d_1(1)$ and $d_1(2)$. It is obvious that $d_1(1) = 0$. Consider the following nondeterministic automaton $A = (\{1, 2\}, \{a, b, c\}, \delta)$: (1) $1a^A = \{1, 2\}$ and $2a^A = \{2\}$. (2) $1b^A = \emptyset$ and $2b^A = \{1, 2\}$. (3) $1c^A = \{1\}$ and $2c^A = \emptyset$.

Then abc is a shortest D_1-directing word of A. Since $d_1(2) \leq 2^2 - 1 = 3$, we have $d_1(2) = 3$.

Now we consider the value $d_3(n)$. Before dealing with the value $d_3(n)$, we define a nondeterministic automaton of partial function type.

A nondeterministic automaton $A = (S, X, \delta)$ is said to be *of partial function type* if $|sa^A| \leq 1$ for any $s \in S$ and any $a \in X$. Then we have:

Remark 1. Let A be a nondeterministic automaton of partial function type. Then $D_3(A) = D_1(A)$.

Let $A = (S, X, \delta)$ be a D_3-directable automaton of partial function type. Consider the following procedure $\mathcal{P}$: Let $u \in D_3(A)$. Assume that $u = u_1 u_2 u_3$ where $u_1, u_3 \in X^*$, $u_2 \in X^+$ and $Su_1^A = S(u_1 u_2)^A$. Then procedure $\mathcal{P}$ can be applied as $u \Rightarrow^{\mathcal{P}} u_1 u_3$.

Then we have the following result.

Lemma 1. *In the above procedure, we have $u_1 u_3 \in D_3(A)$.*

Proof   Let $A = (S, X, \delta)$ be a nondeterministic automaton of partial function type. Moreover, let $u = u_1 u_2 u_3$ where $u_1, u_3 \in X^*$, $u_2 \in X^+$ and $Su_1^A = S(u_1 u_2)^A$. Since $u \in D_3(A)$, there exists $s_0 \in S$ such that $su^A = \{s_0\}$ for any

$s \in S$. From the assumptions that $Su_1{}^{\mathcal{A}} = S(u_1u_2)^{\mathcal{A}}$ and $\mathcal{A}$ is a nondeterministic automaton of partial function type, it follows that $su^{\mathcal{A}} = s(u_1u_2u_3)^{\mathcal{A}} = s(u_1u_3)^{\mathcal{A}} = \{s_0\}$ for any $s \in S$. By Remark 1, this means that $u_1u_3 \in D_3(\mathcal{A})$.

Let $\mathcal{A} = (S, X, \delta)$ be a D_3-directable automaton of partial function type and let $a_1a_2 \cdots a_r \in D_3(\mathcal{A})$ such that $sa_1{}^{\mathcal{A}} = ta_1{}^{\mathcal{A}}$ for some $s, t \in S, s \neq t$.

Assume that $v \in D_3(\mathcal{A}), v = v_1v_2v_3, v_1, v_3 \in X^*, v_2 \in X^+, |Sv_1{}^{\mathcal{A}}| = |S(v_1v_2)^{\mathcal{A}}|$ and $\{s, t\} \subseteq Sv_1{}^{\mathcal{A}}$. Then procedure $\mathfrak{Q}_{(s,t)}$ can be applied as $v \Rightarrow^{\mathfrak{Q}_{(s,t)}} v_1a_1a_2 \cdots a_r$.

Then we have the following results.

Lemma 2. *In the above procedure, we have $v_1a_1a_2 \cdots a_r \in D_3(\mathcal{A})$ and $|Sv_1{}^{\mathcal{A}}| > |Sv_1a_1{}^{\mathcal{A}}|$.*

Proof   Let $s \in S$. Since $v = v_1v_2v_3 \in D_3(\mathcal{A})$, we have $sv_1{}^{\mathcal{A}} \neq \emptyset$, actually $|sv_1{}^{\mathcal{A}}| = 1$. Notice that $\exists s_r \in S, \forall t \in S, t(a_1a_2 \cdots a_r)^{\mathcal{A}} = \{s_r\}$. Therefore, $s(v_1a_1a_2 \cdots a_r)^{\mathcal{A}} = (sv_1{}^{\mathcal{A}})(a_1a_2 \cdots a_r)^{\mathcal{A}} = \{s_r\}$ and hence $v_1a_1a_2 \cdots a_r \in D_3(\mathcal{A})$. Since $\mathcal{A}$ is of partial function type and $\{s, t\} \subseteq Sv_1{}^{\mathcal{A}}, |Sv_1{}^{\mathcal{A}}| \geq |Sv_1a_1{}^{\mathcal{A}}| + 1$. This completes the proof of the lemma.

Lemma 3. *Let $\mathcal{A} = (S, X, \delta)$ be a D_3-directable automaton such that $|S| = n$ and $d_3(\mathcal{A}) = d_3(n)$. Then there exists a nondeterministic automaton $\mathcal{B} = (S, Y, \gamma)$ of partial function type such that $d_3(\mathcal{B}) = d_3(n)$.*

Proof   Let $u = a_1a_2 \cdots a_r \in D_3(\mathcal{A})$ with $|u| = d_3(\mathcal{A})$. Since $u \in D_3(\mathcal{A})$, there are $s_r \in S$ and a sequence of partial functions of S into S, $\rho_1, \rho_2, \ldots, \rho_r$ such that $s(a_1a_2 \cdots a_i)^{\mathcal{A}} \supseteq \rho_i(\rho_{i-1}(\cdots(\rho_1(s))\cdots))$ for any $s \in S$ and any $i = 1, 2, \ldots, r$. Furthermore, $\rho_r(\rho_{r-1}(\cdots(\rho_1(s))\cdots)) = \{s_r\}$ for any $s \in S$. Now we define the automaton of partial function type $\mathcal{B} = (S, Y, \gamma)$ as follows: (1) $Y = \{b_i \mid i = 1, 2, \ldots, r\}$. Remark that $b_1, b_2, \ldots, b_r$ are distinct symbols. (2) $sb_i{}^{\mathcal{B}} = \rho_i(s)$ for any $s \in S$ and any $i = 1, 2, \ldots, r$.

Then $\mathcal{B}$ is a nondeterministic automaton of partial function type. Moreover, it is obvious that $b_1b_2 \cdots b_r \in D_3(\mathcal{B})$. Suppose that $b_{i_1}b_{i_2} \cdots b_{i_k} \in D_3(\mathcal{B})$ where $i_1, i_2, \ldots, i_k \in \{1, 2, \ldots, r\}$. Then we have $a_{i_1}a_{i_2} \cdots a_{i_k} \in D_3(\mathcal{A})$. Therefore, $k \geq r$ and $r = d_3(\mathcal{B})$. This completes the proof of the lemma.

We are now ready to determine an upper bound for $d_3(n)$.

Proposition 7. *For any $n \geq 3$, $d_3(n) \leq \sum_{k=2}^{n-1} \binom{n}{k} - \sum_{k=0}^{n-2} \binom{n-2}{k} + n - 1$.*

Proof   By Lemma 3, there exists a nondeterministic automaton of partial function type $\mathcal{A} = (S, X, \delta)$ such that $|S| = n$ and $d_3(n) = d_3(\mathcal{A})$. Let $u = a_1a_2 \cdots a_r \in D_3(\mathcal{A})$ with $r = d_3(n)$ and let $S_i = S(a_1a_2 \cdots a_i)^{\mathcal{A}}$ for $i = 1, 2, \ldots, r$. Since $\mathcal{A}$ is of partial function type and $r = d_3(n) = d_3(\mathcal{A}), |S| > |S_1| \geq |S_2| \geq \cdots \geq |S_{r-1}| > |S_r| = 1$. Let $S_r = \{s_r\}$. By Lemma 1, $S, S_1, S_2, \ldots,$

S_{r-1} and S_r are distinct. Moreover, since $|S| > |S_1|$, there exist $s_0, s_1 \in S$ such that $s_0 \neq s_1$ and $s_0 a_1{}^A = s_1 a_1{}^A$. Therefore, we can apply procedure $\mathcal{Q}_{(s_0,s_1)}$ to $a_1 a_2 \cdots a_r$ if necessary and we can get $a_1 a_2 \cdots a_r \Rightarrow^{\mathcal{Q}_{(s_0,s_1)}} v_1 a_1 a_2 \cdots a_r$. Now we apply procedure $\mathcal{P}$ to $v_1 a_1 a_2 \cdots a_r$ as many times as possible until we cannot apply procedure $\mathcal{P}$ anymore. Hence we can obtain $w \in D_3(\mathcal{A})$ with $|w| \leq 2^{|S|} - |S|$. Then we apply procedure $\mathcal{Q}_{(s_0,s_1)}$ to w. We will continue the same process until we cannot apply either procedure $\mathcal{P}$ nor $\mathcal{Q}_{(s_0,s_1)}$. Notice that this process will be terminated after a finite number of applications of procedures $\mathcal{P}$ and $\mathcal{Q}_{(s_0,s_1)}$. Let $w = c_1 c_2 \cdots c_s, c_i \in X, i = 1, 2, \ldots, s$ be the last D_3-directing word of $\mathcal{A}$ which was obtained by the above process. Let $T_i = S(c_1 c_2 \cdots c_i)^A$ for any $i = 1, 2, \ldots, s$. Then $T_i \neq T_j$ for any $i, j = 1, 2, \ldots, s$ with $i < j$ and $\{T_1, T_2, \ldots, T_s\}$ contains at most $n - 2$ elements $T_i, i = 1, 2, \ldots, s$ with $T_i \supseteq \{s_0, s_1\}$. Since $|\{T \subseteq S \mid \{s_0, s_1\} \subseteq T\}| = \sum_{k=0}^{n-2} \binom{n-2}{k}$ and by the above observation (includig Lemma 2), we have $d_3(n) \leq \sum_{k=2}^{n-1} \binom{n}{k} - \sum_{k=0}^{n-2} \binom{n-2}{k} + n - 1$.

For the lower bound for $d_3(n)$, we have the following new result.

Proposition 8. Let $n \geq 3$. Then $d_3(n) \geq 2^m + 1$ if $n = 2m$ ($d_3(n) \geq 3 \cdot 2^{m-1} + 1$ if $n = 2m + 1$).

Proof   Let $n \geq 3$ and let $S = \{1, 2, \ldots, n\}$. Moreover, let $S_1 = \{1, 2\}$, let $S_2 = \{3, 4\}, \ldots$, let $S_{m-1} = \{2m - 3, 2m - 2\}$ and let $S_m = \{2m - 1, 2m\}$ if $n = 2m$ ($S_m = \{2m - 1, 2m, 2m + 1\}$ if $n = 2m + 1$).

We define the following D_3-directable automaton $\mathcal{A} = (S, X, \delta)$:

(1) $\{T_1, T_2, \ldots, T_k\} = \{\{n_1, n_2, \ldots, n_m\} \mid (n_1, n_2, \ldots, n_m) \in S_1 \times S_2 \times \cdots \times S_m\}$ where $k = 2^m$ if $n = 2m$ ($k = 3 \cdot 2^{m-1}$ if $n = 2m+1$). (2) $T_1 = \{1, 3, 5, \ldots, 2m - 1\}$. (3) $X = \{a, b_1, b_2, \ldots, b_{k-2}, b_{k-1}, c\}$. (4) $1a^A = 2a^A = \{1\}, 3a^A = 4a^A = \{3\}, \ldots, (2m - 3)a^A = (2m - 2)a^A = \{2m - 3\}$ and $(2m - 1)a^A = (2m)a^A = \{2m - 1\}$ if $n = 2m$ ($(2m - 1)a^A = (2m)a^A = (2m + 1)a^A = \{2m - 1\}$ if $n = 2m + 1$). (5) Let $i = 1, 2, \ldots, k - 1$. By ρ_i, we denote a bijection of T_i onto T_{i+1}. Then $t b_i{}^A = \rho_i(t)$ for any $t \in T_i$ and $t b_i{}^A = \emptyset$, otherwise. (6) $t c^A = \{1\}$ for any $t \in T_k$ and $t c^A = \emptyset$, otherwise.

Then it can be easily verified that $a b_1 b_2 \cdots b_{k-1} c$ is a unique shortest D_3-directing word of $\mathcal{A}$. Therefore, $d_3(n) \geq 2^m + 1$ if $n = 2m$ ($d_3(n) \geq 3 \cdot 2^{m-1} + 1$ if $n = 2m + 1$).

Now we consider the values $cd_2(n)$ and $d_2(n)$. The lower bound is due to [1] and the upper bound is followed by [5].

Proposition 9. For $n \geq 2$, $2^n - n - 1 \leq cd_2(n) \leq d_2(n) < 1 + (2^n - 2)\binom{2^n}{2}$. Remark that $cd_2(1) = d_2(1) = 0$.

Finally, we provide a result on the value of $cd_3(n)$. The result is due to [2] and [5].

Proposition 10. *Let $n \geq 1$. Then $(n-1)^2 \leq cd_3(n) \leq 1 + (n-2)\binom{n}{2}$.*

3   Commutative Nondeterministic Directable Automata

In this section, we will deal with commutative nondeterministic automata and related languages alongside the same line as that of the previous section.

A nondeterministic automaton $\mathcal{A} = (S, X, \delta)$ is said to be *commutative* if $s(ab)^{\mathcal{A}} = s(ba)^{\mathcal{A}}$ holds for any $s \in S$ and any $a, b \in X$. nondeterministic automata@commutative

By $\mathcal{L}'^{X}_{D}, \mathcal{L}'^{X}_{CND(i)}$ and $\mathcal{L}'^{X}_{ND(j)}, i, j = 1, 2, 3$, we denote the classes of regular languages of directing words of deterministic commutative automata, of D_i-directing words of complete commutative nondeterministic automata, and of D_j-directing words of commutative nondeterministic automata, respectively.

Then we have the following inclusion relations among these classes (see Figure 2).

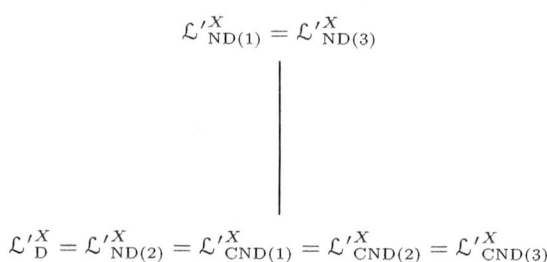

Fig. 2. Commutative case

Now we will consider the shortest directing words of commutative nondeterministic automata. The results in this section are due to [4].

Let $i = 1, 2, 3$ and let $n \geq 1$. Then $cd_{com(i)}(n)$ denotes the value $max\{d_i(\mathcal{A}) \mid \mathcal{A} = (S, X, \delta) : \text{commutative}, \mathcal{A} \in \mathbf{CDir}(i) \text{ and } |S| = n\}$.

Notice that in the definitions of $d_{com(i)}(n)$ and $cd_{com(i)}(n)$, X ranges over all finite nonempty alphabets.

Proposition 11. *For any $n \geq 1, d_{com(1)}(n) = cd_{com(1)}(n) = n - 1$.*

Proposition 12. *Let $n \geq 2$. Then $(n-1)^2 + 1 \leq cd_{com(2)}(n) = d_{com(2)}(n) \leq 2^n - 2$. For $n = 1$, $cd_{com(2)}(1) = d_{com(2)}(1) = 0$.*

Proposition 13. *Let $n \geq 2$. Then $n^2 - 3n + 3 \leq cd_{com(3)}(n) = d_{com(3)}(n) \leq 1 + (n-2)\binom{n}{2}$. For $n = 1, cd_{com(3)}(1) = d_{com(3)}(1) = 0$.*

As for more detailed information on deterministic and nondeterministic directable automata, refer to [6].

Acknowledgement The authors would like to thank Dr. Cs. Imreh for his valuable comments.

References

1. H.V. Burkhard, Zum Längenproblem homogener experimente an determinierten und nicht-deterministischen automaten, Elektronische Informationsverarbeitung und Kybernetik, EIK 12 (1976), 301-306.
2. J. Černý, Poznámka k homogénym experimentom s konečinými automatami, Matematicko-fysikalny Časopis SAV 14 (1964), 208-215.
3. B. Imreh and M. Ito, On some special classes of regular languages, in *Jewels are Forever* (edited by J. Karhumäki et al.) (1999) (Springer, New York), 25-34.
4. B. Imreh, M. Ito and M. Steinby, On commutative directable nondeterministic automata, in *Grammars and Automata for Strings: From Mathematics and Computer Science to Biology, and Back* (edited by C. Martin-Vide et al.) (2003) (Taylor and Francis, London), 141-150.
5. B. Imreh and M. Steinby, Directable nondeterministic automata, Acta Cybernetica 14 (1999), 105-115.
6. M. Ito, *Algebraic Theory of Automata and Languages*, World Scientific (Singapore), 2004.
7. J.-E. Pin, Sur les mots synchronisants dans un automata fini, Elektronische Informationsverarbeitung und Kybernetik, EIK 14 (1978), 297-303.
8. J.-E. Pin, Sur un cas particulier de la conjecture de Cerny, Automata, Lecture Notes in Computer Science 62 (Springer) (1979), 345-352.
9. I. Rystsov, Exact linear bound for the length of reset words in commutative automata, Publicationes Mathematicae of Debrecen 48 (1996), 405-409.
10. I. Rystsov, Reset words for commutative and solvable automata, Theoretical Computer Science 172 (1997), 273-279

Rectangles and Squares Recognized by Two-Dimensional Automata

Jarkko Kari* and Cristopher Moore

[1] Mathematics Department, FIN-20014 University of Turku, Finland, and
Department of Computer Science, University of Iowa, Iowa City IA 52242 USA
`jjkari@cs.uiowa.edu`
[2] Computer Science Department and Department of Physics and Astronomy,
University of New Mexico, Albuquerque NM 87131, and the Santa Fe Institute,
1399 Hyde Park Road, Santa Fe NM 87501,
`moore@cs.unm.edu`

Abstract. We consider sets of rectangles and squares recognized by deterministic and non-deterministic two-dimensional finite-state automata. We show that sets of squares recognized by DFAs from the inside can be as sparse as any recursively enumerable set. We also show that NFAs can only recognize sets of rectangles from the outside that correspond to simple regular languages.

1 Introduction

Two-dimensional languages, or *picture languages,* are an interesting generalization of the standard languages of computer science. Rather than one-dimensional strings, we consider two-dimensional arrays of symbols over a finite alphabet. These arrays can then be accepted or rejected by various types of automata, and this gives rise to different language classes. Such classes may be of interest as formal models of image recognition, or simply as mathematical objects in their own right.

Much of this work has focused on two- or more-dimensional generalizations of regular languages. We can define the regular languages as those recognized by finite-state automata that can move in one direction or both directions on the input, and which are deterministic or non-deterministic. We can also consider finite complement languages, in which some finite set of subwords is forbidden, and then project onto a smaller alphabet with some alphabetic homomorphism. In one dimension, these are all equivalent in their computational power.

In two dimensions, we can consider *4-way finite-state automata,* which at each step can read a symbol of the array, change their internal state, and move up, down, left or right to a neighboring symbol. These can be deterministic or non-deterministic, and DFAs and NFAs of this kind were introduced by Blum and Hewitt [1]. Similarly, we can forbid a finite number of subblocks and then project onto a smaller alphabet, obtaining a class of picture languages which

---

* Research supported by NSF Grant CCR 97-33101

J. Karhumäki et al. (Eds.): Theory Is Forever (Salomaa Festschrift), LNCS 3113, pp. 134–144, 2004.
© Springer-Verlag Berlin Heidelberg 2004

we call *homomorphisms of local lattice languages,* or h(LLL)s [9]. (These are also called the *recognizable* languages [3] or the languages recognizable by *non-deterministic on-line tesselation acceptors* [5].) While DFAs, NFAs and h(LLL)s are equivalent in one dimension, in two or more they become distinct:

$$\text{DFA} \subset \text{NFA} \subset \text{h(LLL)}$$

where these inclusions are strict. Reviews of these classes are given in [9, 4, 7, 12], and a bibliography of papers in the subject is maintained by Borchert at [2].

A fair amount is known about the closure properties of these classes as well. The DFA, NFA, and h(LLL) languages are all closed under intersection and union using straightforward constructions.

The situation for complement is somewhat more complicated. DFAs are closed under complement by an argument of Sipser [13] which allows us to re-move the danger that a DFA might loop forever and never halt. We construct a new DFA that starts in the final halt state, which we can assume without loss of generality is in the lower right-hand corner. Then this DFA does a depth-first search backwards, attempting to reach the initial state of the original DFA. (We use a similar construction in Section 2 below.) This gives a loop-free DFA which accepts if and only if the original DFA accepts, and since a loop-free DFA al-ways halts, we can then switch accepting and rejecting states to recognize the complement of the original language.

It is also known that h(LLL)s are not closed under complement, and in [8] we proved that NFAs are not closed under complement either. In the process, we proved in [8] also that NFAs are more powerful than DFAs, even for rectangles of a single symbol. We note that recognition of rectangles by NFAs, in the context of "recognizable functions," is studied in [6], where some upper bounds are shown.

In this paper we consider recognizing picture languages over a single letter alphabet, that is, recognizing squares and rectangles. We start by investigat-ing sets of squares recognized by DFAs, and we show that DFAs can recognize squares whose sizes belong to sets which are as sparse as any recursively enu-merable set. This resolves an open question raised in [9] about how sparse these sets can be.

We then consider the question of what sets of rectangles can be recognized by an NFA or DFA from the outside, i.e. by an automaton which is not allowed to enter the rectangle but can roam the plane outside it. We show that any such set corresponds to a simple regular language, and therefore that such an automaton can be simulated by a DFA moving along the rectangle's boundary.

2   Square Recognition

Our model of four-way finite automaton is sensitive to the borders of the input rectangle. In other words, the automaton knows which of the neighboring squares are inside and which are outside of the rectangle. Based on this information and the current state of the automaton the transition rule specifies the new state of the automaton and the direction *Left, Right, Up* or *Down* of movement. The automaton is not allowed to move in a direction that takes the automaton to the other side of the border.

In this section we consider the recognition problem from the inside of the rectangle. In this case the automaton is initially located inside the rectangle, at the lower left corner. The rectangle is accepted if and only if the automaton is able to eventually reach its accepting state. A DFA has a deterministic transition rule, while in NFA there may be several choices of the next move. The automaton recognizes set $S \subset \mathbb{N}^2$ where

$$S = \{(w, h) \mid \text{ the automaton accepts the rectangle of size } w \times h \}.$$

A DFA can easily check if a given rectangle is a square by moving diagonally through the rectangle. Since the class of sets recognized by DFAs or NFAs are closed under intersection, the set of squares recognized by a DFA or NFA is recognizable by an automaton of the same type. A natural question to ask is what sets of squares can be recognized by NFA or DFA. We say that a set $S \subset \mathbb{N}$ is *square-recognizable* by an NFA (DFA) if the set of squares $\{(n, n) \mid n \in S\}$ is recognizable by an NFA (DFA).

Example 1. [9] The set $\{2^n \mid n \in \mathbb{N}\}$ is square-recognized by a DFA that uses knight's moves to divide the distance to a corner by two, as in Figure 1.

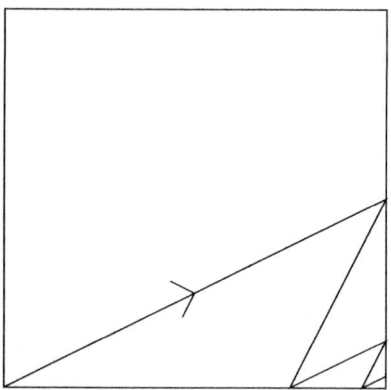

Fig. 1. Using knight's moves to recognize $2^n \times 2^n$ squares

Example 2. A slightly more complex construction shows that the set $\{2^{2^n} \mid n \in \mathbb{N}\}$ can be square-recognized. In this construction a signal of rational speed 3/4 is repeatedly used to move the automaton from position 4^n into position 3^n, in the same style as the knight moves was used in the previous example. That is, we multiply the automaton's position on an edge by 3/4 until the result is no longer an integer. Then signals of speed 2/3 are iterated to move the automaton from position 3^n into position 2^n. This process accomplishes the following changes in the automaton's position:

$$2^{2^n} = 4^{2^{n-1}} \longrightarrow 3^{2^{n-1}} \longrightarrow 2^{2^{n-1}}.$$

The process is iterated until position $2 = 2^{2^0}$ is reached, in which case the square is accepted. If it ends at position 1 instead, we reject. Note that we can check the position mod 2, 3, or 4 at any time by using a finite number of states. At the very beginning we can verify that the square size is a power of 2, using the DFA from our first example.

We see from these examples that rather "sparse" sets of squares can be recognized, even by deterministic finite automata. In what follows we prove that these sets can be made as sparse as any recursive (or recursively enumerable) set. This result is optimal because any square-recognized set has to be recursive.

Theorem 1. *Let $\{a_1, a_2, \ldots\}$ be any recursively enumerable set of positive integers. There exists a 2D DFA that square-recognizes a set $\{b_1, b_2, \ldots\}$ such that $b_i > a_i$ for all $i = 1, 2, \ldots$*

Proof. We exploit classical results by Minsky [11] and a very natural correspondence between automata inside rectangles and 2-counter machines without an input tape. A 2-counter machine without an input tape consists of a deterministic finite state automaton and two infinite registers, called counters. The counters store one non-negative integer each. The machine can detect when either counter is zero. It changes its state according to a deterministic transition rule. The new state only depends on the old state and the zero vs. non-zero status of the two counters. The transition rule also specifies how to change the counters: They may be incremented, decremented (only if they are non-zero) or kept unchanged. The automaton accepts iff it reaches a specified accepting state. We define the set recognized by such a device as the set of integers k such that the machine enters the accepting state when started with counter values 0 and k.

It follows from Minsky's construction in [11] that for every recursively enumerable set X of positive integers there exists a 2-counter machine A that accepts the set

$$Y = \{2^n \mid n \in X\}.$$

Moreover, we may assume that the 2-counter machine A never loops, i.e., never returns back to the same state and counter values it had before. This assumption means no loss of generality because we can start Minsky's construction from a Turing machine that does not loop. Such Turing machine exists for every recursively enumerable set. (It may, for example, count on the tape the number of instructions it executes.)

Notice that a 2-counter machine can be interpreted as a 2D DFA that operates on the (infinite) positive quadrant $\{(x, y) \mid x, y \geq 0\}$ of the plane and has the same finite states as the 2-counter machine. The values of the counters correspond to the coordinates of the position of the automaton. Increments and decrements of the counters correspond to movements of the DFA on the plane. Counter values zero are the positive axeses $(x, 0)$ and $(0, y)$, which can be detected by the DFA.

An accepting computation of a 2-counter machine can therefore be simulated by a 2D DFA inside a sufficiently large square. The square has to be at least as large as the largest counter value used during the accepting computation. Using this observation we prove that if Y is any set of positive integers accepted by a 2-counter machine A that does not loop then some 2D DFA square-recognizes the set

$$Z = \{k \in \mathbb{N} \mid k \text{ is the largest counter value during the accepting}$$
$$\text{computation by } A \text{ of some } i \in Y \}.$$

It is very easy to construct a non-deterministic 2D automaton that square-recognizes set Z: The NFA moves along the lower edge of the square and uses the non-determinism to guess the starting position of the 2-counter machine A. Then it executes A until A accepts, halts otherwise or tries to move outside the square. Moreover, the NFA memorizes in the finite control unit if it ever touched the right or upper edge of the square, indicating that a counter value equal to the size of the square was used. The square is accepted if and only if A accepts after touching the right or the upper edge of the square. Note that the forward simulation part of this process is deterministic. Non-determinism is only needed to find the correct input to the 2-counter machine.

To accept set Z deterministically we use the determinism of the 2-counter machine to depth-first-search the predecessor tree of a given configuration C to see if it is on a computation path for some valid input. This is done by trying all possible predecessors of a configuration in a predetermined order, and recursively processing all these predecessors. When no predecessors exist the DFA can backtrack using the determinism of the 2-counter machine, and continue with the next branch of the predecessor tree. Note that the predecessor tree – restricted inside a square – is finite and does not contain any loops so the depth-first-search process will eventually terminate. It finds a valid starting configuration that leads to the given configuration C if it exists. If no valid input configuration exists the process may identify an incorrect starting configuration if the depth-first-search backtracked beyond C. In any case, if a potential start configuration is found, a forward simulation of the 2-counter machine is done to see if it accepts after touching the right or the upper edge of the square.

The depth-first-search process described above is done on all configurations C that are close to the upper left or the lower right corners of the input square. If the two counter machine A has s states then any accepting computation with maximum counter value k must enter some cell within distance s of one of the corners: otherwise counter value k is present in a loop that takes the automaton away from the axeses, and the computation cannot be accepting.

There are a finite number of configurations C within distance s from the corners, so the 2D DFA can easily try all of them one after the other.

Now we are ready to conclude the proof. If $X = \{a_1, a_2, \ldots\}$ is any r.e. set of positive integers then there exists a 2-counter machine A that does not loop and accepts $Y = \{2^{a_1}, 2^{a_2}, \ldots\}$. The construction above provides a 2D DFA that square accepts the set $Z = \{b_1, b_2, \ldots\}$ where b_i is the maximum counter value during the accepting computation of 2^{a_i} by A. Clearly $b_i \geq 2^{a_i} > a_i$.

Another way of stating the previous theorem is

Corollary 1. *Let $f : \mathbb{N} \longrightarrow \mathbb{N}$ be any computable function. Then some 2D DFA square accepts the range of some function $g : \mathbb{N} \longrightarrow \mathbb{N}$ that satisfies $g(n) \geq f(n)$ for all $n \in \mathbb{N}$.*

It is interesting to note that there seem to be no tools whatsoever, at the present time, to prove that a set S cannot be square-recognized by a DFA. To inspire the creation of such tools, we make the following conjecture:

Conjecture 1. No DFA or NFA in two dimensions can recognize the set of squares of prime size.

In contrast, the reader can easily show that there is a two-dimensional DFA that recognizes squares of size 11^p for p prime. This works by first moving p from the 11-register to the 2-register, that is, by converting 11^p to 2^p as in Example 2 above. It then copies p into to the 3-register, and uses the 5- and 7-registers to maintain a number m by which it divides p, with a loop that passes m back and forth between the 5- and 7-registers while decrementing the 3-register. After each division, it increments m and recopies p from the 2-register to the 3-register. We halt if m exceeds $\lfloor p/4 \rfloor$, which we can check with a similar loop; this suffices for all $p > 9$. Then since $2^p 3^p 7^{p/4} \leq 11^p$ all this can take place inside a square of size 11^p.

In addition, in [9] DFAs are given that recognize cubes of prime size, and tesseracts of perfect size, in three and four dimensions respectively. These use a billiard-ball-like computation to check whether two integers are mutually prime; thus they simulate counter machines which branch both when the counter is zero and when it reaches its maximum, i.e. the size of the cube.

We also conjecture the following:

Conjecture 2. NFAs are more powerful than DFAs on squares of one symbol.

Since we have proved in [8] that NFAs are more powerful than DFAs on rectangles of one symbol, it would be rather strange if they were equivalent on squares.

3   Recognizing Rectangles from Outside

Let us consider next the "dual" problem of recognizing rectangles using a finite automaton that operates outside the rectangle and is not allowed to penetrate inside. Initially the automaton is located outside one of the corners of the rectangle, and the rectangle is accepted if the automaton is able to reach its accepting state. Not surprisingly, such devices turn out to be less powerful than automata that operate inside rectangles. We prove that if $S \subseteq \mathbb{N}^2$ is the set of rectangles recognized by an NFA from outside, then the language

$$L_S = \{0^i 1^j \mid (i, j) \in S\}$$

is regular. In other words, the same set of rectangles can be recognized by a DFA that moves east and south along the edges of the rectangle from the upper left corner to the lower right corner. A similar result was shown for DFAs which are allowed to make excursions into the plane outside the rectangle by Milgram [10].

Lemma 1. *Let $S \subseteq \mathbb{N}^2$ be recognized from the outside by an NFA A. Then there exist positive integers t_H and p_H, called the horizontal transient and the period, respectively, such that for every $w > t_H$ and every h*

$$(w, h) \in S \Longrightarrow (w + p_H, h) \in S$$

where the numbers t_H and p_H are independent of the height h.

Proof. We generalize an argument from [9] for rectangles of height 1. Consider first computation paths P by A on unmarked plane that start in state q at position $(0, 0)$ and end in state r on the same row $y = 0$. Assume that P is entirely above that row, i.e. all intermediate positions (x, y) of P have $y > 0$. Let us prove the following fact: there must exist positive integers t and p such that, if P ends in position $(x, 0)$ with $x > t$ then there is a computation path P' from the same initial position $(0, 0)$ and state q, into position $(x + p, 0)$ and state r, such that also P' is entirely above the row $y = 0$. In other words, the possible horizontal positions where A returns to the initial row $y = 0$ in state r are eventually periodic.

To prove this, notice that sequences I of instructions executed by A during possible P's form a context-free language, accepted by a pushdown automaton that uses its stack to count its vertical position and its finite state control to simulate the states of A. Recall that a Parikh vector of a word is the integer vector whose elements are the counts of different letters in the word. It is well known that in the case of a context-free language the Parikh vectors of the words in the language form a semi-linear set, that is, a finite union of linear sets $\{\bar{u} + \sum_{i=0}^{k} a_i \bar{v}_i \mid a_i = 0, 1, \ldots\}$ for integer vectors $\bar{u}$ and $\bar{v}_i$, $i = 0, 1, \ldots, k$.

Any linear transformation keeps linear sets linear; therefore, the possible values of $n_r(I) - n_l(I)$ form a semilinear set of integers, where $n_r(I)$ and $n_l(I)$ indicate the numbers of right- and left-moving instructions in the word I, respectively. Because $n_r(I) - n_l(I)$ is the total horizontal motion caused by executing I, we conclude that possible horizontal positions at the end of P form a semi-linear set $\cup_j \{t_j + ip_j \mid i = 0, 1, \ldots\}$. By choosing t to be the maximum of all the t_j's and p as the lowest common multiple of the positive p_j's, we obtain numbers that satisfy the following claim: if $x > t$ is the horizontal position at the end of some P, then $x = t_j + ip_j$ for some numbers $i \geq 1$ and j with $p_j > 0$. Because p_j divides p we see that $x + p$ is in the same linear set $\{t_j + ip_j \mid i = 0, 1, \ldots\}$ as x is.

As proved above, the transients t and the periods p exist for all start states q and end states r. We can find a common transient and period for all start and end states by taking the lowest common multiple of the periods p and the largest of the transients t.

U

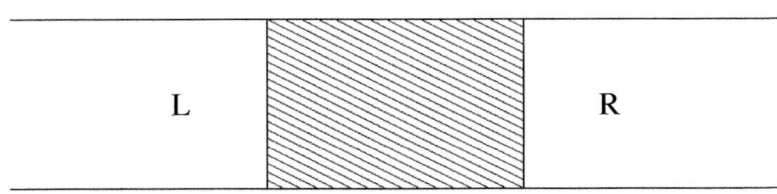

L   R

D

Fig. 2. Four parts of the plane

Now we are ready to proceed with the proof of the Lemma. Let us divide the plane outside a rectangle into four parts as indicated in Figure 2. Parts L and R are semi-infinite strips to the left and to the right of the rectangle. Consider an accepting computation C for a $w \times h$ rectangle and let us divide C into segments at the transitions between the four parts. More precisely, a segment will be a part of C that is entirely inside one of the four regions, and the automaton touches the border between two regions only at the beginning and end of each segment.

If the width w of the rectangle is increased, all segments of C can be executed unchanged except segments that take the automaton from L to R or from R to L through areas U or D.

Let us study a segment that takes the automaton from L to R through U. Let us divide it into subsegments that start and end on the same row as the beginning and end of the entire segment, so that inside the subsegments all positions are strictly above that row. See the illustration in Figure 3. Notice that the subsegments start and end in positions that touch the upper edge of the rectangle. The only exceptions are the start and end positions of the first and last subsegment, which lie above L and R respectively.

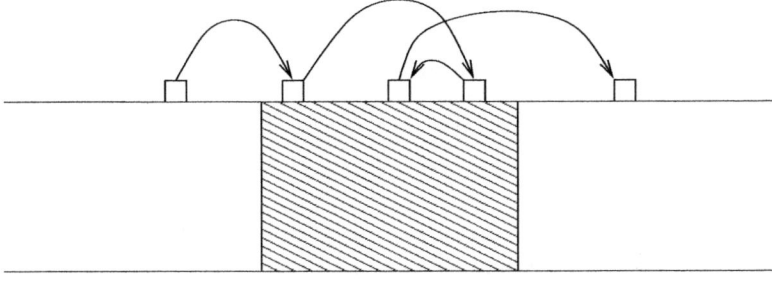

Fig. 3. A sample segment under consideration

If any of the subsegments moves the automaton more than t cells to the right where t is the number from the first part of the proof, then according to the first part that subsegment can be replaced by another one that moves the automaton $k \cdot p$ additional cells to the right, for any $k = 1, 2, \ldots$. This allows us to 'pump' this segment of the computation farther to the right.

If, on the other hand, every subsegment moves A either to the left or at most t cells to the right, we can still pump the segment farther right as follows. Let s be the number of states in A. If the width w of the rectangle is at least $(s+1)t$ then the automaton must touch the upper edge of the rectangle at an increasing series of times at positions $x_1 < x_2 < x_3 < \cdots < x_{s+1}$, all satisfying $x_{i+1} - x_i \leq t$. It follows from the pigeonhole principle that the automaton must be in the same state when it touches the rectangle in two different positions x_i and x_j for some $i < j$. The computation path between positions x_i and x_j can be repeated arbitrarily many times, leading to an additional movement by $k \cdot (x_j - x_i)$ cells to the right, for any chosen $k = 1, 2, \ldots$. Because $x_j - x_i \leq st$ we conclude that an additional movement to the right by any multiple of $(st)!$ cells is possible.

From the two cases above it follows that if the width of the rectangle is at least $(s+1)t$ then path P can be modified to cross an additional distance of the lowest common multiple of $(st)!$ and p. Therefore, the width of the rectangle can be increased by $\operatorname{lcm}(p, (st)!)$.

A similar argument can be made for segments that cross from R to L, or between R and L through the lower half D. By taking the maximum of the transient lengths and the lowest common multiple of the periods we obtain numbers t_H and p_H that satisfy the lemma. The numbers are independent of the height of the rectangle.

The previous lemma has, naturally, a vertical counterpart:

Lemma 2. *Let $S \subseteq \mathbb{N}^2$ be recognized from outside by an NFA A. Then there exist positive integers t_V and p_V such that for every $h > t_V$ and every w*

$$(w, h) \in S \implies (w, h + p_V) \in S$$

where the numbers t_V and p_V are independent of the width w.

Finally, we use the following technical result:

Lemma 3. *Assume that a language $L \subseteq 0^*1^*$ satisfies the following monotonicity conditions: there exist positive integers n and m such that*

$$0^i 1^j \in L, i \geq n \implies (\forall x \geq i)\, 0^x 1^j \in L, \ and$$
$$0^i 1^j \in L, j \geq m \implies (\forall y \geq j)\, 0^i 1^y \in L.$$

Then L is regular.

Proof. For every fixed k the language

$$A_k = L \cap 0^k 1^*$$

is regular: if for some $j \geq m$ we have $0^k 1^j \in L$, then A_k is the union of $0^k 1^j 1^*$
and a finite number of words that are shorter than $0^k 1^j$. If no such j exists then
A_k is finite and therefore regular. Analogously, languages

$$B_k = L \cap 0^* 1^k$$

are regular.

If L contains some word $0^i 1^j$ with $i \geq n$ and $j \geq m$ then

$$\begin{aligned} L = {} & A_0 \cup A_1 \cup \cdots \cup A_{i-1} \\ & \cup B_0 \cup B_1 \cup \cdots \cup B_{j-1} \\ & \cup 0^i 0^* 1^j 1^* \end{aligned}$$

is regular since it is a finite union of regular languages. If no such word $0^i 1^j$ is
in L then

$$\begin{aligned} L = {} & A_0 \cup A_1 \cup \cdots \cup A_{n-1} \\ & \cup B_0 \cup B_1 \cup \cdots \cup B_{m-1} \end{aligned}$$

is a union of a finite number of sets of the form A_k and B_k, and is therefore
regular.

Now we have all the necessary tools to prove the main result of the section:

Theorem 2. *Suppose an NFA recognizes from outside a set $S \subseteq \mathbb{N}^2$ of rectangles. Then the language*
$$L_S = \{0^i 1^j \mid (i,j) \in S\}$$
is regular.

Proof. Let t_H and p_H be the horizontal transient and the period from Lemma 1,
and let t_V and p_V be their vertical counterparts from Lemma 2. For every $a =
0, 1, \ldots, p_H - 1$ and $b = 0, 1, \ldots, p_V - 1$ let us define

$$L_{a,b} = L_S \cap \{0^i 1^j \mid i \equiv a \pmod{p_H}, j \equiv b \pmod{p_V}\}.$$

Because L_S is the union of languages $L_{a,b}$ it is enough to prove that every $L_{a,b}$
is regular.

But it follows from Lemmas 1 and 2 that the language

$$L'_{a,b} = \{0^{(i-a)/p_H} 1^{(j-b)/p_V} \mid 0^i 1^j \in L_{a,b}\}$$

satisfies the monotonicity condition of Lemma 3 and is therefore regular. If h is
the homomorphism that replaces 0 with p_H 0's and 1 with p_V 1's then

$$L_{a,b} = 0^a h(L'_{a,b}) 1^b.$$

Thus $L_{a,b}$ is regular.

Acknowledgments. JK thanks the Santa Fe Institute for an enjoyable visit
where this work began. We also thank Bernd Borchert and Juraj Hromkovič for
helpful conversations.

References

1. M. Blum and C. Hewitt (1967), Automata on a 2-dimensional tape. *8th IEEE Symp. on Switching and Automata Theory* 155–160.
2. B. Borchert, http://math.uni-heidelberg.de/logic/bb/2dpapers.html
3. D. Giammarresi and A. Restivo (1992) Recognizable picture languages. *Int. J. of Pattern Recognition and Artificial Intelligence* **6(2-3)** 241-256.
4. D. Giammarresi and A. Restivo (1996) Two-dimensional languages. In G. Rosenberg and A. Salomaa, Eds., *Handbook of Formal Languages,* Volume III, pp. 215–267. Springer-Verlag.
5. K. Inoue and A. Nakamura (1977) Some properties of two-dimensional on-line tesselation acceptors. *Information Sciences* **13** 95–121.
6. K. Inoue and A. Nakamura (1979) Two-dimensional finite automata and unacceptable functions. *Int. J. Comput. Math. A* **7** 207–213.
7. K. Inoue and I. Takanami (1991) A survey of two-dimensional automata theory. *Information Sciences* **55** 99–121.
8. J. Kari and C. Moore (2001) New results on alternating and non-deterministic two-dimensional finite automata. *"Proceedings of STACS'2001, 18th Annual Symposium on Theoretical Aspects of Computer Science".* Lecture Notes in computer Science **2010**, 396–406, Springer-Verlag.
9. K. Lindgren, C. Moore and M.G. Nordahl (1998) Complexity of two-dimensional patterns. *Journal of Statistical Physics* **91** 909–951.
10. D.L. Milgram (1976) A region crossing problem for array-bounded automata. *Information and Control* **31** 147–152.
11. M. Minsky (1967) *Computation: Finite and Infinite Machines.* Prentice-Hall.
12. A. Rosenfeld (1979) *Picture Languages: Formal Models for Picture Recognition.* Academic Press.
13. M. Sipser (1980) Halting space-bounded computations. *Theoretical Computer Science* **10** 335-338.

Substitution on Trajectories

Lila Kari[1], Stavros Konstantinidis[2], and Petr Sosík[1,3]

[1] Department of Computer Science, The University of Western Ontario, London,
ON, Canada, N6A 5B7
{lila,sosik}@csd.uwo.ca
[2] Dept. of Mathematics and Computing Science, Saint Mary's University, Halifax,
Nova Scotia, B3H 3C3 Canada
s.konstantinidis@stmarys.ca
[3] Institute of Computer Science, Silesian University, Opava, Czech Republic

Abstract. The word substitutions are binary word operations which
can be basically interpreted as a deletion followed by insertion, with some
restrictions applied. Besides being itself an interesting topic in formal
language theory, they have been naturally applied to modelling noisy
channels. We introduce the concept of *substitution on trajectories* which
generalizes a class of substitution operations. Within this framework, we
study their closure properties and decision questions related to language
equations. We also discuss applications of substitution on trajectories in
modelling complex channels and a cryptanalysis problem.

1 Introduction

There are two basic forms of the word substitution operation. The *substitution
in α by β* means to substitute certain letters of the word α by the letters of β.
The *substitution in α of β* means to substitute the letters of β within α by other
letters, provided that β is scattered within α. In both cases the overall length of
α is not changed. Also, we assume that a letter must not be substituted by the
same letter.

These two operations are closely related and, indeed, we prove in Section 4
that they are mutual left inverses. Their motivation comes from coding theory
where they have been used to model certain noisy channels [8]. The natural idea
is to assume that during a transfer through a noisy channel, some letters of the
transferred word can de distorted — replaced by different letters. This can be
modelled by a substitution operation extended to sets of words. This approach
also allows one to take into account that certain substitutions are more likely
than others. Hence the algebraic, closure and other properties of the substitution
operation are of interest, to study how a set of messages (=language) can change
when transferred through a noisy channel.

In this paper we generalize the idea of substitution using the syntactical
constraints — *trajectories*. The *shuffle on trajectories* as a generalization of sequential insertion has been studied since 1996 [16, 17]. Recently also its inverse
— the *deletion on trajectories* has been introduced [1, 10]. A *trajectory* acts as a

J. Karhumäki et al. (Eds.): Theory Is Forever (Salomaa Festschrift), LNCS 3113, pp. 145–158, 2004.

syntactical condition, restricting the positions of letters within the word where an operation places its effect. Hence the shuffle and deletion on trajectories can be understood as meta-operations, defining a whole class of insertion/deletion operations due to the set of trajectories at hand. This idea turned out to be fruitful, with several interesting consequences and applications [1–4, 11, 14, 15].

We give a basic description of these operations in Section 3. Then in Section 4 we introduce on a similar basis the *substitution and difference on trajectories*. From the point of view of noisy channels, the application of trajectories allows one to restrict positions of errors within words, their frequency etc. We then study the closure properties of substitution on trajectories in Section 5 and basic decision questions connected with them in Section 6. In Section 7 we discuss a few applications of the substitution on trajectories in modelling complex noisy channels and a cryptanalysis problem. In the former case, the channels involved permit only substitution errors. This restriction allows us to improve the time complexity of the problem of whether a given regular language is error-detecting with respect to a given channel [13].

2   Definitions

An *alphabet* is a finite and nonempty set of symbols. In the sequel we shall use a fixed alphabet Σ. Σ is assumed to be non-singleton, if not stated otherwise. The set of all words (over Σ) is denoted by Σ^*. This set includes the *empty word* λ. The length of a word w is denoted by $|w|$. $|w|_x$ denotes the number of occurrences of x within u, for $w, x \in \Sigma^*$.

For a nonnegative integer n and a word w, we use w^n to denote the word that consists of n concatenated copies of w. The *Hamming distance* $H(u, v)$ between two words u and v of the same length is the number of corresponding positions in which u and v differ. For example, $H(abba, aaaa) = 2$.

A language L is a set of words, or equivalently a subset of Σ^*. A language is said to be λ-free if it does not contain the empty word. For a language L, we write L_λ to denote $L \cup \{\lambda\}$. If n is a nonnegative integer, we write L^n for the language consisting of all words of the form $w_1 \cdots w_n$ such that each w_i is in L. We also write L^* for the language $L^0 \cup L^1 \cup L^2 \cup \cdots$ and L^+ for the language $L^* - \{\lambda\}$. The notation L^c represents the complement of the language L, that is, $L^c = \Sigma^* - L$. For the classes of regular, context-free, and context sensitive languages, we use the notations REG, CF and CS, respectively.

A nondeterministic finite automaton with λ productions (or transitions), a λ-NFA for short, is a quintuple $A = (S, \Sigma, s_0, F, P)$ such that S is the finite and nonempty set of states, s_0 is the start state, F is the set of final states, and P is the set of productions of the form $sx \to t$, where s and t are states in S, and x is either a symbol in Σ or the empty word. If there is no production with $x = \lambda$, the automaton is called an *NFA*. If for every two productions of the form $sx_1 \to t_1$ and $sx_2 \to t_2$ of an NFA we have that $x_1 \neq x_2$ then the automaton is called a *DFA* (deterministic finite automaton). The language accepted by the automaton A is denoted by $L(A)$. The *size* $|A|$ of the automaton A is the number $|S| + |P|$.

A *finite transducer* (in standard form) is a sextuple $T = (S, \Sigma, \Sigma', s_0, F, P)$ such that Σ' is the output alphabet, the components S, s_0, F are as in the case of λ-NFAs, and the set P consists of productions of the form $sx \to yt$ where s and t are states in S, $x \in \Sigma \cup \{\lambda\}$ and $y \in \Sigma' \cup \{\lambda\}$. If x is nonempty for every production then the transducer is called a *gsm* (generalized sequential machine). If, in addition, y is nonempty for every production then the transducer is called a λ-*free gsm*. The *relation realized by* the transducer T is denoted by $R(T)$. The size $|T|$ of the transducer T (in standard form) is $|S| + |P|$. We refer the reader to [18] for further details on automata and formal languages.

A *binary word operation* is a mapping $\diamondsuit : \Sigma^* \times \Sigma^* \to 2^{\Sigma^*}$, where 2^{Σ^*} is the set of all subsets of Σ^*. The *characteristic relation* of $\diamondsuit$ is

$$C_\diamondsuit = \{(w, u, v) : w \in u \diamondsuit v\}.$$

For any languages X and Y, we define

$$X \diamondsuit Y = \bigcup_{u \in X, v \in Y} u \diamondsuit v. \tag{1}$$

It should be noted that every subset B of $\Sigma^* \times \Sigma^* \times \Sigma^*$ defines a unique binary word operation whose characteristic relation is exactly B. For an operation $\diamondsuit$ we define its left inverse $\diamondsuit^l$ as

$$w \in (x \diamondsuit v) \text{ iff } x \in (w \diamondsuit^l v), \text{ for all } v, x, w \in \Sigma^*,$$

and the right inverse $\diamondsuit^r$ of $\diamondsuit$ as

$$w \in (u \diamondsuit y) \text{ iff } y \in (u \diamondsuit^r w), \text{ for all } u, y, w \in \Sigma^*.$$

Moreover, the word operation $\diamondsuit'$ defined by $u \diamondsuit' v = v \diamondsuit u$ is called reversed $\diamondsuit$. It should be clear that, for every binary operation $\diamondsuit$, the triple (w, u, v) is in $C_\diamondsuit$ if and only if (u, w, v) is in $C_{\diamondsuit^l}$ if and only if (v, u, w) is in $C_{\diamondsuit^r}$ if and only if (w, v, u) is in $C_{\diamondsuit'}$. If x and y are symbols in $\{l, r,'\}$, the notation $\diamondsuit^{xy}$ represents the operation $(\diamondsuit^x)^y$. Using the above observations, one can establish identities between operations of the form $\diamondsuit^{xy}$.

Lemma 1. *(i)* $\diamondsuit^{ll} = \diamondsuit^{rr} = \diamondsuit'' = \diamondsuit$,
(ii) $\diamondsuit^{rl} = \diamondsuit^{r'} = \diamondsuit^{lr}$,
(iii) $\diamondsuit^{'r} = \diamondsuit^{l'} = \diamondsuit^{rl}$.

Bellow we list several binary word operations together with their left and right inverses [6, 7].

Catenation: [4] $u \cdot v = \{uv\}$, with $\cdot^l = \longrightarrow_{\text{rq}}$ and $\cdot^r = \longrightarrow_{\text{lq}}$.
Left quotient: $u \longrightarrow_{\text{lq}} v = \{w\}$ if $u = vw$, with $\longrightarrow_{\text{lq}}^l = \cdot'$ and $\longrightarrow_{\text{lq}}^r = \cdot$.
Right quotient: $u \longrightarrow_{\text{rq}} v = \{w\}$ if $u = wv$, with $\longrightarrow_{\text{rq}}^l = \cdot$ and $\longrightarrow_{\text{rq}}^r = \longrightarrow_{\text{lq}}'$.
Shuffle (or scattered insertion): $u \sqcup\!\sqcup v = \{u_1 v_1 \cdots u_k v_k u_{k+1} \mid k \geq 1,$
    $u = u_1 \cdots u_k u_{k+1}, v = v_1 \cdots v_k\}$, with $\sqcup\!\sqcup^l = \rightsquigarrow$ and $\sqcup\!\sqcup^r = \rightsquigarrow'$.
Scattered deletion: $u \rightsquigarrow v = \{u_1 \cdots u_k u_{k+1} \mid k \geq 1, u = u_1 v_1 \cdots u_k v_k u_{k+1}, v = v_1 \cdots v_k\}$, with $\rightsquigarrow^l = \sqcup\!\sqcup$ and $\rightsquigarrow^r = \rightsquigarrow$.

---

[4] We shall also write uv for $u \cdot v$.

3   Shuffle and Deletion on Trajectories

The above insertion and deletion operations can be naturally generalized using the concept of *trajectories*. A trajectory defines an order in which the operation is applied to the letters of its arguments. Notice that this restriction is purely syntactical, as the content of the arguments has no influence on this order. Formally, a trajectory is a string over the *trajectory alphabet* $V = \{0, 1\}$. The following definitions are due to [1, 16, 10].

Let Σ be an alphabet and let t be a trajectory, $t \in V^*$. Let α, β be two words over Σ.

Definition 1. *The shuffle of α with β on the trajectory t, denoted by $\alpha \sqcup\!\sqcup_t \beta$, is defined as follows:*

$$\alpha \sqcup\!\sqcup_t \beta = \{\alpha_1 \beta_1 \ldots \alpha_k \beta_k \mid \alpha = \alpha_1 \ldots \alpha_k,\ \beta = \beta_1 \ldots \beta_k,\ t = 0^{i_1} 1^{j_1} \ldots 0^{i_k} 1^{j_k},\ \text{where} \\ |\alpha_m| = i_m \text{ and } |\beta_m| = j_m \text{ for all } m,\ 1 \le m \le k\}.$$

Definition 2. *The deletion of β from α on trajectory t is the following binary word operation:*

$$\alpha \leadsto_t \beta = \{\alpha_1 \ldots \alpha_k \mid \alpha = \alpha_1 \beta_1 \ldots \alpha_k \beta_k,\ \beta = \beta_1 \ldots \beta_k,\ t = 0^{i_1} 1^{j_1} \ldots 0^{i_k} 1^{j_k},\ \text{where} \\ |\alpha_m| = i_m \text{ and } |\beta_m| = j_m \text{ for all } m,\ 1 \le m \le k\}.$$

Observe that due to the above definition, if $|\alpha| \neq |t|$ or $|\beta| \neq |t|_1$, then $\alpha \leadsto_t \beta = \emptyset$.

A *set of trajectories* is any set $T \subseteq V^*$. We extend the shuffle and deletion to sets of trajectories as follows:

$$\alpha \sqcup\!\sqcup_T \beta = \bigcup_{t \in T} \alpha \sqcup\!\sqcup_t \beta, \qquad \alpha \leadsto_T \beta = \bigcup_{t \in T} \alpha \leadsto_t \beta. \tag{2}$$

The operations $\sqcup\!\sqcup_T$ and $\leadsto_T$ generalize to languages due to (1).

Example 1. The following binary word operations can be expressed via shuffle on trajectories using certain sets of trajectories.

1. Let $T = 0^*1^*$, then $\sqcup\!\sqcup_T = \cdot$, the catenation operation, and $\leadsto_T = \longrightarrow_{\text{rq}}$, the right quotient.
2. For $T = 1^*0^*$ we have $\sqcup\!\sqcup_T = \cdot'$, the anti-catenation, and $\leadsto_T = \longrightarrow_{\text{lq}}$, the left quotient.
3. Let $T = \{0, 1\}^*$, then $\leadsto_T = \sqcup\!\sqcup$, the shuffle, and $\leadsto_T = \leadsto$, the scattered deletion.

We refer to [1, 16, 10] for further elementary results concerning shuffle and deletion on trajectories.

4   Substitution on Trajectories

Based on the previously studied concepts of the insertion and deletion on trajectories, we consider a generalization of three natural binary word operations which are used to model certain noisy channels [8]. Generally, *channel* [13] is a binary relation $\gamma \subseteq \Sigma^* \times \Sigma^*$ such that (u, u) is in γ for every word u in the input domain of γ – this domain is the set $\{u \mid (u, v) \in \gamma \text{ for some word } v\}$. The fact that (u, v) is in γ means that the word v can be received from u via the channel γ. In [8], certain channels with insertion, deletion and substitution errors are characterized via word operations. For instance, the channel with exactly m insertion errors is the set of all pairs (u, v) such that $v \in u \sqcup \Sigma^m$, and analogously for deletion errors. The following definitions allow one to characterize channels with substitution errors.

Definition 3. *If $u, v \in \Sigma^*$ then we define the substitution in u by v as*

$$u \bowtie v = \{u_1 v_1 u_2 v_2 \ldots u_k v_k u_{k+1} \mid k \geq 0, \ u = u_1 a_1 u_2 a_2 \ldots u_k a_k u_{k+1}, \ v = v_1 \ldots v_k,$$
$$a_i, v_i \in \Sigma, 1 \leq i \leq k, \ a_i \neq v_i, \forall i, 1 \leq i \leq k\}.$$

The case $k = 0$ corresponds to $v = \lambda$ when no substitution is performed.

Definition 4. *If $u, v \in \Sigma^*$ then we define the substitution in u of v as*

$$u \triangle v = \{u_1 a_1 u_2 a_2 \ldots u_k a_k u_{k+1} \mid k \geq 0, \ u = u_1 v_1 u_2 v_2 \ldots u_k v_k u_{k+1}, \ v = v_1 \ldots v_k,$$
$$a_i, v_i \in \Sigma, 1 \leq i \leq k, \ a_i \neq v_i, \forall i, 1 \leq i \leq k\}.$$

Definition 5. *Let $u, v \in \Sigma^*$, $|u| = |v|$, let $H(u, v)$ be the Hamming distance of u and v. We define*

$$u \triangleright v = \{v_1 v_2 \ldots v_k \mid k = H(u, v), \ u = u_1 a_1 \ldots u_k a_k u_{k+1}, \ v = u_1 v_1 \ldots u_k v_k u_{k+1},$$
$$a_i, v_i \in \Sigma, 1 \leq i \leq k, \ a_i \neq v_i, \forall i, 1 \leq i \leq k\}.$$

The above definitions are due to [8], where it is also shown that the left- and the right-inverse of $\bowtie$ are $\triangle$ and $\triangleright$, respectively. Given two binary word operations $\diamondsuit_1, \diamondsuit_2$, their composition $(\diamondsuit_1 \diamondsuit_2)$ is defined as

$$w \in u(\diamondsuit_1 \diamondsuit_2)v \iff w \in (u \diamondsuit_1 v_1) \diamondsuit_2 v_2, \quad v = v_1 v_2,$$

for all $u, v, w \in \Sigma^*$. Then it is among others shown that:

(i) The channel with at most m substitution and insertion errors is equal to
$\{(u, v) \mid v \in u(\triangle \sqcup)(\Sigma^0 \cup \cdots \cup \Sigma^m)\}$.
(i) The channel with at most m substitution and deletion errors is equal to
$\{(u, v) \mid v \in u(\leadsto \bowtie)(\Sigma^0 \cup \cdots \cup \Sigma^m)\}$.

Moreover, further consequences including composition of channels, inversion of channels etc. are derived. The above substitution operations can be generalized using trajectories as follows.

Definition 6. *For a trajectory $t \in V^*$ and $u, v \in \Sigma^*$ we define the* substitution *in u by v on trajectory t as*

$$u \bowtie_t v = \{u_1 v_1 u_2 v_2 \ldots u_k v_k u_{k+1} \mid k \geq 0, \ u = u_1 a_1 \ldots u_k a_k u_{k+1}, \ v = v_1 \ldots v_k,$$
$$t = 0^{j_1} 10^{j_2} 1 \ldots 0^{j_k} 10^{j_{k+1}}, \ a_i, v_i \in \Sigma, 1 \leq i \leq k, \ a_i \neq v_i, \forall i, 1 \leq i \leq k,$$
$$j_i = |u_i|, 1 \leq i \leq k+1\}.$$

Definition 7. *For a trajectory $t \in V^*$ and $u, v \in \Sigma^*$ we define the* substitution *in u of v on trajectory t as*

$$u \triangle_t v = \{u_1 a_1 u_2 a_2 \ldots u_k a_k u_{k+1} \mid k \geq 0, \ u = u_1 v_1 \ldots u_k v_k u_{k+1}, \ v = v_1 \ldots v_k,$$
$$t = 0^{j_1} 10^{j_2} 1 \ldots 0^{j_k} 10^{j_{k+1}}, \ a_i, v_i \in \Sigma, 1 \leq i \leq k, \ a_i \neq v_i, \forall i, 1 \leq i \leq k,$$
$$j_i = |u_i|, 1 \leq i \leq k+1\}.$$

Definition 8. *For a trajectory $t \in V^*$ and $u, v \in \Sigma^*$ we define the* right difference *of u and v on trajectory t as*

$$u \triangleright_t v = \{v_1 v_2 \ldots v_k \mid k \geq 0, \ u = u_1 a_1 \ldots u_k a_k u_{k+1}, \ v = u_1 v_1 \ldots u_k v_k u_{k+1},$$
$$t = 0^{j_1} 10^{j_2} 1 \ldots 0^{j_k} 10^{j_{k+1}}, \ a_i, v_i \in \Sigma, 1 \leq i \leq k, \ a_i \neq v_i, \forall i, 1 \leq i \leq k,$$
$$j_i = |u_i|, 1 \leq i \leq k+1\}.$$

These operations can be generalized to sets of trajectories in the natural way:

$$u \bowtie_T v = \bigcup_{t \in T} u \bowtie_t v, \quad u \triangle_T v = \bigcup_{t \in T} u \triangle_t v \ \text{ and } \ u \triangleright_T v = \bigcup_{t \in T} u \triangleright_t v.$$

Example 2. Let $T = V^*$, i.e. the set T contains all the possible trajectories. Then $\bowtie_T = \bowtie$, $\triangle_T = \triangle$ and $\triangleright_T = \triangleright$.

One can observe that similarly as in [8], the above defined substitution on trajectories could be used to characterize channels where errors occur in certain parts of words only, or with a certain frequency and so on. Due to the fact that the trajectory is a *syntactic* restriction, only such channels can be modelled where the occurrence of errors may depend on the *length* of the transferred message, but not on its *content*. In the sequel we study various properties of the above defined substitution operations.

Lemma 2. *For a set of trajectories T and words $u, v \in \Sigma^*$, the following holds:*

(i) $\bowtie_T^l = \triangle_T$ and $\bowtie_T^r = \triangleright_T$,
(ii) $\triangle_T^l = \bowtie_T$ and $\triangle_T^r = \triangleright_T'$,
(iii) $\triangleright_T^l = \triangle_T'$ and $\triangleright_T^r = \bowtie_T$.

Proof. (i) Consider the characteristic relation $C_{\bowtie_t}$ of the operation $\bowtie_t$. Observe that $(w, u, v) \in C_{\bowtie_t}$ iff $(u, w, v) \in C_{\bowtie_t^l}$ iff $(v, u, w) \in C_{\bowtie_t^r}$. Then the statements

$\bowtie_t^l = \triangle_t$ and $\bowtie_t^r = \rhd_t$, $t \in T$, follow directly by careful reading the definitions of $\bowtie_t$, $\triangle_t$ and $\rhd_t$. Now observe that

$$u \bowtie_T^l v = \bigcup_{t \in T} u \bowtie_t^l v = \bigcup_{t \in T} u \triangle_t v = u \triangle_T v.$$

The proof for $\bowtie_T^r$ is analogous.

(ii) Due to Lemma 1, $\bowtie_T^l = \triangle_T$ implies $\triangle_T^l = \bowtie_T$ and $\bowtie_T^r = \rhd_T$ implies $\triangle_T^r = \bowtie_T^{lr} = \bowtie_T^{r'} = \rhd_T^l$.

(iii) Similarly, $\bowtie_T^r = \rhd_T$ implies $\rhd_T^l = \bowtie_T$, and consequently $\rhd_T^l = \bowtie_T^{rl} = \bowtie_T^{l'} = \triangle_T'$. □

5 Closure Properties

Before addressing the closure properties of substitution, we show first that any (not necessarily recursively enumerable) language over a two letter alphabet can be obtained as a result of substitution.

Lemma 3. *For an arbitrary language $L \subseteq \{a, b\}^*$ there exists a set of trajectories T such that*

(i) $L = a^* \bowtie_T b^*$,
(ii) $L = a^* \triangle_T a^*$.

Proof. Let $T = \phi(L)$, $\phi : \{a, b\}^* \longrightarrow V^*$ being a coding morphism such that $\phi(a) = 0$, $\phi(b) = 1$. The statements follow easily by definition. □

Similarly as in the case of shuffle and deletion on trajectories [1, 16, 10], the substitution on trajectories can be characterized by simpler language operations.

Lemma 4. *Let $\Diamond_T$ be any of the operations $\bowtie_T$, $\triangle_T$, $\rhd_T$. Then there exists a finite substitution h_1, morphisms h_2, g and a regular language R such that for all languages $L_1, L_2 \subseteq \Sigma^*$, and for all sets of trajectories $T \subseteq V^*$,*

$$L_1 \Diamond_T L_2 = g((h_1(L_1) \sqcup\!\!\sqcup h_2(L_2) \sqcup\!\!\sqcup T) \cap R). \tag{3}$$

Proof. Let $\Sigma_i = \{a_i \mid a \in \Sigma\}$, for $i = 1, 2, 3$, be copies of Σ such that $\Sigma, \Sigma_1, \Sigma_2, \Sigma_3$ and V are pairwise disjoint alphabets. For a letter $a \in \Sigma$, we denote by a_i the corresponding letter from Σ_i, $i = 1, 2, 3$.

Let further $h_1 : \Sigma \longrightarrow (\Sigma_1 \cup \Sigma_3)$ be a finite substitution and let $h_2 : \Sigma \longrightarrow \Sigma^2$ and $g : (\Sigma_1 \cup \Sigma^2 \cup \Sigma^3 \cup V) \longrightarrow \Sigma$ be morphisms.

(i) If $\Diamond_T = \bowtie_T$, then define $h_1(a) = \{a_1, a_3\}$, $h_2(a) = a_2$ for each $a \in \Sigma$. Let

$$R = (\Sigma_1 \cdot \{0\} \cup \{a_3 b_2 1 \mid a, b \in \Sigma, a \neq b\})^*.$$

Let further $g(a_1) = a$, $g(a_2) = a$ for all $a_1 \in \Sigma_1$, $a_2 \in \Sigma_2$, and $g(x) = \lambda$ for all $x \in \Sigma_3 \cup V$. Then one can easily verify that (3) holds true.

(ii) If $\Diamond_T = \triangle_T$, then let $h_1(a) = \{a_1\} \cup \{a_3\} \cdot \Sigma_1$, $h_2(a) = a_2$ for each $a \in \Sigma$. Let further

$$R = (\Sigma_1 \cdot \{0\} \cup \{a_3 a_2 b_1 1 \mid a, b \in \Sigma, a \neq b\})^*,$$

and $g(a_1) = a$ for all $a_1 \in \Sigma_1$, $g(x) = \lambda$ for all $x \in \Sigma_2 \cup \Sigma_3 \cup V$.

(iii) If $\Diamond_T = \rhd_T$, then define $h_1(a) = a_1$, $h_2(a) = \{a_2, a_3\}$ for each $a \in \Sigma$. Let

$$R = (\{a_1 a_2 0 \mid a \in \Sigma\} \cup \{a_1 b_3 1 \mid a, b \in \Sigma, a \neq b\})^*,$$

and $g(a_3) = a$ for all $a_3 \in \Sigma_3$, $g(x) = \lambda$ for all $x \in \Sigma_1 \cup \Sigma_2 \cup V$.

□

The previous lemmata allow us to make statements about closure properties of the substitution operations now.

Theorem 1. *For a set of trajectories $T \subseteq V^*$, the following three statements are equivalent.*

(i) T is a regular language.
(ii) $L_1 \bowtie_T L_2$ is a regular language for all $L_1, L_2 \subseteq \Sigma^*$.
(iii) $L_1 \triangle_T L_2$ is a regular language for all $L_1, L_2 \subseteq \Sigma^*$.

Proof. The implications (i) $\Rightarrow$ (ii) and (i) $\Rightarrow$ (iii) follow by Lemma 4 due to the closure of the class of regular languages with respect to shuffle, finite substitution, morphisms and intersection.

To show the implication (ii) $\Rightarrow$ (i), assume that $L_1 \bowtie_T L_2$ is a regular language for all $L_1, L_2 \subseteq \Sigma^*$. Let $a, b \in \Sigma$ without loss of generality, then also $L = a^* \bowtie_T b^*$ is a regular language, and $T = \phi^{-1}(L)$, ϕ being the coding defined in the proof of Lemma 3. Consequently, T is regular. The implication (iii) $\Rightarrow$ (i) can be shown analogously.

□

Theorem 2. *For all regular set of trajectories $T \subseteq V^*$ and regular languages $L_1, L_2 \subseteq \Sigma^*$, $L_1 \rhd_T L_2$ is a regular language.*

Proof. The same as the proof of Theorem 1, (i) $\Rightarrow$ (ii).

□

Theorem 3. *Let $\Diamond_T$ be any of the operations $\bowtie_T$, $\triangle_T$, $\rhd_T$.*

(i) *Let any two of the languages L_1, L_2, T be regular and the third one be context-free. Then $L_1 \Diamond_T L_2$ is a context-free language.*
(ii) *Let any two of the languages L_1, L_2, T be context-free and the third one be regular. Then $L_1 \Diamond_T L_2$ is a non-context-free language for some triples (L_1, L_2, T).*

Proof.   (i) Follows by Lemmata 4, and by closure of the class of context-free languages with respect to finite substitution, shuffle, morphisms and intersection with regular languages.

(ii) Consider the alphabet $\Sigma = \{a, b, c, d\}$.
   1. Let $\Diamond_T = \bowtie_T$.

(1) Consider $L_1 = \{a^n db^{2n} \mid n > 0\}$, $L_2 = \{a^m c^m \mid m > 0\}$ and $T = V^*$, then $(L_1 \bowtie_T L_2) \cap a^* da^* c^* = a^n da^n c^n$.

(2) Consider $L_1 = \{a^n b^{2n} \mid n > 0\}$, $L_2 = c^+$ and $T = \{0^{2m} 1^m \mid m > 0\}$, then $L_1 \bowtie_T L_2 = a^n b^n c^n$.

(3) Consider $L_1 = a^+$, $L_2 = \{b^n c^n \mid n > 0\}$ and $T = \{0^m 1^{2m} \mid m > 0\}$, then $L_1 \bowtie_T L_2 = a^n b^n c^n$.

2. Let $\Diamond_T = \triangle_T$. Consider:

   (1) $L_1 = \{a^n ba^k ba^l \mid k + l + 1 = 2n > 0\}$, $L_2 = \{a^m ba^{m+1} \mid m > 0\}$ and $T = 0^+ 1^+$,

   (2) $L_1 = \{a^n b^n a^* \mid n > 0\}$, $L_2 = a^+$ and $T = 0^{2m+1} 1^m$,

   (3) $L_1 = a^+ ba^+$, $L_2 = \{a^n ba^n \mid n > 0\}$ and $T = \{0^m 1^{2m+1} \mid m > 0\}$,

   then in all three cases $(L_1 \triangle_T L_2) \cap a^* b^* ab^* = a^n b^n ab^n$.

3. Let $\Diamond_T = \triangleright_T$. Consider:

   (1) $L_1 = \{c^{2m} dc^m a^* \mid m > 0\}$, $L_2 = \{a^n b^n da^* \mid n > 0\}$ and $T = V^+$,

   (2) $L_1 = \{b^n a^n db^+ a^* \mid n > 0\}$, $L_2 = a^+ b^+ da^+$ and $T = \{1^{2m} 01^m 0^* \mid m > 0\}$,

   (3) $L_1 = c^+ dc^+ a^*$, $L_2 = \{a^n b^n da^* \mid n > 0\}$ and $T = \{1^{2m} 01^m 0^* \mid m > 0\}$,

   then in all three cases $(L_1 \triangleright_T L_2) \cap \{a, b\}^* = a^n b^n a^n$.

In all the above cases we have shown that $L_1 \Diamond_T L_2$ is a non-context-free language.

$\square$

6   Decision Problems

In this section we study three elementary types of decision problems for language equations of the form $L_1 \Diamond_T L_2 = R$, where $\Diamond_T$ is one of the operations $\bowtie_T$, $\triangle_T$, $\triangleright_T$. These problems, studied already for various binary word operations in [7, 6, 1, 10, 5] and others, are stated as follows. First, given L_1, L_2 and R, one asks whether the above equation holds true. Second, the existence of a solution L_1 to the equation is questioned, when L_1 is unknown (the left operand problem). Third, the same problem is stated for the right operand L_2. All these problems have their variants when one of L_1, L_2 (the unknown language in the case of the operand problems) consists of a single word.

We focus now on the case when L_1, L_2 and T are all regular languages. Then $L_1 \Diamond_T L_2$ is also a regular language by Theorems 1, 2, $\Diamond_T$ being any of the operations $\bowtie_T$, $\triangle_T$, $\triangleright_T$. Immediately we obtain the following result.

Theorem 4. *The following problems are both decidable if the operation $\Diamond_T$ is one of $\bowtie_T$, $\triangle_T$, $\triangleright_T$, T being a regular set of trajectories:*

(i) For given regular languages L_1, L_2, R, is $L_1 \Diamond_T L_2 = R$?

(ii) For given regular languages L_1, R and a word $w \in \Sigma^*$, is $L_1 \Diamond_T w = R$?

Also the decidability of the left and the right operand problems for languages are straightforward consequences of the results in Section 5 and some previously known facts about language equations [7].

Theorem 5. *Let $\diamond_T$ be one of the operations $\bowtie_T$, $\triangle_T$, $\triangleright_T$. The problem "Does there exist a solution X to the equation $X \diamond_T L = R$?" (left-operand problem) is decidable for regular languages L, R and a regular set of trajectories T.*

Proof. Due to [7], if a solution to the equation $X \diamond_T L = R$ exists, then also $X_{\max} = (R^c \diamond_T^l L)^c$ is also a solution, $\diamond_T$ being an invertible binary word operation. In fact, $X_{\max}$ is the maximum (with respect to the subset relation) of all the sets X such that $X \diamond_T L \subseteq R$. We can conclude that a solution X exists iff

$$(R^c \diamond_T^l L)^c \diamond_T L = R. \tag{4}$$

holds. Observe that if $\diamond_T$ is one of $\bowtie_T$, $\triangle_T$, $\triangleright_T$, then $\diamond_T^l$ is $\triangle_T$, $\bowtie_T$ or $\triangle_T'$, respectively, by Lemma 2. Hence the left side of the equation (4) represents an effectively constructible regular language by Theorems 1, 2. Consequently, the validity of (4) is decidable and moreover the maximal solution $X_{\max} = (R^c \diamond_T^l L)^c$ can be effectively found if one exists. □

Theorem 6. *Let $\diamond_T$ be one of the operations $\bowtie_T$, $\triangle_T$, $\triangleright_T$. The problem "Does there exist a solution X to the equation $L \diamond_T X = R$?" (right-operand problem) is decidable for regular languages L, R and a regular set of trajectories T.*

Proof. Similarly as in the proof of Theorem 5, a maximal solution to the equation $L \diamond_T X = R$ is $X_{\max} = (L \diamond_T^r R^c)^c$ for a binary word operation $\diamond_T$, see [7]. Hence a solution X exists iff

$$L \diamond_T (L \diamond_T^r R^c)^c = R \tag{5}$$

By Lemma 2, if $\diamond_T$ is one of $\bowtie_T$, $\triangle_T$, $\triangleright_T$, then $\diamond_T^r$ is $\triangleright_T$, $\triangleright_T'$ or $\bowtie_T$, respectively. Again the validity of (5) is effectively decidable by Theorems 1, 2, and, moreover, an eventual maximal solution $X_{\max} = (L \diamond_T^r R^c)^c$ can be effectively found. □

The situation is a bit different in the case when the existence of a singleton solution to the left or the right operand problem is questioned. Another proof technique takes place.

Theorem 7. *Let $\diamond_T$ be one of the operations $\bowtie_T$, $\triangle_T$, $\triangleright_T$. The problem "Does there exist a word w such that $w \diamond_T L = R$?" is decidable for regular languages L, R and a regular set of trajectories T.*

Proof. Assume that $\diamond_T$ is one of $\bowtie_T$, $\triangle_T$, $\triangleright_T$. Observe first that if $y \in w \diamond_T x$ for some $w, x, y \in \Sigma^*$, then $|y| \leq |w|$. Therefore, if R is infinite, then there cannot exist a solution w of a finite length satisfying $w \diamond_T L = R$. Hence for an infinite R the problem is trivial.

Assume now that R is finite. As shown in [7], the regular set $X_{\max} = (R^c \diamond_T^l L)^c$ is the maximal set with the property $X \diamond_T L \subseteq R$. Hence w is a solution of $w \diamond_T L = R$ iff

(i) $w \diamond_T L \subseteq R$, i.e. $w \in X_{\max}$, and
(ii) $w \diamond_T L \not\subseteq R$.

Moreover, (ii) is satisfied iff $w \Diamond_T L \not\subseteq R_1$ for all $R_1 \subset R$, and hence $w \notin (R_1^c \Diamond_T^l L)^c$. Hence we can conclude that the set S of all singleton solutions to the equation $w \Diamond_T L = R$ can be expressed as

$$S = (R^c \Diamond_T^l L)^c - \bigcup_{R_1 \subset R} (R_1^c \Diamond_T^l L)^c.$$

Since we assume that R is finite, the set S is regular and effectively constructible by Lemma 2, Theorems 1, 2 and the closure of REG under finite union and complement. Hence it is also decidable whether S is empty or not, and eventually all its elements can be effectively listed.   □

Theorem 8. *Let $\Diamond_T$ be one of the operations $\bowtie_T$, $\triangle_T$, $\triangleright_T$. The problem "Does there exist a word w such that $L \Diamond_T w = R$?" is decidable for regular languages L, R and a regular set of trajectories T.*

Proof. Assume first that $\Diamond_T$ is one of $\bowtie_T$, $\triangle_T$. Observe that if $y \in x \Diamond_T w$ for some $w, x, y \in \Sigma^*$, then $|y| \geq |w|$. Therefore, if a solution w to the equation $L \Diamond_T w = R$ exists, then $|w| \leq k$, where $k = \min\{|y| \mid y \in R\}$. Hence, to verify whether a solution exists or not, it suffices to test all the words from $\Sigma^0 \cup \Sigma^1 \cup \cdots \cup \Sigma^k$.

Focus now on the operation $\triangleright_T$. Analogously to the case of Theorem 7, we can deduce that there is no word w satisfying $L \triangleright_T w = R$, if R is infinite. Furthermore, the set $X_{\max} = (L \triangleright_T^r R^c)^c = (L \bowtie_T R^c)^c$ is the maximal set with the property $L \triangleright_T X \subseteq R$. The same arguments as in the proof of Theorem 7 allow one to express the set of all singleton solutions as

$$S = (L \bowtie_T R^c)^c - \bigcup_{R_1 \subset R} (L \bowtie_T R_1^c)^c.$$

For a finite R, the set S is regular and effectively constructible, hence we can decide whether it contains at least one solution.   □

We add that in the above cases of the left and the right operand problems, if there exists a solution, then at least one can be effectively found. Moreover, in the case of their singleton variants, *all* the singleton solutions can be effectively enumerated.

7   Applications

In this section we discuss a few applications of the substitution-on-trajectories operation in modelling certain noisy channels and a cryptanalysis problem. In the former case, we revisit a decidability question involving the property of error-detection.

For positive integers m and l, with $m < l$, consider the SID channel [12] that permits at most m substitution errors in any l (or less) consecutive symbols of any input message. Using the operation $\bowtie_T$, this channel is defined as the set of pairs of words (u, v) such that u is in $v \bowtie_T \Sigma^*$, where T is the set of all trajectories t such that, for any subword s of t, if $|s| \leq l$ then $|s|_1 \leq m$. In general, following the notation of [8], for any trajectory set T we shall denote by $[\bowtie_T \Sigma^*]$ the channel $\{(u, v) \mid v \in u \bowtie_T \Sigma^*\}$. In the context of noisy channels, the concept of error-detection is fundamental [13]. A language L is called *error-detecting for* a channel γ, if γ cannot transform a word in L_λ to another word in L_λ; that is, if $u, v \in L_\lambda$ and $(u, v) \in \gamma$ then $u = v$. Here L_λ is the language $L \cup \{\lambda\}$. The empty word in this definition is needed in case the channel permits symbols to be inserted into, or deleted from, messages – see [13] for details. In our case, where only substitution errors are permitted, the above definition remains valid if we replace L_λ with L.

In [13] it is shown that, given a rational relation γ and a regular language L, we can decide in polynomial time whether L is error-detecting for γ. Here we take advantage of the fact that the channels $[\bowtie_T \Sigma^*]$ permit only substitution errors and improve the time complexity of the above result.

Theorem 9. *The following problem is decidable in time $O(|A|^2 |T|)$.*

   Input: NFA A over Σ and NFA T over $\{0, 1\}$.

   Output: Y/N, depending on whether $L(A)$ is error-detecting for $[\bowtie_T \Sigma^*]$.

Proof. In [9] it is shown that given an NFA A, one can construct the NFA A^σ, in time $O(|A|^2)$, such that the alphabet of A^σ is $E = \Sigma \times \Sigma$ and the language accepted by A^σ consists of all the words of the form $(x_1, y_1) \cdots (x_n, y_n)$, with each $(x_i, y_i) \in E$, such that $x_1 \cdots x_n \neq y_1 \cdots y_n$ and the words $x_1 \cdots x_n$ and $y_1 \cdots y_n$ are in $L(A)$. Let ϕ be the morphism of E into $\{0, 1\}$ such that $\phi(x, y) = 0$ iff $x = y$. One can verify that $L(A)$ is error-detecting for $[\bowtie_T \Sigma^*]$ iff the language $\phi(L(A^\sigma)) \cap L(T)$ is empty. Using this observation, the required algorithm consists of the following steps: (i) Construct the NFA A^σ from A. (ii) Construct the NFA $\phi(A^\sigma)$ by simply replacing each transition $s(x, y) \rightarrow t$ of A^c with $s\phi(x, y) \rightarrow t$. (iii) Use a product construction on $\phi(A^\sigma)$ and T to obtain an NFA B accepting $\phi(L(A^\sigma)) \cap L(T)$. (iv) Perform a depth first search algorithm on the graph of B to test whether there is a path from the start state to a final state. □

We close this section with a cryptanalysis application of the operation $\bowtie_T$. Let V be a set of candidate binary messages (words over $\{0, 1\}$) and let K be a set of possible binary keys. An unknown message v in V is encrypted as $v \oplus t$, where t is an unknown key in K, and $\oplus$ is the exclusive-OR logic operation. Let e be an observed encrypted message and let T be a set of possible guesses for t, with $T \subseteq K$. We want to find the subset X of V for which $X \oplus T = e$, that is, the possible original messages that can be encrypted as e using the keys we have guessed in T. In general T can be infinite and given, for instance, by a regular expression describing the possible pattern of the key. We can model this problem using the following observation whose proof is based on the definitions of the operations $\bowtie_T$ and $\oplus$, and is left to the reader.

Lemma 5. *For every word v and trajectory t, $v \bowtie_T \Sigma^* = \{v \oplus t\}$.*

By the above lemma, we have that the equation $X \oplus T = e$ is equivalent to $X \bowtie_T \Sigma^* = e$. By Theorem 5, we can decide whether there is a solution for this equation and, in this case, find the maximal solution $X_{\max}$. In particular, $X_{\max} = (e^c \triangle_T \Sigma^*)^c$. Hence, one needs to compute the set $M \cap X_{\max}$. Most likely, for a general T, this problem is intractable. On the other hand, this method provides an alternate way to approach the problem.

References

1. M. Domaratzki, *Deletion Along Trajectories.* Tech. report 464-2003, School of Computing, Queen's University, 2003, and accepted for publication.
2. M. Domaratzki, *Splicing on Routes versus Shuffle and Deletion Along Trajectories.* Tech. report 2003-471, School of Computing, Queen's University, 2003.
3. M. Domaratzki, *Decidability of Trajectory-Based Equations.* Tech. report 2003-472, School of Computing, Queen's University, 2003.
4. M. Domaratzki, A. Mateescu, K. Salomaa, S. Yu, Deletion on Trajectories and Commutative Closure. In T. Harju and J. Karhumaki, eds., *WORDS'03: 4th International Conference on Combinatorics on Words.* TUCS General Publication No. 27, Aug. 2003, 309–319.
5. M. Ito, L. Kari, G. Thierrin, Shuffle and scattered deletion closure of languages. *Theoretical Computer Science* **245** (2000), 115–133.
6. L. Kari, On insertion and deletion in formal languages, *PhD thesis*, University of Turku, Finland, 1991.
7. L. Kari, On language equations with invertible operations, *Theoretical Computer Science* **132** (1994), 129–150.
8. L. Kari, S. Konstantinidis, Language equations, maximality and error detection. Submitted.
9. L. Kari, S. Konstantinidis, S. Perron, G. Wozniak, J. Xu, *Finite-state error/edit-systems and difference measures for languages and words.* Dept. of Math. and Computing Sci. Tech. Report No. 2003-01, Saint Mary's University, Canada, 2003.
10. L. Kari, P. Sosík, *Language deletion on trajectories.* Dept. of Computer Science technical report No. 606, University of Western Ontario, London, 2003, and submitted for publication.
11. L. Kari, S. Konstantinidis, P. Sosík, *On Properties of Bond-Free DNA Languages.* Dept. of Computer Science Tech. Report No. 609, Univ. of Western Ontario, 2003, and submitted for publication.
12. S. Konstantinidis, An algebra of discrete channels that involve combinations of three basic error types. *Information and Computation* **167** (2001), 120–131.
13. S. Konstantinidis, Transducers and the properties of error detection, error correction and finite-delay decodability. *J. Universal Comp. Science* **8** (2002), 278–291.
14. C. Martin-Víde, A. Mateescu, G. Rozenberg, A. Salomaa, *Contexts on Trajectories*, TUCS Technical Report No. 214, Turku Centre for Computer Science, 1998.
15. A. Mateescu, A. Salomaa, Nondeterministic trajectories. *Formal and Natural Computing: Essays Dedicated to Grzegorz Rozenberg, LNCS* **2300** (2002), 96-106.
16. A. Mateescu, G. Rozenberg, A. Salomaa, Shuffle on trajectories: syntactic constraints, *Theoretical Computer Science* **197** (1998), 1–56.

17. A. Mateescu, G. Rozenberg, A. Salomaa, *Shuffle on Trajectories: Syntactic Constraints*, TUCS Technical Report No. 41, Turku Centre for Computer Science, 1996.
18. G. Rozenberg, A. Salomaa (eds.), *Handbook of Formal Languages*, Springer-Verlag, Berlin, 1997.

Recombination Systems

Mikko Koivisto, Pasi Rastas, and Esko Ukkonen*

Department of Computer Science, P.O. Box 26 (Teollisuuskatu 23)
FIN-00014 University of Helsinki, Finland
{mikko.koivisto, pasi.rastas, esko.ukkonen}@cs.helsinki.fi

Abstract. We study biological recombination from language-theoretic and machine learning point of view. Two generative systems to model recombinations are introduced and polynomial-time algorithms for their language membership, parsing and equivalence problems are described. Another polynomial-time algorithm is given for finding a small model for a given set of recombinants.

1 Introduction

Recombination is one of the main mechanisms producing genetic variation. Simply stated, recombination refers to the process in which the DNA molecules of a father chromosome and a mother chromosome get entangled and then split off to produce the DNA of the child chromosome, composed of segments taken alternately from the father DNA and the mother DNA (Fig 1) [1].

Fig. 1. Recombination

Combinatorial structures created by iterated recombinations have attracted lots of interest recently. The discovery of so-called haplotype blocks [3, 5] has also inspired the development of new efficient algorithms for the analysis of structural regularities of the DNA, from various perspectives; e.g. [9, 4]. Some methods for genetic mapping such as the recent approach of [7] also model recombinations.

In this paper we study recombination from language-theoretic and machine learning point of view. Two simple systems are introduced to generate recombinants starting from certain founding strings. Membership, parsing and equivalence problems for these systems turn out in general easy. More interesting and also much harder is the problem of inverting recombinations: given a sample set of recombinants we want to construct a smallest possible system generating a language that contains the sample.

* Supported by the Academy of Finland under grant 201560.

J. Karhumäki et al. (Eds.): Theory Is Forever (Salomaa Festschrift), LNCS 3113, pp. 159–169, 2004.

The paper is organized as follows. Section 2 introduces simple recombination systems. Such a system is specified just by giving a set of strings, the "founders" of a population. Section 3 introduces another system, called the fragmentation model, in which the strings that can be used as segments of recombinants are listed explicitly. Language membership, parsing and equivalence problems for these two systems are polynomial-time solvable, by well-known techniques from finite automata [6] and string matching [2]. In Section 4 we consider a machine learning type of problem of constructing a good fragmentation model for a sample set of recombinants. We give a polynomial-time algorithm that finds a smallest model in a special case. Also in the general case the algorithm seems useful although the result is not necessarily minimal.

2   Simple Recombination Systems

A *recombination* is an operation that takes two strings u and v of equal length n and produces a new string w, also of length n, called a *recombinant* of u and w, such that

$$w = xy$$

where x is a prefix of u and y is a suffix of v or x is a prefix of v and y is a suffix of u. The recombinant w is said to have a *cross-over* at location $|x|$. For simplicity we assume that a recombinant may have only one cross-over. As x or y may be the empty string, u and v themselves are recombinants of u and v.

Let A be a set of strings of length n. The set of strings generated from A in one recombination step is denoted

$$\mathcal{R}(A) = \{w \mid w \text{ is a recombinant of some } u, v \in A\}.$$

Let Σ be a finite alphabet. A *simple $m \times n$ recombination system* in Σ is defined by a set $F \subseteq \Sigma^n$ consisting of m strings of length n in Σ. The strings in F are called the *founders* of the system. System F generates new sequences by iterating the recombination operation. The generative process has a natural division into generations giving the corresponding languages $G_0(F), G_1(F), \ldots$ as follows:

$$G_0(F) = F$$
$$G_1(F) = \mathcal{R}(F)$$

$$\vdots$$

$$G_i(F) = \mathcal{R}\Big(G_{i-1}(F)\Big).$$

As $G_0(F) \subseteq G_1(F) \subseteq \cdots \subseteq G_i(F) \subseteq \cdots \subseteq \Sigma^n$ there must be j such that after the jth generation nothing new can be produced, that is, $G_{j'}(F) = G_j(F)$ for all $j' \geq j$. We call $L(F) = G_j(F)$ the *full recombinant language* of system F.

Example 1. Let $\Sigma = \{0,1\}$, $n = 4$, and consider 2×4 system $F = \{0000, 0111\}$. Then $G_1(F) = \{0000, 0111, 0100, 0110, 0011, 0001\}$ and $G_2(F) = \{0000, 0111, 0100, 0110, 0011, 0001, 0101, 0010\}$. Language $G_2(F)$ consists of all strings in Σ^4 that start with 0. This is also the full language $L(F)$.

It should be obvious that w is in $L(F)$ if and only if w can be written as

$$w = \alpha_1 \alpha_2 \cdots \alpha_p \tag{1}$$

for some non-empty strings $\alpha_i \in \Sigma^+$ such that each α_i occurs in some founder string $f_j \in F$ at the same location as in w. That is, we have $f_j = \gamma \alpha_i \delta$ for some γ such that $|\gamma| = |\alpha_1 \cdots \alpha_{i-1}|$. Each decomposition (1) of w into *fragments* α_i is called a *parse* of w with respect to F.

String w may have several different parses. Two of them are of special interest. First, if w has some parse (1) then it also has a parse such that $|\alpha_i| = 1$ for all $i = 1, 2, \ldots, p$ and $p = n$. We then note that a string $w_1 w_2 \cdots w_n$, where $w_i \in \Sigma$, belongs to $L(F)$ if and only if for each w_i there is some $f_j \in F$ whose ith symbol is w_i. Let us denote by Σ_i the symbols in Σ that occur at the ith location of some string in F. We call Σ_i the *local alphabet* of F at i. Summarized we get the following simple result.

Theorem 1. $L(F) = \Sigma_1 \Sigma_2 \cdots \Sigma_n$ □

This immediately gives a language equivalence test for recombination systems. Let E and F be two recombination systems of length n, and let $\Pi_1, \Pi_2, \ldots, \Pi_n$ be the local alphabets of E and $\Sigma_1, \Sigma_2, \ldots, \Sigma_n$ the local alphabets of F. Then $L(E) = L(F)$ if and only if $\Pi_i = \Sigma_i$ for all $i = 1, 2, \ldots, n$. So, for example systems $\{0000, 1111\}$ and $\{0101, 1010\}$ are equivalent as all local alphabets are equal to $\{0, 1\}$.

The simplicity of the equivalence test also indicates that the sequential structure of the founders has totally disappeared in $L(F)$. Therefore it is more interesting to look at strings that have a parse consisting of a small number of fragments α_i. This leads us to define the canonical parses.

Let $w \in L(F)$. Then a parse $w = \alpha_1 \alpha_2 \cdots \alpha_p$ of w with respect to F is *canonical*, if

1. p is smallest possible; and
2. among parses of w with p fragments, each $|\alpha_1 \alpha_2 \cdots \alpha_i|$, $1 \le i < p$, is largest possible.

A canonical parse of w is easily seen unique. It can be found by the following *greedy parsing algorithm*. First find the longest prefix of w that is also a prefix of some string in F. This prefix is fragment α_1 of the canonical parse. Then remove $|\alpha_1|$ symbols long prefix from w and from all members of F. Repeat the same steps to find longest prefix that becomes α_2, and so on, until the entire w has been processed or it turns out that parsing can not be continued to the end of w, in which case $w \notin L(F)$.

We will use the number $p - 1$ of the cross-overs in the canonical parse as a distance measure for strings: the *recombination distance* $\rho(w, F)$ between w

and F is $p - 1$, the smallest possible number of cross-overs in a parse of w with respect to F. Note that $\rho(w, F) \leq n - 1$ for all $w \in L(F)$. If $w \notin L(F)$, we let $\rho(w, F) = \infty$.

The greedy parsing algorithm finds $\rho(w, F)$. The algorithm works without any preprocessing of F and can be organized to run in time $O(mn)$, i.e. linear in the total length of the strings in F. We now describe a preprocessing of F which constructs a collection of trie structures such that canonical parsing of any string with respect of F can be done in optimal time $O(n)$.

In the canonical parsing by the greedy method one has to find the longest prefix of the current suffix of w that is common with the corresponding suffix of some founder $f \in F$. Let $w^i = w_i w_{i+1} \cdots w_n$ and $f_j^i = f_{ji} f_{ji+1} \cdots f_{jn}$ denote the ith suffixes of w and the founders, and let T^i denote the trie representing strings $f_1^i, f_2^i, \cdots, f_m^i$. The longest common prefix, that will become the first fragment of the parse, can be found by traversing the path of T^1 for w^1 until a symbol of w^1 is encountered, say w_h, that is not present in T^1 (or w^1 ends). The scanned prefix is the fragment α_1 of the canonical parse. We needed $O(|\alpha_1|)$ time to find it this way. The parsing continues by next traversing the path of T^h for w^h, giving α_2 in time $O(|\alpha_2|)$, and so on.

To make this work we need the tries $T^1, T^2, \ldots, T^n$. A straightforward construction of a trie for $m = |F|$ strings of length n takes time $O(mn)$ assuming that $|\Sigma|$ is constant. Hence the total time for all tries would be $O(mn^2)$. We next describe a suffix-tree based technique for constructing these tries in time $O(mn)$.

The suffix-tree of a string x is a (compacted) trie representing all the suffixes of x. The size of the tree is $O(|x|)$, and it can be constructed in time $O(|x|)$ by several alternative algorithms [2]. To get the tries T^h that form our parser for F we first augment the founder strings with explicit location indices, such that founder string $f_i = f_{i1} f_{i2} \cdots f_{in}$ becomes $\hat{f}_i = (f_{i1}, 1)(f_{i2}, 2) \cdots (f_{in}, n)$. Now construct the suffix-tree T for string $\hat{f} = \hat{f}_1 \hat{f}_2 \cdots \hat{f}_m$. Then trie T^h consists of the subtrees of T representing suffixes that start with symbols (a, h), where $a \in \Sigma$. Hence tries T^h can be extracted from T in one scan through the edges that are adjacent to the root.

This construction can be performed in $O(mn)$ time, i.e., linear time in the length of $\hat{f}$ although we have formally used alphabet of non-constant size $|\Sigma|n$. This is because the non-root nodes of T may only have $|\Sigma|$ branches and hence the branching degree at such nodes does not depend on n. While the root node can have $O(|\Sigma|n)$ branches, the dependency on n can be made constant by direct indexing (or bucketing) on the second component of a symbol.

Finally note that the tries T^h extracted from T are of compacted form, i.e., the non-branching nodes of the trie are represented only implicitly. The edges of a compacted trie correspond to strings (instead of single symbols), represented by pairs of pointers to the original strings in F. In our greedy parsing algorithm such tries can be used as well, without significant overhead.

Theorem 2. *Given an $m \times n$ recombination system F, a greedy parser for F can be constructed in time $O(mn)$. For any string w, the parser computes in time $O(n)$ the canonical parse of w with respect to F and the recombination distance $\rho(w, F)$.* □

Canonical parsing is not the only possible use of the parser of Theorem 2. All possible parses of w can be generated if, instead of greedily traversing the current trie as far as possible, the parsing jumps to the next trie at any point on the way. It is also possible to check whether or not w has a parse with given cross-over points: Then the parsing should jump to the next trie exactly on these points. The parses can also be utilized to find a string w with largest possible distance $\rho(w, F)$.

3   Generalized Recombination Systems and Fragmentation Models

Parsing a string as introduced in the previous section means decomposing the string into fragments taken from the founders. The available fragments were implicitly defined by the founders: any substring of a founder can be used in a parse.

We now go one step further and introduce models in which the available fragments are listed explicitly.

A *fragmentation model* of length n in alphabet Σ is a state-transition system $M = (S, Q, \Sigma, n)$ consisting of a finite set S of the *states* and a set Q of *transitions*. Each $s \in S$ is a pair (i, v), where i is an integer $1 \leq i \leq n$, and string $v \in \Sigma^*$ is the *fragment* of the state such that $|v| \leq n - i + 1$. We call $b(s) = i$ the *begin location* and $b(s) = i + |v|$ the *end location* of s. A state s is a *start state* if $b(s) = 1$ and an *end state* if $e(s) = n + 1$. The transition set Q is any subset of $S \times S$ such that if $(r, s) \in Q$ then $e(r) = b(s)$, that is, the location intervals covered by r and s should be next to each other.

The language $L(M)$ of M consists of all strings generated along the transition paths from a start state to an end state. More formally, $e \in L(M)$ if and only if there are states $s_1, s_2, \ldots s_p$ such that $(s_i, s_{i+1}) \in Q$ for $1 \leq i < p$, s is a start state and s_p is an end state, and $w = v_1 v_2 \cdots v_p$ where v_i is the fragment of state s_i. Note that all $w \in L(M)$ are of length n. Also note that fragmentation models are a subclass of finite-state automata. Hence for example their language equivalence is solvable by standard methods [6].

Example 2. A simple $m \times n$ recombination system F of the previous section consisting of m founders $f_j = f_{j1} f_{j2} \cdots f_{jn}$ can be represented as a fragmentation model $M = (S, Q, \Sigma, n)$ as follows: set S consists of all states (i, v) where $1 \leq i \leq n$ and $v = f_{ji} f_{ji+1} \cdots f_{jh}$ for some $1 \leq j \leq m$ and $i \leq h \leq n$. The transition (r, s) is included into Q for all r, s such that $e(r) = b(s)$. Note that M is much larger than F. It has $O(mn^2)$ states and $O(m^2 n^3)$ transitions, and the fragments of the states have total length $O(mn^3)$. □

Each transition path gives for the generated string a parse into fragments. Different parses for the same string can be efficiently enumerated and analyzed for example by using dynamic programming combined with breadth-first traversal of the transition graph of M. We describe next an algorithm for finding a parse with smallest number of fragments, i.e., a shortest path through M that generates the string to be parsed.

Let w be the string to be parsed. We say that state (i, v) of M *matches* w if $w = xvy$ where $|x| = i - 1$. We associate with each state s a counter $c(s)$ whose value will be the length of a shortest path to s that will generate the prefix of length $b(s) - 1$ of w. Variable P will be used to store the length of a shortest parse. The parsing algorithm is as follows:

1. Let $s_1, s_2 \ldots s_t$ be the states of M ordered according to increasing value of $e(s_j)$
2. Initialize the counters

$$P \leftarrow \infty$$
$$c(s_j) \leftarrow \begin{cases} 0 & \text{, if } s_j \text{ is a start state} \\ \infty & \text{, otherwise} \end{cases}$$

3. **for** $j \leftarrow 1, 2, \ldots, t$ **do**
       if $c(s_j) < \infty$ and s_j matches w **then**
           if s_j is an end state **then**
               $P \leftarrow \min(P, c(s_j) + 1)$
           else
               for all s_k such that $(s_j, s_k) \in Q$
                   $c(s_k) \leftarrow \min(c(s_k), c(s_j) + 1)$

The algorithm can be implemented such that the running time is linear in the size of M. We also observe that testing whether or not some states of M and w match can be done very fast by first constructing Aho-Corasick multi-pattern matching automaton [2] for the fragments of the states and then scanning w with this automaton.

4   Model Reconstruction Problems

The language membership and equivalence well as parsing problems for recombination systems turned out solvable by fast algorithms, not unexpectedly as we are dealing with a limited subclass of the regular languages. We now discuss much harder problems concerning inversion of recombinations.

Given a set D of strings of length n we want to find a model that could have generated D. This question was addressed in [8] in the case of simple recombination systems. For example, an algorithm was given in [8] that constructs an $m \times n$ recombination system F such that $D \subseteq L(F)$ and the average recombination distance of the elements of D from F is minimized. Here we will consider the problem of finding fragmentation models for D. The fragments of such a model can be thought to represent the "conserved" substrings of D.

The goodness of a fragmentation model M for set D can be evaluated using various criteria. A possibility is to consider probabilistic generalizations of fragmentation models and apply model selection criteria such as the Minimum Description Length principle. We will resort to combinatorial approach and consider the following parsimony criterion: find a fragmentation model $M = (S, Q, \Sigma, n)$ such that $D \subseteq L(M)$ and the number of states in Q is smallest possible. We call this the *minimal fragmentation model reconstruction problem*.

Example 3. Let D consist of strings

$$
\begin{array}{l}
0\ 0\ 0\ 0\ 1\ 1 \\
0\ 0\ 0\ 1\ 1\ 1 \\
0\ 0\ 1\ 0\ 1\ 1 \\
0\ 0\ 0\ 0\ 0\ 0 \\
0\ 0\ 0\ 1\ 0\ 0 \\
1\ 1\ 1\ 0\ 1\ 1 \\
1\ 1\ 0\ 0\ 0\ 0 \\
1\ 1\ 0\ 1\ 0\ 0 \\
1\ 1\ 1\ 0\ 0\ 0
\end{array}
\tag{2}
$$

By taking the strings in D as such (and nothing else) as the fragments we get a fragmentation model which generates exactly D and has 9 states. However, the model depicted in Fig 2 has only 7 states. It generates a language that properly contains D. □

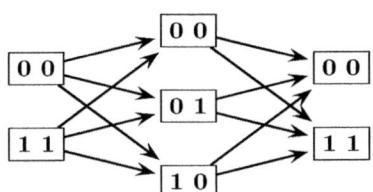

Fig. 2. A fragmentation model (begin locations of states not explicitly shown)

We rephrase now the minimal fragmentation model reconstruction in terms of certain tilings of D. Let us refer to the m strings in D by $1, 2, \ldots, m$; the ith string is $d_{i1}d_{i2}\cdots d_{in}$. Then any triple $\tau = (A, h, k)$, where $A \subseteq \{1, 2, \ldots, m\}$ and h and k are integers such that $1 \leq h \leq k \leq m$, is a *tile* of D. Set A is the *row-set* of τ. The tile τ *covers* all substrings $d_{ih}d_{ih+1}\cdots d_{ik}$ where $i \in A$. The tile is *uniform* if all substrings it covers are equal, i.e., $d_{ih}\cdots d_{ik} = d_{jh}\cdots d_{jk}$ for all $i, j \in A$. A set T of tiles of D is a *uniform tiling* of D if the tiles in T are uniform and disjoint and cover D, i.e., for each d_{ij} there is exactly one tile in T that covers d_{ij}.

A fragmentation model M such that $D \subseteq L(M)$ induces a uniform tiling $T(M)$ of D as follows. Fix for each string $d \in D$ a path of M that spells out d. For any state $s = (i, v)$ of M, let A be the set of strings in D whose path goes through s. Then add the tile $(A, i, i + |v| - 1)$ to $T(M)$. It should be clear that $T(M)$ is a uniform tiling of D. Note that there are several different tilings $T(M)$ if some $d \in D$ is ambiguous with respect to M, i.e., if M has more than one path for d.

On the other hand, given a uniform tiling T of D, one may construct a fragmentation model $M(T)$ as follows. For each tile $(A, h, k) \in T$, add to $M(T)$ a state $s = (h, v)$ where $v = d_{ih} \cdots d_{ik}$ for some $i \in A$. Also add a transition (s, s') to $M(T)$ if the tiles (A, h, k) and (A', h', k') in T that correspond to s and s' are such that row-set intersection $A \cap A'$ is nonempty and $k + 1 = h'$.

As the number of states of $M(T)$ equals the number of tiles in T, and the number of tiles in $T(M)$ is at most the number of states of M, we get the following result.

Proposition 1. *The number of states of the smallest fragmentation model M such that $D \subseteq L(M)$ equals the number of tiles in the smallest uniform tiling of D.*

To solve the minimal fragmentation model reconstruction we will construct small uniform tilings for D. We will proceed in two main steps. First a rather simple dynamic programming algorithm is given to find optimal tilings in a subclass called the column-structured tilings. In the second step we apply certain local transformations to further improve the solution.

A uniform tiling of D is *column-structured* if the tiles cover D in columns: for each two tiles (A, h, k) and (A', h', k'), if $h = h'$ then $k = k'$. The corresponding class of fragmentation models (models whose fragments with the same begin location are of equal length) is also called column-structured models. If a column-structured tiling is smallest possible, then the number of tiles in each column should obviously be minimal. Such minimal tiling for a column is easy to find as follows. Consider set $D(h, k)$ consisting of strings $d_{ih} \cdots d_{ik}$ for $1 \leq i \leq m$. Let $A_1, A_2, \ldots A_p$ be the partition of $\{1, 2, \ldots, m\}$ such that i and j belong to the same class A_r if and only if $d_{ih} \cdots d_{ik} = d_{jh} \cdots d_{jk}$. Then the tiling $\big((A_1, h, k), \ldots, (A_p, h, k)\big)$ of $D(h, k)$ is uniform and has the smallest possible number of tiles among tilings whose tiles are from h to k. We denote this tiling by $t(h, k)$ and its size p by $\sigma(h, k)$.

Let $S(j)$ be the size of smallest column-structured tiling of $D(1, j)$. Then $S(j)$ can be evaluated for $j = 0, 1, \ldots, n$ from

$$\begin{cases} S(0) = 0 \\ S(j) = \min_{i < j} \Big(S(i) + \sigma(i + 1, j) \Big) \end{cases} \tag{3}$$

and $S(n)$ gives the size of smallest column-structured uniform tiling of entire D. The usual trace-back of dynamic programming can be used to find the end locations $j_1, j_2, \ldots, j_q = n$ of the corresponding columns. Then the smallest tiling itself is $t(1, j_1) \cup t(j_1 + 1, j_2) \cup \cdots \cup t(j_{q-1} + 1, n)$.

Evaluation of (3) takes time $O(n^2)$ plus the time for evaluating tables σ and t which can be accomplished in time $O(n^2m)$ using straightforward trie-based techniques. We have obtained the following theorem.

Theorem 3. *Minimal column-structured fragmentation model for D can be constructed in time $O(n^2m)$ where n is the length and m the number of strings in D.*

Example 4. The fragmentation model in Fig 2 for the set (2) of Example 3 is column-structured and minimal. If string 011010 is added to (3), then algorithm (3) will give the column-structured model in Fig 3(a). However, the model in Fig 3(b) is smaller.   □

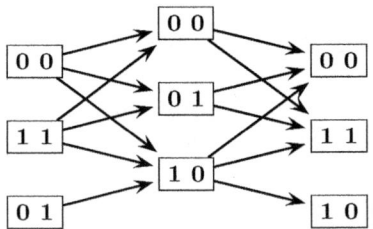
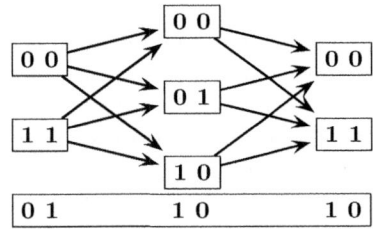

Fig. 3. (a) A column-structured fragmentation model   (b) A smaller model

The tilings given by the column-wise approach can further be improved by applying local transformations. The transformations use the following basic step. Assume that our current tiling has adjacent tiles $(A, h, k - 1)$ and (B, k, r). We may replace these tiles by the tiles

$$(A \cap B, h, r),$$
$$(A \setminus B, h, k - 1),$$
$$(B \setminus A, k, r),$$

and the tiling stays uniform and covers still the entire D. The replacement operation has no effect if row-set $A \cap B$ is empty. Otherwise it changes the structure of the tiling. If $A = B$, the number of tiles is reduced by one; if $A \subseteq B$ or $B \subseteq A$, the number stays the same; and if $A \cap B$, $A \setminus B$ and $B \setminus A$ are all non-empty, the number increases by one.

Given any tiling T we can improve it by the following iterative reduction rule: apply the above local transformation on any pair of tiles $(A, h, k - 1)$ and (B, k, r) such that $A \subseteq B$ or $B \subseteq A$ (i.e., transformation does not increase the number of tiles). Repeat this until the local transformation is not any more applicable. It is easy to see that the process stops in $O(mn)$ iterations. Note that the seemingly useless transformation steps that do not make the number of tiles smaller are indirectly helpful: they make the tiles narrower (and longer) and hence may create possibility for true size reduction in the next step.

There are two possible ways to include the reduction step into algorithm (3). On can apply it only on the final result of (3). This would, for example, improve the tiling in Fig 3(a) into that in Fig 3(b). Other possibility is to apply the reduction rule also on the intermediate tiling obtained for each $D(1, j)$ during algorithm (3) and to use the reduced tiling in subsequent computation. Sometimes this strategy will give better results than the previous one.

There is also another local transformation that makes the tiles longer and narrower without reducing the number of tiles. This transformation eliminates certain loop-like structures from the tiling, defined as follows. The *inclusion graph* of a tiling T at j is a bipartite graph which has as its nodes all tiles (A, h, k) and (A', h', k') such that $k = j - 1$ and $h' = j$ and as its (undirected) arcs all $\left((A, h, j - 1), (A', j, k') \right)$ such that row-set intersection $A \cap A'$ is not empty. A connected component of this graph is a *simple loop* if it contains as many nodes as arcs. In a simple loop every node has degree 2 (i.e., two arcs are adjacent to a node). The number of tiles in a simple loop equals the number of their non-empty pairwise row-set intersections. But this means that applying our local transformation on all such pairs will keep the number of tiles unchanged. Hence the loop-removal transformation can safely be added to the local transformations one should apply to make the tiling smaller.

Summarized, we get an optimization algorithm that combines dynamic programming and local transformations. It finds a local optimum with respect to the local transformations. Running-time is polynomial in the size of D.

5   Conclusion

We introduced two simple language–generating systems inspired by the recombination mechanism of the nature. For the model reconstruction problem we delineated some initial results while many questions remained open, most notably the complexity status and approximability of the minimal model reconstruction. Probabilistic generalizations of our models are another interesting direction for further study.

References

1. Creighton H. and McClintock B.: A correlation of cytological and genetical crossing-over in Zea mays. *PNAS* **17** (1931), 492-497
2. Crochemore, M. and Rytter, W.: *Jewels of Stringology.* World Scientific 2002
3. Daly, M., Rioux, J., Schaffner, *et al.*: High-resolution haplotype structure in the human genome. *Nature Genetics* **29** (2001), 229–232
4. Koivisto, M., Perola, M., Varilo, *et al.*: An MDL method for finding haplotype blocks and for estimating the strength of haplotype block boundaries. In: *Pacific Symposium on Biocomputing (PSB2003)*, pp. 502–513. World Scientific 2003
5. Patil, N., Berno, A. and Hinds, D.A. *et al.*: Blocks of limited haplotype diversity revealed by high-resolution scanning of human chromosome 21. *Science* **294** (2001), 1719–1723
6. Salomaa, A.: *Jewels of Formal Language Theory.* Computer Science Press 1981
7. Sevon, P., Ollikainen, V. and Toivonen, H.T.T.: Tree pattern mining for gene mapping. *Information Sciences* (to appear)

8. Ukkonen, E.: Finding founder sequences from a set of recombinants. In: *Algorithms in Bioinformatics (WABI 2002)*, LNCS 2452, pp. 277–286. Springer 2002
9. Zhang, K., Deng, M., Chen, T., *et al.*: A dynamic programming algorithm for haplotype block partitioning. *PNAS* **99** (2002), 7335–7339

Algebraic Aspects of Parikh Matrices

Alexandru Mateescu

Faculty of Mathematics, University of Bucharest
Academiei, 14, Bucharest, Romania
alexmate@pcnet.ro

Abstract. This paper contains algebraic aspects of Parikh matrices. We present new, but also some old results, concerning this topic. It is proved that in some cases the set of Parikh matrices is a noncommutative semiring with a unit element. Also we prove that the set of Parikh matrices is closed under the operation of shuffle on trajectories and thus it is closed under many other operations. It is presented also the notion of extended Parikh matrix that it is an extension of the notion of the Parikh matrix. The paper contains also a number of open problems.

1 Introduction

The Parikh vector is an important notion in the theory of formal languages. This notion was introduced in [11]. One of the important results concerning this notion is that the image by the Parikh mapping of a context-free language is always a *semilinear* set. (For details and ramifications, see [14].) The basic idea behind Parikh vectors is that properties of words are expressed as *numerical* properties of vectors. However, much information is lost in the transition from a word to a vector.

In this paper we introduce a sharpening of the Parikh vector, where somewhat more information is preserved than in the original Parikh vector. The new notion is based on a certain type of matrices. All other entries above the main diagonal contain information about the *order* of letters in the original word. All matrices are triangular, with 1's on the main diagonal and 0's below it.

We introduce also the notion of *extended Parikh matrix*.

Two words with the same Parikh matrix always have the same Parikh vector, but the converse is not true. The exact meaning of the entries in a Parikh matrix is given below in Theorem 1. Our second main result, Theorem 2, shows an interesting interconnection between the inverse of a Parikh matrix and the Parikh matrix of the mirror image.

We remind some basic notations and definitions. The set of all nonnegative integers is denoted by N. Let Σ be an alphabet. The set of all words over Σ is Σ^* and the empty word is λ. If $w \in \Sigma^*$ then $|w|$ denotes the length of w.

We very often use "ordered" alphabets. An ordered alphabet is an alphabet $\Sigma = \{a_1, a_2, \ldots a_k\}$ with a relation of order ("<") on it. If we have $a_1 < a_2 < \cdots < a_k$, then we use the notation

$$\Sigma = \{a_1 < a_2 < \cdots < a_k\}.$$

J. Karhumäki et al. (Eds.): Theory Is Forever (Salomaa Festschrift), LNCS 3113, pp. 170–180, 2004.
© Springer-Verlag Berlin Heidelberg 2004

Let $a \in \Sigma$ be a letter. The number of occurrences of a in a word $w \in \Sigma^*$ is denoted by $|w|_a$. Let u, v be words over Σ. The word u is a *scattered subword* of v if there exists a word t such that $v \in u \sqcup t$, where $\sqcup$ denotes the shuffle operation. We now introduce a *notation* very important in our subsequent considerations.

If $u, v \in \Sigma^*$, then the number of occurrences of u in v as a scattered subword is denoted by $|v|_{scatt-u}$. For instance,

$$|acbbb|_{scatt-ab} = 3, \quad |acbabb|_{scatt-ab} = 5 \text{ and } |aabbbc|_{scatt-abc} = 6.$$

Thus, partially overlapping occurrences of a word as a scattered subword are counted as distinct occurrences. The number $|v|_{scatt-u}$ is denoted as a binomial coefficient in [13].

Let $\Sigma = \{a_1 < a_2 < \cdots < a_k\}$ be an ordered alphabet. The *Parikh vector* $\Psi : \Sigma^* \to N^k$, is defined by

$$\Psi(w) = (|w|_{a_1}, |w|_{a_2}, \ldots, |w|_{a_k}).$$

The *Parikh vector* of w is $(|w|_{a_1}, |w|_{a_2}, \ldots, |w|_{a_k})$. Note that the Parikh vector is also a mapping Ψ that is a morphism from the monoid $(\Sigma^*, \cdot, \lambda)$ to the monoid $(N^k, +, (0, 0, \ldots, 0))$.

The *mirror image* of a word $w \in \Sigma^*$, denoted $mi(w)$, is defined as: $mi(\lambda) = \lambda$ and $mi(b_1 b_2 \ldots b_n) = b_n \ldots b_2 b_1$, where $b_i \in \Sigma$, $1 \leq i \leq n$.

The reader is referred to [12] as a comprehensive treatment on formal languages and diverse background material. The most fundamental applications and interconnections of Parikh vectors with language theory are presented in [14].

2   Parikh Matrices and Extended Parikh Matrices

In this paper we consider "triangle" matrices. A *triangle matrix* is a square matrix $M = (m_{i,j})_{1 \leq i,j \leq k}$, such that $m_{i,j} \in N$, for all $1 \leq i, j \leq k$, $m_{i,j} = 0$, for all $1 \leq j < i \leq k$, and, moreover, $m_{i,i} = 1$, for all $1 \leq i \leq k$.

The set of all triangle matrices is denoted by $\mathfrak{M}$. The set of all triangle matrices of dimension $k \geq 1$ is denoted by $\mathfrak{M}_k$. Clearly, $\mathfrak{M}_k$ constitutes a monoid under matrix multiplication.

Now we introduce the main notion of this paper.

Definition 1. *Let $\Sigma = \{a_1 < a_2 < \cdots < a_k\}$ be an ordered alphabet, where $k \geq 1$. The* Parikh matrix, *denoted Ψ_{M_k}, is the morphism:*

$$\Psi_{M_k} : \Sigma^* \to \mathfrak{M}_{k+1},$$

defined by the condition: if $\Psi_{M_k}(a_q) = (m_{i,j})_{1 \leq i,j \leq (k+1)}$, then for each $1 \leq i \leq (k+1)$, $m_{i,i} = 1$, $m_{q,q+1} = 1$, all other elements of the matrix $\Psi_{M_k}(a_q)$ being 0.

The notion of Parikh matrix was introduced in [6] and studied in [7,8].

The Parikh matrix is not injective. One of the major open problems is to characterize non-injectivity, that is, to provide some natural conditions for two words to possess the same Parikh matrix. This problem is closely linked with the fundamental problem about the information content of a Parikh matrix: how much does the Parikh matrix tell about a word?

It was brought to our attention by one of the referees that the term *Parikh matrix* was used in [10] for the growth matrix of a morphism (or a D0L system).

Now we present the notion of extended Parikh matrix. This important notion was introduced in [15].

Definition 2. *Let Σ be an alphabet and $u = b_1 \ldots b_{|u|}$ be a word in Σ^* ($b_i \in \Sigma$ for all $1 \leq i \leq |u|$). The* extended Parikh matrix *induced by the word u over the alphabet Σ (shortly, the u-Parikh matrix), denoted $\Psi_{\Sigma,u}$, is the monoid morphism*

$$\Psi_{\Sigma,u} : (\Sigma^*, \cdot, \lambda) \rightarrow (\mathcal{M}_{|u|+1}, \cdot, I_{|u|+1}),$$

defined by the condition: if $a \in \Sigma$ and $\Psi_{\Sigma,u}(a) = (m_{i,j})_{1 \leq i,j \leq (|u|+1)}$, then:

$$m_{i,j} = \begin{cases} 1 & \text{if } j = i \\ \delta_{b_i,a} & \text{if } j = i+1 \\ 0 & \text{otherwise} \end{cases}$$

where $\delta_{a,b}$ is the Kronecker Symbol *regarding letters, that is*

$$\delta_{a,b} = \begin{cases} 1, \text{ if } a = b \\ 0, \text{ if } a \neq b \end{cases}$$

In the notation $\Psi_{\Sigma,u}$, Σ has to be mentioned because u can be considered over any alphabet which contains its letters, and we need to know the context we are working in.

If Σ is known, then we will use the notation Ψ_u for $\Psi_{\Sigma,u}$, especially in proofs, for reasons of simplicity.

It is clear that if the symbol $a \in \Sigma$ doesn't occur in u, then $\Psi_{\Sigma,u}(a) = I_{|u|+1}$.

Let $u \in \Sigma^*$. We say that $M \in \mathcal{M}_{|u|+1}$ is an *extended Parikh matrix induced by u* if there exists a word $w \in \Sigma^*$ such that $M = \Psi_{\Sigma,u}(w)$. Generally, we say that $M \in \mathcal{M}_{k+1}$ is a *Parikh matrix induced by a word* if there exists a word $u \in \Sigma^*$ such that $|u| = k$ and M is a Parikh matrix induced by u.

It's easy to see that the Parikh matrix can be obtained as a particular case of this definition, when u contains all the symbols in Σ only once. The ordering of the alphabet is then given by the order in which the symbols appear in u.

For example, if $\Sigma = \{b_1 < b_2 < \ldots < b_k\}$ is an ordered alphabet and $u \in \Sigma^*$, $u = b_1 b_2 \ldots b_k$, then it can be easily seen that for all $a \in \Sigma$, $\Psi_{\Sigma,k}(a) = \Psi_{\Sigma,u}(a)$. It follows that for all words $w \in \Sigma^*$,

$$\Psi_{\Sigma,k}(w) = \Psi_{\Sigma,u}(w).$$

Similarly, it follows that for all words $w \in \Sigma^*$,

$$\Psi_{\Sigma_\circ,k}(w) = \Psi_{\Sigma,mi(u)}(w).$$

Let us now give an example of an u-Parikh matrix computation. Let $\Sigma = \{a, b\}$ and $u = aba$. We will compute $\Psi_{\Sigma,u}(abba)$.

We have that $\Psi_{\Sigma,u}(abba) = \Psi_{\Sigma,u}(a)\Psi_{\Sigma,u}(b)\Psi_{\Sigma,u}(b)\Psi_{\Sigma,u}(a)$, which leads to:

$$\Psi_{\Sigma,u}(abba) = \begin{pmatrix} 1 & 1 & 0 & 0 \\ 0 & 1 & 0 & 0 \\ 0 & 0 & 1 & 1 \\ 0 & 0 & 0 & 1 \end{pmatrix} \begin{pmatrix} 1 & 0 & 0 & 0 \\ 0 & 1 & 1 & 0 \\ 0 & 0 & 1 & 0 \\ 0 & 0 & 0 & 1 \end{pmatrix} \begin{pmatrix} 1 & 0 & 0 & 0 \\ 0 & 1 & 1 & 0 \\ 0 & 0 & 1 & 0 \\ 0 & 0 & 0 & 1 \end{pmatrix} \begin{pmatrix} 1 & 1 & 0 & 0 \\ 0 & 1 & 0 & 0 \\ 0 & 0 & 1 & 1 \\ 0 & 0 & 0 & 1 \end{pmatrix} = \begin{pmatrix} 1 & 2 & 2 & 2 \\ 0 & 1 & 2 & 2 \\ 0 & 0 & 1 & 2 \\ 0 & 0 & 0 & 1 \end{pmatrix}$$

3   About the Entries of a Parikh Matrix

In this section we characterize the entries of the Parikh matrix. We first introduce some notation that will be applied in our first theorem. Recall also the notation $|v|_{scatt-u}$ defined in Section 1.

Consider the ordered alphabet $\Sigma = \{a_1 < a_2 < \cdots < a_k\}$, where $k \geq 1$. We denote by $a_{i,j}$ the word $a_i a_{i+1} \ldots a_j$, where $1 \leq i \leq j \leq k$.

We are now ready to prove the basic property of the Parikh matrix.

Theorem 1. *Let $\Sigma = \{a_1 < a_2 < \cdots < a_k\}$ be an ordered alphabet, where $k \geq 1$, and assume that $w \in \Sigma^*$. The matrix $\Psi_{M_k}(w) = (m_{i,j})_{1 \leq i,j \leq (k+1)}$, has the following properties:*

(i) $m_{i,j} = 0$, *for all* $1 \leq j < i \leq (k+1)$,
(ii) $m_{i,i} = 1$, *for all* $1 \leq i \leq (k+1)$,
(iii) $m_{i,j+1} = |w|_{scatt-a_{i,j}}$, *for all* $1 \leq i \leq j \leq k$.

Corollary 1. *The matrix $\Psi_{M_k}(w)$ has as the second diagonal (i.e., the vector $(m_{1,2}, m_{2,3}, \ldots, m_{k,k+1})$) the Parikh vector of w, i.e., $(m_{1,2}, m_{2,3}, \ldots, m_{k,k+1}) = (|w|_{a_1}, |w|_{a_2}, \ldots, |w|_{a_k})$.*

Comment The above results are not true for the extended Parikh matrix. One can easily find all this information as for instance the Parikh vector by a simple method.

As already pointed out, the Parikh matrix gives more information about a word than the classical Parikh vector, although the Parikh matrix is still not injective. Injectivity would of course mean that the information given by Parikh matrices is *complete*. This would be more than one can reasonably hope for: one cannot expect that words could be expressed as matrices in this fashion, which would give all information in a simple numerical form.

So far very little is known about *sets of Parikh matrices associated to languages belonging to a fixed family* such as the families in the Chomsky hierarchy. The following remark shows that the semilinearity result of context-free languages does not carry over to sets of matrices. Results concerning semilinearity and Parikh matrices are in [2]. The notions of slender and Parikh slender languages are studied in [3–5]. Results about the injectivity of Parikh mapping can be found in [1].

Remark 1. Consider the ordered alphabet $\{a < b\}$ and the context-free language $L = \{a^n b^n | n \geq 1\}$. Clearly,

$$\Psi_{M_2}(a^n b^n) = \begin{pmatrix} 1 & n & n^2 \\ 0 & 1 & n \\ 0 & 0 & 1 \end{pmatrix}$$

Hence $\Psi_{M_2}(L)$ cannot be a semilinear set (for any reasonable extension of the definition of semilinearity to matrices).

Clearly, every triangle matrix is not a Parikh matrix of some word. For instance, the matrix

$$\begin{pmatrix} 1 & 2 & 7 \\ 0 & 1 & 3 \\ 0 & 0 & 1 \end{pmatrix}$$

is not a Parikh matrix. This follows because ab occurs as a scattered subword at most 6 times in a word with the Parikh vector $(2,3)$.

The product of the entries in the Parikh vector constitutes an upper bound for the entry $m_{1,k+1}$. Thus, the size of the entry $m_{1,3}$ in Remark 1 is maximal. Whether or not a given triangle matrix is a Parikh matrix is clearly a decidable question.

4   On the Inverse of Parikh Matrices

We investigate interrelations between the inverse of a Parikh matrix associated to a word w and the Parikh matrix of $mi(w)$, the mirror image of w. Clearly, the set of all triangle matrices of order $k \geq 2$ with integer entries is a noncommutative group with respect to multiplication, the unit element being the unit matrix of order k. Consequently, for each Parikh matrix A, there exists the inverse matrix A^{-1}.

Definition 3. *Let* $\Sigma = \{a_1 < a_2 < \cdots < a_k\}$ *be an ordered alphabet and let* $w \in \Sigma^*$ *be a word. Assume that the Parikh matrix of w is* $\Psi_{M_k}(w) = (m_{i,j})_{1 \leq i,j \leq k+1}$*. The alternate Parikh matrix of w, denoted* $\overline{\Psi}_{M_k}(w)$*, is the matrix* $(m'_{i,j})_{1 \leq i,j \leq k+1}$*, where* $m'_{i,j} = (-1)^{i+j} m_{i,j}$*, for all* $1 \leq i, j \leq k+1$*.*

Observe that the mapping $\overline{\Psi}_{M_k}(w)$ is a morphism of Σ^*. For the Parikh vector Ψ and for every word w, $\Psi(w) = \Psi(mi(w))$. However, for the Parikh matrix the situation is completely different. The next theorem reveals the interrelation between the inverse of the Parikh matrix of a word w and the alternate Parikh matrix of the mirror image of w.

Theorem 2. *Let* $\Sigma = \{a_1 < a_2 < \cdots < a_k\}$ *be an ordered alphabet and let* $w \in \Sigma^*$ *be a word. Then:*

$$[\Psi_{M_k}(w)]^{-1} = \overline{\Psi}_{M_k}(mi(w)).$$

Observe that Theorem 2 provides a very simple method to compute the inverse of a Parikh matrix. One can also apply it directly to matrices: inverses of matrices of a certain type can be computed in this way.

As an example, consider the ordered alphabet $\Sigma = \{a < b < c\}$ and assume that $w = cbbaa$. Then

$$\Psi_{M_3}(cbbaa) = \begin{pmatrix} 1 & 2 & 0 & 0 \\ 0 & 1 & 2 & 0 \\ 0 & 0 & 1 & 1 \\ 0 & 0 & 0 & 1 \end{pmatrix}$$

Since $mi(cbbaa) = aabbc$, we have by Theorem 2:

$$[\Psi_{M_3}(cbbaa)]^{-1} = \overline{\Psi}_{M_3}(aabbc) =$$

$$= \begin{pmatrix} 1 & -2 & 4 & -4 \\ 0 & 1 & -2 & 2 \\ 0 & 0 & 1 & -1 \\ 0 & 0 & 0 & 1 \end{pmatrix}$$

A special relation between $|w|_{scatt-a_{i,j}}$ and $|mi(w)|_{scatt-a_{i,j}}$ is obtained in the next corollary. In the statement the last vertical bars stand for the absolute value.

Corollary 2. *Let $\Sigma = \{a_1 < a_2 < \cdots < a_k\}$ be an ordered alphabet and let $w \in \Sigma^*$ be a word. Assume that the Parikh matrix of w is $\Psi_{M_k}(w) = (m_{i,j})_{1 \leq i,j \leq k+1}$, and that $[\Psi_{M_k}(w)]^{-1} = (m'_{i,j})_{1 \leq i,j \leq k+1}$. Then $|mi(w)|_{scatt-a_{i,j}} = |(m'_{i,j+1})|$ for all $1 \leq i,j \leq k$.*

We consider now another method to compute the inverse of a Parikh matrix. We begin with some further definitions and notations.

Let $(A, <)$ be an ordered set. The *dual order* of the order $<$, denoted $<^\circ$, is defined as:

$$a <^\circ b \text{ iff } b < a.$$

Let $\Sigma = \{a_1 < a_2 < \cdots < a_k\}$ be an ordered alphabet. The *dual ordered alphabet*, denoted $\Sigma_\circ$, is $\Sigma_\circ = \{a_k < a_{k-1} < \cdots < a_1\}$.

Consider the ordered alphabet $\Sigma = \{a_1 < a_2 < \cdots < a_k\}$ and let $w \in \Sigma^*$ be a word. The Parikh matrix associated to w with respect to the dual order on Σ is denoted by $\Psi_{M_k,\circ}(w)$.

Let $v = (v_1, v_2, \ldots, v_n)$ be a vector. The *reverse* of v, denoted $v^{(rev)}$, is the vector $v^{(rev)} = (v_n, v_{n-1}, \ldots, v_1)$.

Now we introduce the notion of a reverse of a triangle matrix. Let $M = (m_{i,j})_{1 \leq i,j \leq n}$ be a triangle matrix. The *reverse* of M, denoted $M^{(rev)}$, is the matrix $M^{(rev)} = (m'_{i,j})_{1 \leq i,j \leq n}$, where $m'_{i,j} = m_{n+1-j,n+1-i}$, for all $1 \leq i < j \leq n$. (The entries on and below the main diagonal are the same in M and $M^{(rev)}$.)

Note that $M^{(rev)}$ is also a triangle matrix. An easy way to obtain $M^{(rev)}$ is to reverse in M all diagonals that are parallel to the main diagonal. For instance,

$$\text{If } M = \begin{pmatrix} 1\,2\,3\,7 \\ 0\,1\,4\,5 \\ 0\,0\,1\,6 \\ 0\,0\,0\,1 \end{pmatrix} \text{ then } M^{(rev)} = \begin{pmatrix} 1\,6\,5\,7 \\ 0\,1\,4\,3 \\ 0\,0\,1\,2 \\ 0\,0\,0\,1 \end{pmatrix}$$

The reader can easily verify the following proposition. (Observe that Definition 3 can be immediately extended to concern arbitrary matrices A.)

Proposition 1. *Let* A, B *be two triangle matrices of the same dimension. Then*

(i) $[A^{(rev)}]^{(rev)} = A.$
(ii) $(AB)^{(rev)} = B^{(rev)} A^{(rev)}.$
(iii) $\overline{\overline{A}} = A.$
(iv) $\overline{AB} = \overline{A}\,\overline{B}.$

The next theorem gives another method of computing the inverse of a Parikh matrix.

Theorem 3. *Let* $\Sigma = \{a_1 < a_2 < \cdots < a_k\}$ *be an ordered alphabet and let* $w \in \Sigma^*$ *be a word. Then*

$$[\Psi_{M_k}(w)]^{-1} = [\overline{\Psi}_{M_k,\circ}(w)]^{(rev)}.$$

The above Theorem 3 provides a simpler method to compute the inverse of a Parikh matrix. Here we have to reverse a matrix that is of a fixed size $(card(\Sigma) + 1)$, whereas in the case of Theorem 2 we have to reverse the word w that can be arbitrarily long.

From Theorems 2 and 3 we deduce:

Corollary 3. *Let* $\Sigma = \{a_1 < a_2 < \cdots < a_k\}$ *be an ordered alphabet and let* $w \in \Sigma^*$ *be a word. Then:*

$$\Psi_{M_k}(mi(w)) = \Psi_{M_k,\circ}(w)^{(rev)}.$$

The subsequent final observation concerning the functions introduced is rather obvious. Consider the following four functions from $\mathcal{M}_k$ to $\mathcal{M}_k$: the identity I, the mapping $-$ of A to $\overline{A}$, the mapping (rev) of A to $A^{(rev)}$ and the mapping $\overline{(rev)}$ of A to $\overline{A}^{(rev)}$. Then these four functions together with the operation of composition constitute a group and, moreover, this is the well-known Four-Group of Klein.

Comment Note that the above two methods to compute the inverse of a Parikh matrix does not work in the case of the extended Parikh matrix.

5   Some Algebraic Properties

Next theorem is a direct consequence of the definition of Parikh matrix.

Theorem 4. *The entries* $m_{i,j+1}$, $1 \leq i < j \leq k$ *in a Parikh matrix* $\Psi_{M_k}(w)$ *satisfy the inequality*

$$m_{i,j+1} \leq m_{i,j} \cdot m_{i+1,j+1}.$$

Concerning the minors of a Parikh matrix we can prove that:

Theorem 5. *The value of each minor of an arbitrary Parikh matrix is a non-negative integer.*

Comment The above result is true also for extended Parikh matrices.

The following general inequality is a consequence of (extended) Parikh matrices.

Theorem 6. *The inequality* $|w|_{xyz}|w|_y \leq |w|_{xy}|w|_{yz}$ *holds for arbitrary words* w, x, y, z.

A generalization of the above inequality is:

Theorem 7. *Consider an integer t and let $w, x, y_1, y_2, \ldots y_t, z$ be arbitrary words. Then*

$$|w|_{y_1} \cdots |w|_{y_t}|w|_{xy_1 \ldots y_t z} \leq |w|_{xy_1}|w|_{y_1 y_2} \cdots |w|_{y_{t-1} y_t}|w|_{y_t z}.$$

Next result concerns equality between sums of terms of the form $|w|_x$.

Theorem 8. *The equality*

$$|w|_{x_1} + |w|_{x_2} + \ldots |w|_{x_n} = |w|_{y_1} + |w|_{y_2} + \ldots |w|_{y_m}$$

is decidable for all words $w, x_1, x_2, \ldots x_n, y_1, y_2, \ldots y_m$.

Open Problem It is not known the decidability of the inequality:

$$|w|_{x_1} + |w|_{x_2} + \ldots |w|_{x_n} \leq |w|_{y_1} + |w|_{y_2} + \ldots |w|_{y_m}$$

where $w, x_1, x_2, \ldots x_n, y_1, y_2, \ldots y_m$ are arbitrary words.

6   Algebraic Structures and Other Operations

Let k be a positive integer and denote by $\mathcal{PM}_k$ the set of all Parikh matrices of dimension k. The set of all Parikh matrices is denoted by $\mathcal{PM}$.

We define a special type of sum between Parikh matrices, denoted by $\oplus$.

If A and B are Parikh matrices then the sum $A \oplus B = C$ where C is obtained as the usual sum of matrices except that all elements on the main diagonal of C have by definition value 1.

Theorem 9. *Let Σ be an alphabet with $card(\Sigma) = k \leq 2$. Then, if A and B are Parikh matrices from $\mathcal{PM}_k$, then $A \oplus B$ is also a Parikh matrix.*

Proof For $k = 1$ the result is trivial. Assume now that $k = 2$ and let x be a preimage of A and y a preimage of B. Assume that $pq = c_{1,3}$ We define $z = b^t a^p b^q a^r$, where $t + q = |x|_b + |y|_b$ and $p + r = |x|_a + |y|_a$ and $pq = c_{1,3}$. One can verify that z is a preimage of C.

As a consequence we obtain:

Theorem 10. *Both $(\mathcal{PM}, \cdot, \oplus, 0, I_1)$ and $(\mathcal{PM}, \cdot, \oplus, 0, I_1)$ are semirings.*

Comment. The above results are not true if $card(\Sigma) \geq 3$ For instance if $\Sigma = \{a, b, c\}$ and consider $x = abc$ and $y = b$, then the partial sum of the corresponding Parikh matrices is not a Parikh matrix.

Open Problem For the time being we don't know under what conditions the partial sum of two Parikh matrices continue to be a Parikh matrix.

The last part of this section is dedicated to closure properties of $\mathcal{PM}_k$ and of $\mathcal{PM}$ at certain operations. We start recalling the operation of *shuffle on trajectories*. This notion was defined in [9].

Consider the alphabet $V = \{r, u\}$. We say that r and u are *versors* in the plane: r stands for the *right* direction, whereas u stands for the *up* direction.

Definition 4. *A trajectory is an element $t \in V^*$.*

We will consider also sets T of trajectories, $T \subseteq V^*$.

Let Σ be an alphabet and let t be a (finite) trajectory, let d be a versor, $d \in V$, let α, β be two (finite) words over Σ.

Definition 5. *The shuffle of α with β on the trajectory dt, denoted $\alpha \sqcup\!\sqcup_{dt} \beta$, is recursively defined as follows:*

if $\alpha = ax$ and $\beta = by$, where $a, b \in \Sigma$ and $x, y \in \Sigma^*$, then:

$$ax \sqcup\!\sqcup_{dt} by = \begin{cases} a(x \sqcup\!\sqcup_t by), & \text{if } d = r, \\ b(ax \sqcup\!\sqcup_t y), & \text{if } d = u. \end{cases}$$

if $\alpha = ax$ and $\beta = \lambda$, where $a \in \Sigma$ and $x \in \Sigma^*$, then:

$$ax \sqcup\!\sqcup_{dt} \lambda = \begin{cases} a(x \sqcup\!\sqcup_t \lambda), & \text{if } d = r, \\ \emptyset, & \text{if } d = u. \end{cases}$$

if $\alpha = \lambda$ and $\beta = by$, where $b \in \Sigma$ and $y \in \Sigma^*$, then:

$$\lambda \sqcup_{dt} by = \begin{cases} \emptyset, & \text{if } d = r, \\ b(\lambda \sqcup_t y), & \text{if } d = u. \end{cases}$$

Finally,

$$\lambda \sqcup_t \lambda = \begin{cases} \lambda, & \text{if } t = \lambda, \\ \emptyset, & \text{otherwise.} \end{cases}$$

Comment. Note that if $|\alpha| \neq |t|_r$ or $|\beta| \neq |t|_u$, then $\alpha \sqcup_t \beta = \emptyset$.

If T is a set of trajectories, the *shuffle of α with β on the set T of trajectories,* denoted $\alpha \sqcup_T \beta$, is:

$$\alpha \sqcup_T \beta = \bigcup_{t \in T} \alpha \sqcup_t \beta.$$

The above operation is extended to languages over Σ, if $L_1, L_2 \subseteq \Sigma^*$, then:

$$L_1 \sqcup_T L_2 = \bigcup_{\alpha \in L_1, \beta \in L_2} \alpha \sqcup_T \beta.$$

Consider the following example. Let α and β be the words $\alpha = a_1 a_2 a_3 a_4 a_5 a_6 a_7 a_8$, $\beta = b_1 b_2 b_3 b_4 b_5$ and assume that $t = r^3 u^2 r^3 ururu$. The shuffle of α with β on the trajectory t is:

$$\alpha \sqcup_t \beta = \{a_1 a_2 a_3 b_1 b_2 a_4 a_5 a_6 b_3 a_7 b_4 a_8 b_5\}.$$

Comment. Note that the operation of shuffle on trajectories can be extended for finite strings of matrices, for finite sequences of finite graphs, etc.

For instance in the above example the letters a_i can be square matrices and the catenation to be the multiplication of matrices.

Theorem 11. *Both $\mathcal{PM}_k$ and $\mathcal{PM}$ are closed at the operation of shuffle on trajectories.*

Consequently we obtain that

Theorem 12. *Both $\mathcal{PM}_k$ and $\mathcal{PM}$ are closed at the operations of multiplication, shuffle, insertion, shuffle literal, bicatenation, etc.*

7   Conclusion

We presented the most important results and problems concerning Parikh matrices. A problem area we have not discussed at all in this paper concerns *sets of Parikh matrices* and *families* of such sets, analogous to the family of *semilinear* sets of Parikh vectors.

References

1. Atanasiu, A,. Martín-Vide, C. and Mateescu, A., On the injectivity of the Parikh matrix mapping. Submitted for publication (2000).
2. Harju, T., Ibarra, O., Karhumäki, J. and Salomaa, A., Some decision problems concerning semilinearity and commutation. To appear in *J. Comput. System Sci.* (2002).
3. Honkala, J., On slender languages. *EATCS Bulletin* 64 (1998), 145–152.
4. Honkala, J., On Parikh slender languages and power series. *J. Comput. System Sci.* 52 (1996), 185–190.
5. Ilie, L., Rozenberg, G., and Salomaa, A., A characterization of poly-slender context-free languages. *Theor. Inform. Appl.* 34 (2000), 77–86.
6. Mateescu, Alexandru, Salomaa, Arto, Salomaa, Kai, and Yu, Sheng, *A Sharpening of the Parikh Mapping*, RAIRO - Theoretical Informatics and Applications 35, 551-564 (2001)
7. Mateescu, Alexandru, Salomaa, Arto, Salomaa, Kai and Yu, Sheng, *On an extension of the Parikh mapping*, TUCS Technical Report No 364
8. Mateescu, Alexandru, Salomaa, Arto and Yu, Sheng, *Subword Histories and Parikh Matrices*, TUCS Technical Report No 442, February 2002.
9. A. Mateescu, G. Rozenberg and A. Salomaa, "Shuffle on Trajectories: Syntactic Constraints", Theoretical Computer Science, TCS, Fundamental Study, 197, 1-2, (1998) 1-56.
10. Pansiot, J.J., A decidable property of iterated morphisms. *Springer Lecture Notes in Computer Science* 104 (1981), 152–158.
11. Parikh, R.J., On context-free languages. *J. Assoc. Comput. Mach.*, 13 (1966) 570–581.
12. Rozenberg, G. and Salomaa, A. (eds.), *Handbook of Formal Languages 1-3.* Springer-Verlag, Berlin, Heidelberg, New York (1997).
13. Sakarovitch, J. and Simon, I., Subwords. In M. Lothaire: *Combinatorics on Words*, Addison-Wesley, Reading, Mass. (1983) 105–142.
14. Salomaa, A., *Formal Languages.* Academic Press, New York (1973).
15. Şerbănuţă, Traian-Florin, *Extending Parikh Matrices* Theoretical Computer Science, jauary, 2004.

On Distributed Computing
on Elliptic Curves

Tommi Meskanen[1,2], Ari Renvall[1], and Paula Steinby[1,2]

[1] Department of Mathematics, University of Turku
20014 Turku, Finland
[2] Turku Centre for Computer Science
20520 Turku, Finland

Abstract. Let C be a device performing computations of a crypto-graphic protocol. Assume C to have limited computing power, but to have access to another device A with superior capacities. This setting could occur, for instance, with a smart card C and a mobile phone A. We consider the situation where C is supposed to calculate the basic operation of elliptic curve cryptography: the scalar multiplication of a point P on a curve. We investigate whether C's performance could be improved by means of distributed computation; that is, whether C could exploit A's computing power, without compromising the safety of the procedure. We set up three models of computation, varying the demand for C's trust on A's honesty.

1   Arithmetic on Elliptic Curves

Denote by $\mathbf{F}_q$ the field with q elements. If the field is binary, then the (non-supersingular) elliptic curve $E(\mathbf{F}_q)$ over $\mathbf{F}_q$ is defined as the set of points

$$E(\mathbf{F}_q) = \{(x,y) \mid y^2 + xy = x^3 + ax^2 + b\} \cup \mathcal{O} \qquad (a,b \in \mathbf{F}_q, b \neq 0),$$

where $\mathcal{O}$ is the point at infinity. (In non-binary fields the term ax^2 must be replaced by ax and the condition $4a^3 + 27b^2 \neq 0$ should hold). If the group operation is suitably defined, then $E(\mathbf{F}_q)$ is an Abelian group with $\mathcal{O}$ as the neutral element. It is customary to adopt additive notation, so we denote the group operation by $+$. This addition can be performed using the basic arithmetic operations of the base field. Below we give the equations for a binary field. Let $P = (x,y) \neq \mathcal{O}$ be a point on the curve. Then

$$\begin{aligned}
-P &= (x, x+y) \\
2P &= (u,v) \\
&= (\theta^2 + \theta + a, \theta(x+u) + u + y),
\end{aligned}$$

where $\theta = x + \dfrac{y}{x}$. Moreover, if also $Q = (x',y') \neq \mathcal{O}$, $Q \neq P$ and $Q \neq -P$ (i.e. $x \neq x'$), then

$$\begin{aligned}
P + Q &= (u',v') \\
&= (\theta'^2 + \theta' + a + x + x', \theta'(x+u') + u' + y),
\end{aligned}$$

J. Karhumäki et al. (Eds.): Theory Is Forever (Salomaa Festschrift), LNCS 3113, pp. 181–191, 2004.
© Springer-Verlag Berlin Heidelberg 2004

where $\theta' = \dfrac{y+y'}{x+x'}$.

Elliptic curve cryptosystems are public key cryptosystems [Sal]. In particular they are El Gamal based systems [ElG], hence their security is based on the difficulty of the discrete logarithm problem in the group $(E(\mathbf{F}_q), +)$. More specifically, EC based systems exploit the fact that given points P and kP (for some unknown integer k) on a curve $E(\mathbf{F}_q)$, it is practically impossible to calculate k. The operation of computing $kP = P + \ldots + P$ is called *scalar multiplication* of the point P .

The users of an elliptic curve cryptosystem all share the same domain parameters $(m, f(x), a, b, G, r)$. Of these m and $f(x)$ fix the underlying field and the presentation of field elements, a and b define the curve and G is a point on the curve of order r. The private key of a user is then simply an integer $s < r$, and the corresponding public key is the point $Q = sG$.

For an extensive study on elliptic curves in cryptography we refer to [BSS]. Here, as an example, we describe the ECDSA signature algorithm, which is included in all ECC standards, see for example [P1363,X9.62]. Let s be A's private key and Q the public key. A signature for a message m is then computed as follows.

Signature generation:

1. A computes a representative $h(m)$ of the message, where h is a hash function agreed beforehand (SHA-1, for example).
2. A selects a random integer k, computes the point $P = kG$ and converts the x-coordinate of it to an integer c.
3. A's signature for the message m is the pair (c, d), where $d = k^{-1}(h(m) + sc)$ (mod r).

A verifier B can check the validity of a signature (c, d) for a message m using A's public key $Q = sG$:

Signature verification:

1. B computes the hash value $h(m)$.
2. B computes the point $R = d^{-1}(h(m)G + cQ)$ and converts the x-coordinate to an integer c'
3. If $c = c'$ then B accepts the signature.

It is an easy task to verify that if both A and B follow their algorithms then $P = R$, and thus also $c = c'$:

$$R = d^{-1}(h(m)G + cQ) = d^{-1}(h(m) + sc)G = kG = P.$$

In ECDSA (and in fact in any cryptographic primitive employing elliptic curves), scalar multiplication is the most time consuming operation. This motivates us to concentrate on how to securely and efficiently distribute the computation of kP.

2   Scalar Multiplication of a Point

In the following we introduce our favourite algorithm for scalar multiplication: the *Fixed Base Windowing* algorithm (FBW for short), see [MOV]. Slightly modified, we find it the most suitable for efficient and secure calculation of kP. FBW seems recommendable even in the cases where the base P is not fixed. The FBW algorithm is a constant time algorithm where all point doublings occur in the precomputation, and all point additions during the actual computation. This feature diminishes the chance of a successful *side-channel attack*. These are attacks against specific implementations of some algorithm, which collect information by measuring e.g. time or power consumption of the device when it is processing some secret data. For more information on such attacks, see [Koc,KJJ].

2.1   The FBW Algorithm

The Fixed Base Windowing algorithm is a general exponentiation algorithm, as it can be applied in any group. As the name suggests, it is specially designed for situations where the base is fixed. A significant part of the computation consists of constructing a precomputation table T, the contents of which only depends on the base. If we always have the same base, then it suffices to calculate T just once and store it for further use. But, if necessary, we can also compute T each time separately.

Below we present the basic version of FBW. We adopt the notation from elliptic curves setup.

Let $P = (x, y)$ be a point of order r on some curve $E(\mathbf{F}_q)$, and let k be a positive integer less than r. Our task is to compute the point kP. The idea of the algorithm is to look at the multiplier k through a "window" of fixed width w. For notational convenience we assume that the length of k's binary representation is $n = lw$, and we denote $k = \sum_{i=0}^{l-1} k_i 2^{iw}$, where $0 \leq k_i < 2^w$. Denote also $P_i = 2^{iw}P$ $(i = 0, 1, \ldots, l-1)$; hence $kP = \sum_{i=0}^{l-1} k_i P_i$.

Algorithm 1

```
Q := P; R := 0; B := 0; For i := 0 To l-1 Do (* precomputation *)
  T[i] := Q;
  For j := 1 to w Do Q := 2*Q;
For i := 2^w-1 Downto 1 Do (* computation depending on k *)
  For j := 0 To l-1 Do
    If k[j] = i Then B := B + T[j];
  R := R + B;
Return(R)
```

First part of the algorithm is to generate the precomputation table $T[i] = P_i$. Then the product kP is accumulated to R such that each table element $T[i]$ is included in B when there are k_i additions R := R + B left. This sums up to $R = \sum k_i P_i = kP$. The first part of the algorithm consists of doublings only, whereas there are just additions in the second part. As already mentioned, this is a strength (compared to, for instance, the traditional double-and-add -algorithm) against side-channel attacks.

Let us consider our model of distributed computing: when computing the product kP (with FBW), C takes advantage of A's computing power. We recommend C to use A's assistance with the precomputation only, and the actual computation to be performed solely by C. There are at least two good reasons for this. First, if A is used in computations which depend on the coefficient k, then C has to ask A for something specific. That would bring on a need for duplex data transfer. Second, it seems impossible to build a scheme where C would profit from A's computation without having to reveal any relevant information on k.

3   Models of Computation

The main objective of this article is to analyze the benefits of distributed computation when using elliptic curve cryptography. As scalar multiplication is the basic operation in EC cryptography, and most of the time in any cryptographic EC operation is spent computing kP, it seems natural to concentrate on it.

Let C be a device to calculate kP. Assume that k is known to C only, whereas P is public. Suppose that A is another device with superior computing power, and that A is willing to help C in the computations. For example, C could be a smart card inserted in a mobile phone A.

We consider three different models.

I A does not participate in the computation at all, so C does everything independently. Thus, no data needs to be transferred between A and C.

II A is used during the computation, but all data that A provides is checked by C. No information on k is given to A, so C doesn't need to trust A.

III A participates in the computations, and C trusts the correctness of all data A gives to C. If A really is honest, then the data given by C doesn't reveal A any information on k.

Model I is included to give the basis for relevant comparisons of the benefits of the other two models. In I and II the result of C is never incorrect, but in model III this is not necessarily the case if A cheats. It should also be noted, that we really cannot make any weaker assumptions on A than we do in model III. The same information that an honest A may learn is available also to an eavesdropper listening the traffic between A and C. Moreover, if A learns k, it could as well do all the calculations by itself, making C useless.

In the following we assume that kP is computed using the Fixed Base Windowing algorithm, and that A is used only in the precomputation (if at all). We also assume that $|k| = n = lw$. These choices are motivated by the following reasons:

- Based on our tests, a modification of FBW seems a recommendable choice for calculating kP, as it combines efficiency with suitability to avoid side-channel attacks. Information on different exponentiation algorithms can be found in [Gor,MOV].
- If A helps C during the precomputation, then A can send the required data without C's request. On the other hand, if A helps C in calculations depending on k, then first C needs to request some specific data from A. As a consequence more data needs to be transmitted between the parties. In the smart-card environment data transmission is relatively slow.
- If C's help request depends on k, there is always a risk that some information on k gets revealed. In our experience it seems difficult to use A during computations that depend on k.

We represent the complexity of an algorithm as a triple (s, m, i), where s, m and i stand for the number of squarings, multiplications and inversions of the field elements, respectively. For example, the complexity of both point doubling and ordinary point addition on elliptic curves is $(1, 2, 1)$. (We ignore the costs due to adding field elements, as that is performed by XORing the summands and is extremely fast.)

3.1 Model I

This is the traditional model, where C needs to do all computations by itself. Looking at algorithm 1 it is easy to count the complexity. In the precomputation we need n doublings, so the complexity is $(n, 2n, n) = (lw, 2lw, lw)$. In the actual computation we need $c_k = l + 2^w - 1$ additions. Thus, the total complexity is $(lw + c_k, 2(lw + c_k), lw + c_k)$.

3.2 Model II

Let us now discuss how C could utilize A's computing power during the precomputation. It seems reasonable to reduce the problem to performing a single doubling, since computing the table points consists of consecutive doublings. So, let us look at the situation where A and C need to calculate the point $Q = 2P$. We denote $P = (x, y)$ and $Q = (u, v)$.

As A is not trusted, C must be able to check that every piece of information given by A is correct. Therefore, a precondition for the success of this model is to find operations such that performing them is more costly than verifying the results.

In [Knu] it was shown that, under certain conditions, point halving costs less than point doubling. Thus, one possible idea is that A simply gives C the point $Q = 2P$, and C verifies this by checking that $\frac{1}{2}Q = P$. However, this approach has some problems. An efficient point halving algorithm requires relatively much memory, which is not acceptable in our main application where C is a smart card. More importantly, if C does have the necessary resources for efficient point halving, it would be best exploited by using a so-called "halve-and-add" algorithm (see [Knu]).

In the following we present two ideas, both of which are based on a simple observation: the result of a field inversion is easily verified by performing one field multiplication. Probably in any implementation of a binary field, multiplication costs reasonably less than an inversion. Further, let us point out that using the doubling formula involves computing the value $\theta = x + \frac{y}{x}$. These facts together motivate us to suggest a scheme, where A after computing the value of θ hands it over to C, who first checks its validity and then makes use of it in computing the coordinates of Q.

Algorithm 2

$$
\begin{array}{ll}
A \to C & \theta = x + \frac{y}{x} \\
C & x(\theta + x) = y? \qquad (* \text{ checking } \theta *) \\
C & u = \theta^2 + \theta + a \\
C & v = \theta(x + u) + u + y
\end{array}
$$

Computing the complexity of Algorithm 2 is straightforward. C needs one multiplication to verify that A has given the correct θ; computing u and v requires one squaring and a multiplication, respectively. These sum up to $(1, 2, 0)$ for one doubling and $(lw, 2lw, 0)$ for all precomputation. Altogether, we save time equal to that needed for lw inversions.

Our second algorithm for model II is based on the idea of representing a point as a pair (x, θ) instead of (x, y) (where $\theta = x + \frac{y}{x}$). If $P = (x, y) \sim (x, \theta)$ and $Q = 2P = (u, v) \sim (u, \theta')$ then, by the doubling formula:

$$
\begin{aligned}
u &= \theta^2 + \theta + a \\
\theta' &= u + vu^{-1} = u + (\theta(x + u) + u + y) \cdot u^{-1} \\
&= u + (\theta(x + u) + u + x(\theta + x)) \cdot u^{-1} \\
&= u + (u(\theta + 1) + x^2) \cdot u^{-1} \\
&= \theta + u + 1 + x^2 u^{-1}
\end{aligned}
$$

The complexity of one doubling using this method is $(1, 2, 1)$. Thus, there would be no time savings if C doubled points this way. But if C receives the value θ' from A, then C only has to check its validity by making sure that $x^2 = u(\theta' + \theta + u + 1)$. The cost of the validation operation is $(1, 1, 0)$, and computing u requires one additional squaring.

Algorithm 3

$$
\begin{array}{ll}
C & u = \theta^2 + \theta + a \\
A \to C & \theta' = u + vu^{-1} \\
C & x^2 = u(\theta' + \theta + u + 1)?
\end{array}
$$

In the computation of kP we need the actual y-coordinate of every wth table element. Since $y = x(\theta + x)$, this costs an additional multiplication per each of the l table elements. Altogether, the complexity of precomputation will then be $(2lw, lw + l, 0)$. We notice that algorithm 3 is more efficient than algorithm 2, if w squarings in the base field can be computed faster than $w - 1$ multiplications.

Model III

Analyzing this model is trivial. As A is trusted, A can simply give the whole precomputation table for C, who only needs to do computations depending on k. Thus C's total complexity in this case is $(c_k, 2c_k, c_k)$.

It should be noted, however, that this model is quite dangerous. In the following section we consider the risks.

4   What if A Cheats?

In this section we consider what damage can be done if C trusts a cheating A in our model III above. The obvious consequence is that C's calculations are then incorrect, which might be very harmful as such. But the real danger is that the false result might reveal some secret information.

Assume again that C is trying to compute kP, where k is secret and P is public. Suppose that A gives C a precomputation table T, and C (trusting A) computes the point

$$X_T = \sum_{i=0}^{l-1} k_i T[i]$$

using the FBW algorithm. The point X_T is again public information, so also A learns it. From A's point of view k is a uniformly distributed random variable. Theoretically, the information that X_T reveals about k is $I(X_T, k) = H(X_T) - H(X_T \mid k)$, where $H(Y)$ is the entropy of a random variable Y: $H(Y) = \sum_{y \in Y} p(y) \log_2(p(y))$. In our case X_T is completely determined by k, thus $H(X_T \mid k) = 0$ and $I(X_T \mid k) = H(X_T)$.

For simplicity, suppose that k is an integer selected uniformly from all integers of length $n = lw$: $0 \le k < 2^n$.

Suppose that A has been honest, so the table T is correct. Then $X_T = kP$ and it is easy to verify that $H(X_T) = n$. In other words, X_T completely determines k (as it should). The problem is that to extract the available information A needs to compute the discrete logarithm.

On the other hand, assume that A has given C a table consisting of only one non-zero point $T[i]$. Then $X_T = k_i T[i]$ and $H(X_T) = w$. This means that X_T reveals w bits of information about k. Again, to extract this information A needs to compute the discrete logarithm. But in this case it is trivial, as there are only 2^w alternatives and typically w is very small.

There is a natural generalization of this attack. A can select T to consist of a "small" number of non-zero points. For example, suppose that $T[i_0 + i] = 2^{iw} \cdot T[i_0]$ for some i_0 and $i = 0, \ldots, j-1$ and the rest $T[i] = 0$. Then $H(X_T) = jw$ and, if j is small enough, A can learn the secret bits $k_{i_0}, \ldots, k_{i_0+j-1}$. The optimal choice for j depends on A's computational resources.

The attacks described above can be avoided if C checks that the table elements $T[i]$ are all non-zero. Unfortunately this is of little help, as we can modify the attack in so many ways that C cannot possibly rule out all alternatives. For example, A can choose all the table points as a small multiple of some point R. If $T[i] = iR$, then $X_T = (\sum ik_i)R$. It can be computed that $H(X_T) \approx 11,5$ if $n = 160$ and $w = 4$.

From the above it follows that, unless C is absolutely certain that A is honest, C should avoid our model III.

Let us finally consider a setup somewhere between models II and III. Suppose that, as in algorithm 2, A gives C the values θ, but C does not verify their validity (or alternatively verifies only some of them at random). But not even this idea seems too promising. If A manages to give C two incorrect θ's in succession, then A has a good chance to force the table elements $T[i]$ to any point (α, β) it desires. For example, A could make T to repeat itself by forcing $T[\frac{l}{2}] = T[0]$, whence $T[\frac{l}{2} + 1] = T[1]$ and so on. In this case it is easy to compute that $I(X_T, k) < \frac{n}{2} + w$. In other words, the security provided by k is roughly halved.

Let us describe how A can cheat. Denote by $\varphi(P, \theta)$ the point computed by algorithm 2 without the verification of θ. Hence, if θ is correct, $\varphi(P, \theta) = 2P$. Given a starting point $P_0 = (x, y)$ and a target point $P_2 = (\alpha, \beta)$, A's task is to find such values z and t that $\varphi(P_1, t) = P_2$, where $P_1 = (u, v) = \varphi(P_0, z)$. Necessarily t should satisfy the equation $t^2 + t + a = \alpha$. Such a t is easily computed provided that $\mathrm{Tr}(a + \alpha) = 0$. It remains to select z such that the resulting y-coordinate will be β. Applying the doubling formulae we obtain

$$
\begin{aligned}
\beta &= t(u + \alpha) + \alpha + v \\
&= t(z^2 + z + a + \alpha) + \alpha + \\
&\quad z(x + z^2 + z + a) + z^2 + z + a + y \\
&= z^3 + tz^2 + (t + x + a + 1)z + (t(a + \alpha) + a + y + \alpha).
\end{aligned}
$$

Thus, if A can solve z from the equation above, A's chances for a successful cheating are good. A's task is to solve a polynomial equation of third degree in a binary field. The number of solutions is always either 0, 1 or 3. The following lemma from [BRS] tells us that the equation has at least one solution with probability greater than 50%. In the same article an efficient method to calculate the solution is given.

Lemma. The equation $z^3 + c_1 z^2 + c_2 z + c_3 = 0$ $(c_i \in \mathbf{F}_{2^m})$ has exactly one solution in $\mathbf{F}_{2^m}$, if and only if $\mathrm{Tr}(\frac{(c_2 + c_1^2)^{3/2}}{c_3 + c_1 c_2}) \neq \mathrm{Tr}(1)$.

5   An Example

To get a better understanding of the benefits of distributed computation we consider a concrete example. Currently a 1024 bit RSA modulus is considered to provide adequate security for most applications. The same level of security using elliptic curve cryptography is achieved if the order of the group $E(\mathbf{F}_q)$ is roughly 2^{160}. Then the length of k's binary representation is (approximately) 160 bits, and, from the point of view of FBW's efficiency, the optimal size for the window (w) is 3–5.

Thus, assume that C is calculating a point kP on a curve over a field of size 2^{160}, and that an untrusted A helps C (model II). Suppose that C uses the FBW algorithm with $w = 4$ and $l = 40$. To measure the possible time savings we need to know the relative costs of the basic operations: squaring, multiplication and inversion on the underlying field. Naturally these figures depend heavily on the available resources. We consider two different environments.

If C is a personal computer, then (according to our own implementation) one multiplication takes roughly twice the time required for one squaring, and an inversion costs approximately four multiplications. Then, in model I, the fastest way for C to compute the precomputation table for P is to double P repeatedly using the formula given in the first section. The complexity of it is $(1, 2, 1)$, thus the relative cost for the whole precomputation is $160 + 2 \cdot 320 + 8 \cdot 160 = 2080$.

In model II we shall use algorithm 3, because squaring is cheaper than multiplication. The complexity of one doubling is then $(2, 1, 0)$, and one additional multiplication is needed for the l table points. The cost for precomputation is then $320 + 2 \cdot (160 + 40) = 720$. This means that the time required for precomputation is cut to roughly one third.

To obtain the total saving we should take into account also the computation depending on k. This is performed similarly in both models requiring 55 point additions, and the cost of it is $(55 + 2 \cdot 110 + 8 \cdot 55 = 715)$. Therefore the calculation time for kP is approximately halved in model II.

If C is a smart card with hardwired field multiplication, then squaring and multiplication (which in this case are performed similarly) are really fast compared with inversion. In the following we assume that one inversion takes the same time as 40 multiplications. In this case it pays to use so-called *projective* coordinates instead of the *affine* coordinates (which we have used so far) to present points. This follows, since if computations are performed using projective coordinates, then we can almost completely avoid inversions (with the cost of some additional squarings and multiplications). To understand what projective coordinates are we refer to [BSS], here it suffices to present a table for the complexities of point doubling and point addition using different coordinates. (In the mixed case the other summand is given in affine, the other in projective coordinates.)

	complexity as (s, m, i)		
	affine	mixed	projective
Addition	$(1, 2, 1)$	$(3, 11, 0)$	$(5, 15, 0)$
Doubling	$(1, 2, 1)$	–	$(5, 5, 0)$

As we notice, the complexity of one doubling is now $(5, 5, 0)$. The time required for precomputation is thus $5 \cdot 160 + 5 \cdot 160 = 1600$. The table points are then given in projective coordinates, and the complexity of the remaining 55 point additions is $(15, 5, 0)$ each. Moreover, the result needs to be converted from projective coordinates back to affine coordinates, which requires one inversion, three multiplications and one squaring. This yields to an additional cost of $15 \cdot$

$55 + 5 \cdot 55 + 1 + 3 + 40 \cdot 1 = 1144$, and thus the total cost for computing kP is 2744.

Using model II the most efficient precomputation is performed via affine coordinates and employing algorithm 2. The complexity of one doubling is $(1, 2, 0)$, which gives $160 + 2 \cdot 160 = 480$ for the cost of precomputation. The remaining 55 point additions should again be done using projective coordinates. Of these additions 40 belong to the mixed category, the rest 15 being purely projective. This gives a cost of $(40 \cdot 3 + 15 \cdot 5) + (40 \cdot 11 + 15 \cdot 15 + 3) + 40 \cdot 1 = 904$, as again the final result needs to be converted back to affine coordinates. The total cost is then 1384, which again is about 50% of the cost of model I.

We summarize the results in the table below. It should be remembered that the times are relative, no comparison between the PC setup and smart card setup can be done. Also, these figures give only the maximum possible savings with our methods. We have not included A's computation times and, more importantly, the times required for data transfer between A and C. The main application we have had in mind is a smart card. Unfortunately with the current technology i/o times to and from the card are too slow to exploit these methods.

	precomputation			complete computation		
	compl.	cost	sav.	compl.	cost	sav.
PC						
model I	$(160, 320, 160)$	2080		$(215, 430, 215)$	2795	
model II	$(320, 200, 0)$	720	65%	$(375, 310, 55)$	1435	49%
smart card						
model I	$(800, 800, 0)$	1600		$(1076, 1628, 1)$	2744	
model II	$(160, 320, 0)$	480	70%	$(356, 988, 1)$	1384	50%

6   Conclusion

We have shown that in elliptic curve cryptography a device with restricted computing power can exploit external computational resources without jeopardizing the security. In theory about 50% of the computation times can be saved using the described methods. We have also considered the dangers of trusting the external device.

References

[X9.62] ANSI X9.62 Working Draft, Public Key Cryptography for the Financial Services Industry: The Elliptic Curve Digital Signature Algorithm, September 1998.

[P1363] IEEE P1363/D3 (Draft version 3), Standard Specifications for Public Key Cryptography, May 1998.

[BRS] E. R. Berlekamp, H. Rumsey and G. Solomon, On the Solution of Algebraic Equations over Finite Fields, Information and Control 10 (1967), pp. 553–564

[BSS] I. Blake, G. Seroussi and N. Smart, Elliptic Curves in Cryptography, Cambridge University Press, Cambridge, 1999.

[ElG]   T. ElGamal, A Public-Key Cryptosystem and a Signature Scheme Based on Discrete Logarithms, IEEE Trans. Info. Theory, 31, 469–472, 1985.

[Gor]   M. D. Gordon, A Survey of Fast Exponentiation Methods, Journal of Algorithms 27 (1998), 129–146.

[Knu]   Erik Woodward Knudsen, Elliptic Scalar Multiplication Using Point Halving, Advances in Cryptology - Asiacrypt'99, LNCS 1716, Springer-Verlag (1999), pp. 135–149.

[Koc]   Paul C. Kocher, Timing Attacks on Implementations of Diffie-Hellman, RSA, DSS and Other Systems, Proceedings CRYPTO '96, LNCS 1109, Springer-Verlag, 1996, pp. 104–113.

[KJJ]   P. Kocher, J. Jaffe and B. Jun, Introduction to Differential Power Analysis and Related Attacks, http://www.cryptography.com/dpa/technical/index.html.

[MOV]   A. Menezes, P. van Oorschot and S. Vanstone, Handbook of Applied Cryptography, CRC Press, Boca Raton, Florida, 1996.

[Sal]   Arto Salomaa, Public-key Cryptography, Springer-Verlag, Berlin, 1990.

On the Formal Modelling of Trust in Reputation-Based Systems

Mogens Nielsen[1,2] and Karl Krukow[1,2]

[1] BRICS*, University of Aarhus, Denmark
{krukow, mn}@brics.dk
[2] Supported by SECURE**

Abstract. In a reputation-based trust management system an entity's behaviour determines its reputation which in turn affects other entities interaction with it. We present a mathematical model for trust aimed at global computing environments which, as opposed to many traditional trust management systems, supports the dynamics of reputation-based systems in the sense that trusting relationships are monitored and changes over time depending on the behaviour of the entities involved. The main contribution is the discovery that the notion of event structures, well studied e.g. in the theory of concurrency, can faithfully model the important concepts of *observation* and *outcome* of interactions. In this setting observations are events and an outcome of an interaction is a maximal set of consistent events describing what happened. We also touch upon the problem of transferring trust or behavioural information between contexts, and we propose a generalised definition of morphism of event structures as an information-transfer function.

1   Introduction

In the Global Computing (GC) vision very large numbers of networked, mobile, computational entities interact to fulfill their respective goals. To be successful in such environments, entities (the terms principal, agent and entity are used synonymously) must collaborate, must be capable of operating under only partial information, and security decisions must be made autonomously, as no central authority is feasible.

The classical trust management approach [1], first introduced by Blaze, Feigenbaum and Lacy in [2], was proposed as a solution to the inadequacy of traditional security mechanisms in larger decentralised environments. Roughly, a classical trust management system deals with deciding the so-called compliance checking problem: given a request together with a set of credentials, does the request comply with the local security policy of the provider? The same authors also

---

* Basic Research in Computer Science funded by the Danish National Research Foundation
** Authors are supported by SECURE: Secure Environments for Collaboration among Ubiquitous Roaming Entities, EU FET-GC IST-2001-32486
http://secure.dsg.cs.tcd.ie/

J. Karhumäki et al. (Eds.): Theory Is Forever (Salomaa Festschrift), LNCS 3113, pp. 192–204, 2004.

developed tool-support in the form of PolicyMaker [2,3] and later KeyNote [4] for handling the trust management problem. In his paper [5], Weeks displayed a simple mathematical framework, and showed how this framework would instantiate to various existing trust management systems, including KeyNote, SPKI [6] and some logic based systems (see [5] for details), sometimes even leading to more efficient algorithms for the compliance checking problem. The framework expresses a trust management system as a complete lattice $(D, \leq)$ of possible *authorisations*, a set of *principal names* $\mathcal{P}$, and a language for specifying so-called *licenses*. The lattice elements $d, e \in D$ express the authorisations relevant for a particular system, e.g. access-rights, and $d \leq e$ means that e authorises at least as much as d. An *assertion* is a pair $a = \langle p, l \rangle$ consisting of a principal $p \in \mathcal{P}$, the *issuer*, and a monotone function $l : (\mathcal{P} \rightarrow D) \rightarrow D$, called a *license*. In the simplest case l could be a constant function, say d_0, meaning that p authorises d_0. In the general case the interpretation of a is: given that all principals authorise as specified in the *authorisation map*, $m : \mathcal{P} \rightarrow D$, then p authorises as specified in $l(m)$. This means that a license such as $l(m) = m(A) \vee m(B)$ expresses a policy saying "give the lub of what A says and what B says". Weeks showed that a collection of assertions $L = \langle p_i, l_i \rangle_{i \in I}$ gives rise to a monotone function $L_\lambda : (\mathcal{P} \rightarrow D) \rightarrow \mathcal{P} \rightarrow D$, with the property that a coherent authorisation map representing the authorisations of the involved principals is given by the least fixed point, lfp L_λ.

The ideas on trust management systems seeded a substantial amount of research in the area of security in large distributed systems, but as noted in [7], which serves as a survey on existing systems anno 2000, the current trust management solutions do not adequately deal with the dynamic aspects of trust: a trusting relationship evolves over time and requires monitoring and reevaluation. In [8,9] it was argued that while the idea of having mutually referring licenses resolved by fixed points was good, the Weeks-framework for trust would be too restrictive in GC environments. One reason is that principals often do not have sufficient information to specify precise authorisations for all other principals. In the framework this means that any unknown or only partially known principal is always assigned the bottom authorisation. The proposed solution was to have the set T of "authorisations", here called *trust values*, equipped with *two* orderings, denoted $\preceq$ and $\sqsubseteq$. Here $\preceq$, called the *trust ordering*, corresponds to Weeks' way of ordering by "more privilege", whereas $\sqsubseteq$, called the *information ordering*, introduces a notion of precision or information. The key idea was that the elements of the set should embody also various degrees of uncertainty, and then $d \sqsubseteq e$ reflects that e is more precise or contains more information than d. In the simplest of cases the trust values could be just symbolic, e.g. unknown $\sqsubseteq$ low $\preceq$ high, but they might also have more structure, as will become clear in the following sections. It was shown how least fixed points with respect to the information ordering, leads to a way of distinguishing an unknown principal from a known and distrusted one.

The notion of reputation-based systems (see e.g. [10–12]) also addresses some of these issues. In a reputation-based system, an agent's past behaviour determines together with local security policies how other agents assign privileges to that agent, and more generally affects any decisions concerning that agent. The SECURE project [13, 14] aims at providing a framework for decision-making in GC environments, based on the notion of trust. The formal model for trust deployed is that of [8, 9], and a particular application defines a triple $(T, \sqsubseteq, \preceq)$ of trust values with the two orderings. In this model, trust exists *between principals*, and so for any principals P and Q the trust that P has in Q is modelled as an element of T. As in the Weeks-framework this value is defined in terms of a license issued by P which is called P's *trust policy*. Thus, at any given time the trust-state of the system can be described as function, $m : \mathcal{P} \to \mathcal{P} \to T$, where $\mathcal{P}$ is the set of principals, and the interpretation is that $m(P)(Q)$ describes P's trust in Q. At any time there is a unique trust-state describing how principals trust, and this state is the $\sqsubseteq$-least fixed point of the monotone function induced by the collection of all licenses.

In SECURE, each principal P has its own decision making framework which is invoked when an application needs to make some decision involving another principal. The decision making framework contains three primary components: the risk engine, the trust engine, and the collaboration monitor. At the most abstract level, the collaboration monitor records the behaviour of principals with which P has interacted. This information together with a trust policy defines how P assigns trust values to any other principal. The trust information, in turn, serves as a basis for a risk analysis of any interaction. In fact, with each type of interaction with a principal, say Q, there is a finite set of possible *outcomes* of the interaction. The outcome that occurs is determined by the behaviour of Q. Each of these outcomes has an associated cost[3] which could be represented simply as a number, but could also be a more complex object like a probability distribution on say $\mathbb{R}$. Since the outcome depends on Q, the decision of how to interact is based on the trust in Q. In this set-up it is necessary that the trust value for Q carries enough information that estimation of the likelihood of each of the outcomes is possible. If this estimation is possible, one may start reasoning about risk, e.g. the *expected* cost of an interaction. The rest of this paper describes the model for trust deployed in SECURE applications.

2   An Evidence Based Framework

As we discussed in the previous section the SECURE architecture brings forward the need for a formal model for trust supporting the approximation of likelihood of interaction outcomes, based on previous observations. We now propose a framework supporting this reasoning. We will use the mathematical structures

---

[3] The term cost should be understood more generally as cost or *benefit*. If costs are represented as non-negative numbers, one might represent benefit as negative number.

known as event structures (see [15] for an original reference and the handbook chapter [16] for an extensive reference).

Definition 1 (Event Structure). *An event structure is a triple $(E, \leq, \#)$ consisting of a set E of events which are partially ordered by $\leq$, the necessity relation (or causality relation), and $\#$ is a binary, symmetric, irreflexive relation $\# \subset E \times E$, called the conflict relation. The relations satisfy*

$$\{e' \in E \mid e' \leq e\} \text{ is finite,}$$

$$\text{if } e \# e' \text{ and } e' \leq e'' \text{ then } e \# e''$$

for all $e, e', e'' \in E$. We say that two events are independent *if they are not in either of the two relations.*

As an example, the event structure in Figure 1 could model a small scenario where a principal may ask a bank for the transfer of electronic cash from its bank account to an electronic wallet. After making the request, the principal observes that the request is either rejected or granted. After a successful transaction, the principal could observe that the cash sent in the transaction is forged or perhaps run an authentication algorithm to establish that it is authentic. Also, the principal could observe a withdrawal from its bank account with the present transaction's id, and this withdrawal may or may not be of the correct amount. The two basic relations on event structures have an intuitive meaning in our set

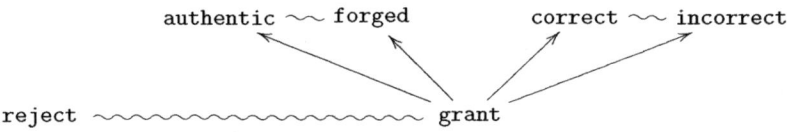

Fig. 1. An event structure describing our example. The curly lines $\sim$ describe the immediate conflict relation and pointed arrows, the causality relation

up. An event may *exclude* the possibility of the occurrence of a number of other events. In our example the occurrence of the event 'transaction rejected' clearly excludes the event 'transaction granted'. The necessity relation is also natural: some events are *only possible* when others have already occurred. In the example structure, 'money forged' only makes sense in a transaction where the transfer of money actually did occur. Whether the e-cash is forged and whether the correct amount is charged are two independent observations that may be observed, in any order, which is modelled as independence in the event structure.

Definition 2 (Configurations of an Event Structure). *Let $ES = (E, \leq, \#)$ be an event structure. Say that a subset of events $x \subseteq E$ is* consistent *if it satisfies the following two properties:*

1. *Conflict free:* for any $e, e' \in x : e \,\#\, e'$ *(i.e.* $(e, e') \notin \#$*)*.
2. *Necessity downwards closed:* for any $e \in x, e' \in E : e' \leq e \Rightarrow e' \in x$.

Define the configurations *of ES, written* $\mathcal{C}_{ES}$*, to be the set of consistent subsets of E. We will define* $\mathcal{C}^0_{ES}$ *to be the* finite *configurations. Define relation* $\rightarrow$ *on* $\mathcal{C}_{ES} \times E \times \mathcal{C}_{ES}$ *by*

$$x \xrightarrow{e} x' \iff e \notin x \text{ and } x' = x \cup \{e\}$$

A (finite) configuration models information regarding the result of one interaction. Note that the *outcomes* of an action corresponds to the maximal configurations (ordered by inclusion) of the event structures, and knowing the outcome corresponds to having complete information. The configurations of our example is given in Figure 2.

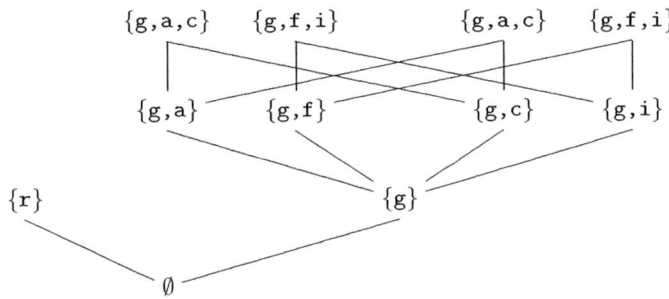

Fig. 2. Configurations of the event structure in Figure 1. The lines indicate inclusion and the events are abbreviated

We can now be more precise about the role of the collaboration monitor in the SECURE framework. Informally, its function is to monitor the behaviour of principals with whom interaction is made. For a particular interaction the possible events that may occur are modelled by an event structure, say *ES*. The information about the outcome of this interaction is simply a configuration, $x \in \mathcal{C}^0_{ES}$.

Definition 3 (Interaction History). *Let* $ES = (E, \leq, \#)$ *be an event structure. Define an* interaction history *in ES to be a finite ordered sequence H of configurations,* $H = x_1 x_2 \cdots x_n \in \mathcal{C}^0_{ES}{}^*$*. The individual components* x_i *in the history H will be called* interactions.

An interaction history in the event structure from Figure 1 could be the sequence $\{g, a, c\}\{g, c\}\{g\}\{r\}$. The concept of interaction histories models one principal's recording of previous interactions with another. When the collaboration monitor learns about the occurrence of an event, e, this information is increased. We define a simple relation expressing this operation.

Definition 4 (Information Relation). *Let $ES = (E, \leq, \#)$ be an event structure and let $H = x_1 \cdots x_n$ and $K = y_1 \cdots y_n$ be interaction histories in ES, $e \in E$ an event, and $i \in \mathbb{N}, 1 \leq i \leq n$ be an index, then define:*

$$H \stackrel{(e,i)}{\Rightarrow} K \iff x_i \stackrel{e}{\to} y_i \text{ and } \forall (1 \leq j \leq n) : j \neq i \Rightarrow x_j = y_j$$

and also let new be a special event $new \notin E$ then

$$H \stackrel{new}{\Rightarrow} H \cdot \emptyset$$

Let $H \Rightarrow K$ denote that either $H \stackrel{new}{\Rightarrow} K$ or there exists $e \in E$, $i \in \mathbb{N}$ so that $H \stackrel{(e,i)}{\Rightarrow} K$, and $\Rightarrow^*$ the reflexive and transitive closure of $\Rightarrow$.

2.1 Evaluating Evidence

We will equate the notion of trust values with "evidence values". That is, values expressing the amount of evidence regarding a particular partial outcome (i.e. a configuration). We will consider the derivation of such values based on interaction histories.

Consider an event structure $ES = (E, \leq, \#)$. A trust value will be a function from $\mathcal{C}_{ES}$ into a domain of *evidence values*. The function applied to a configuration $x \in \mathcal{C}_{ES}$ is then a value reflecting the evidence for x. It will be natural to express this evidence value as a triple of natural numbers $(s, i, c) \in \mathbb{N}^3$. The interpretation is that out of $s + i + c$ interactions, s of these *support* the occurrence of configuration x, c of these *contradict* it, and i are *inconclusive* about x in the sense that they do not support or contradict it.

Definition 5. *Let $ES = (E, \leq, \#)$ be an event structure and let x be a configuration of ES. Define the* effect *of x as a function,* $\mathbf{eff}_x : \mathcal{C}_{ES} \to \mathbb{N}^3$ *by*

$$\mathbf{eff}_x(w) = \begin{cases} (1,0,0) & \text{if } w \subseteq x \\ (0,0,1) & \text{if } x \# w \text{ (i.e. } \exists e \in x, e' \in w : e \# e') \\ (0,1,0) & \text{otherwise} \end{cases}$$

Also for $(s, i, c), (s', i', c') \in \mathbb{N}^3$ define $(s, i, c) + (s', i', c') = (s + s', i + i', c + c')$.

The intuition behind the definition of $\mathbf{eff}_x$ is the following. Think of x as a configuration which has already been observed. We are now considering engaging in another interaction which will end up in some configuration. Thus, we would like to estimate the likelihood of ending up in a particular configuration w, given that the last interaction ended in x. There are exactly three cases for any fixed configuration w: if $w \subseteq x$ then the fact that x occurred last time supports the occurrence of w. If instead $x \# w$ then x contains an event which rules out the configuration w. Finally, if neither of these are the case, i.e. w didn't occur but also wasn't excluded, we say that x is inconclusive about w. There is a strong similarity between this division of configurations in three disjoint classes and the way Jøsang [17] derives his uncertain probabilities in the Dempster-Shafer framework for evidence [18]. We discuss this in the concluding section.

Definition 6. *Let $ES = (E, \leq, \#)$ be an event structure, define the function* $\mathbf{eval} : \mathcal{C}_{ES}^0{}^* \to (\mathcal{C}_{ES} \to \mathbb{N}^3)$:

$$\mathbf{eval}(x_1 x_2 \cdots x_n) = \lambda w. \sum_{i=1}^{n} \mathbf{eff}_{x_i}(w)$$

We would like to note that the functions **eff** and **eval** allow for many useful variations when computing trust values from interaction histories. For example, suppose we want to model a "memory" so that a principal only remembers the last $M + 1 \in \mathbb{N}$ interactions. This could be done by simply taking

$$\mathbf{eval}^M(x_1 x_2 \cdots x_n) = \lambda w. \sum_{i=n-M}^{n} \mathbf{eff}_{x_i}(w) = \mathbf{eval}(x_{n-M} x_{n-M+1} \cdots x_n)$$

One could also imagine older interactions "counting less", which could be modelled by scaling and rounding of the value of, say, the interactions older than a certain boundary.

2.2 Ordering Evidence

Given this intuition we will consider two orderings on evidence values: an *information ordering*, and an ordering expressing "more evidence in favour of", which we call the *trust ordering*.

Information order. The information ordering $\sqsubseteq$ of $\mathbb{N}^3$ is defined as follows:

$$(s, i, c) \sqsubseteq (s', i', c') \iff (s \leq s') \wedge (c \leq c') \wedge (s + i + c \leq s' + i' + c')$$

The rationale is that (s', i', c') represents more information than (s, i, c) if it can be obtained from (s, i, c) by performing some additional number of interactions, or by refining the information about a particular interaction (or both). By refining we mean to change an "inconclusive" to "supporting" or "contradicting". $\widehat{\mathbb{N}^3}$ denotes the completion of $\mathbb{N}^3$ by a greatest element $\top_{\sqsubseteq}$.

Trust order. The trust ordering $\preceq$ of $\mathbb{N}^3$ is defined as:

$$(s, i, c) \preceq (s', i', c') \iff (s \leq s') \wedge (c \geq c') \wedge (s + i + c \leq s' + i' + c')$$

Here (s', i', c') expresses "more evidence in favour of" than (s, i, c) if it contains more supporting evidence, less contradicting evidence, and still at least as many interactions. Intuitively one can obtain (s', i', c') from (s, i, c) by changing contradicting evidence to inconclusive or supporting, changing inconclusive to supporting, or by adding inconclusive or positive events.

Theorem 1. *The structure* $(\widehat{\mathbb{N}^3}, \sqsubseteq)$ *is a complete lattice. The binary join is given by* $(s_0, i_0, c_0) \sqcup (s_1, i_1, c_1) = (\bar{s}, \bar{i}, \bar{c})$ *where* $\bar{s} = \max\{s_0, s_1\}$, $\bar{c} = \max\{c_0, c_1\}$ *and*

$$\bar{i} = \min\{i \in \mathbb{N} \mid \bar{s} + i + \bar{c} \geq \max\{s_0 + i_0 + c_0, s_1 + i_1 + c_1\}\}$$

The join with respect to $\top_{\sqsubseteq}$ is as expected, and the join of any infinite set is $\top_{\sqsubseteq}$. Furthermore, the structure $(\mathbb{N}^3, \preceq)$ is a lattice. The binary $\preceq$-join is given by $(s_0, i_0, c_0) \vee (s_1, i_1, c_1) = (\hat{s}, \hat{i}, \hat{c})$ where $\hat{s} = \max\{s_0, s_1\}$, $\hat{c} = \min\{c_0, c_1\}$ and

$$\hat{i} = \min\{i \in \mathbb{N} \mid \hat{s} + i + \hat{c} \geq \max\{s_0 + i_0 + c_0, s_1 + i_1 + c_1\}\}$$

The meet is obtained dually. Finally, the join and meet functions for the trust order, $\vee, \wedge : \mathbb{N}^3 \times \mathbb{N}^3 \to \mathbb{N}^3$ are monotone with respect to the information order.

In the following we use $\sqsubseteq$ also for the pointwise extension of $\sqsubseteq$ to trust values, i.e. the functions $\mathcal{C}_{ES} \to \widehat{\mathbb{N}^3}$. We can relate the relation $\Rightarrow^*$ on interaction histories with the information relation on trust values.

Proposition 1. *Let ES be an event structure and $H, K \in \mathcal{C}_{ES}^{0}{}^{*}$ interaction histories. Then **eval** is monotonic in the sense that if $H \Rightarrow^* K$ then also* **eval**$(H) \sqsubseteq$ **eval**(K).

Some information is discarded by **eval**, and the following proposition explains what is lost. The function **eval** is injective up to rearranging the order of interactions.

Proposition 2. *Let $ES = (E, \leq, \#)$ be an event structure and $H, K \in \mathcal{C}_{ES}^{0}$ be configurations, $H = x_1 x_2 \cdots x_n$ and $K = y_1 y_2 \cdots y_m$. If* **eval**$(H) =$ **eval**(K) *then $n = m$ and there exists a permutation on n elements $\sigma : [n] \overset{\sim}{\to} [n]$ so that*

$$H = \sigma(K) = y_{\sigma(1)} y_{\sigma(2)} \cdots y_{\sigma(n)}$$

Returning to the SECURE architecture, the risk engine uses trust values to derive estimates on the likelihood of the various outcomes. Our trust values convey sufficient information to enable estimation of probability distributions on the configurations. There are several ways to do this, depending on the application. For example one might derive an opinion $\omega_x = (b_x, u_x, d_x)$ for $x \in \mathcal{C}_{ES}$ in the sense of Jøsang, which gives rise to a probability pdf [17, 12].

3   Trust Policies

As discussed in the introduction, each principal defines a local *trust policy* following the idea from [5]. We give an example of a language for specifying such policies. The syntax is given in Figure 3. A policy is a list of specific policies, terminated by a general policy. The specific policies explicitly name a principal and a corresponding trust expression (τ), whereas the general policy applies to any principal not explicitly listed. In this simple example language, the trust expressions are built up from the basic constructs of "local reference" and "policy reference", and these can then be combined with the various joins and meets we have available. The two types of references are similar in that both refer to a principal P's trust value for a principal Q. The difference is that the local reference refers to P's personal observation on Q, whereas the trust reference instead refers to the value that P would compute using its *policy*.

$$\pi ::= \quad \star : \tau \qquad\qquad\qquad\qquad\qquad\qquad \text{(default policy)}$$
$$| \; p : \tau ; \pi \qquad\qquad\qquad\qquad\qquad (p \in \mathcal{P}, \text{ specific policies})$$
$$\tau ::= \quad p?_{loc} q \qquad\qquad\qquad\quad \text{(local reference to } p, q \in \mathcal{P} \cup \{\star\})$$
$$| \; p?q \qquad\qquad\qquad\qquad \text{(policy reference to } p, q \in \mathcal{P} \cup \{\star\})$$
$$| \; \tau_1 \; \text{binop} \; \tau_2 \qquad\qquad \text{(binary operation binop} \in \{\wedge, \vee, \sqcap, \sqcup\})$$

Fig. 3. An example policy language

The semantics of a policy is interpreted relative to an environment providing for each pair P, Q of principals a trust value which we think of as being P's interaction history with Q evaluated as in the previous section. This serves as the data for the local references. Let $obs : \mathcal{P} \to \mathcal{P} \to \mathcal{C}_{ES} \to \widehat{\mathbb{N}^3}$ be a fixed function representing this. The semantics of a policy π is a function which takes as input the observation data obs, and gives as output a $\sqsubseteq$-monotone function mapping the global current trust state (an element in $GS = \mathcal{P} \to \mathcal{P} \to (\mathcal{C}_{ES} \to \widehat{\mathbb{N}^3}))$, to a local trust state (an element of $LS = \mathcal{P} \to (\mathcal{C}_{ES} \to \widehat{\mathbb{N}^3}))$. We denote this as

$$[\![\pi]\!]^{obs} : GS \to LS$$

The semantic function $[\![\cdot]\!]^{obs}$ is defined by structural induction on the syntax of π in Figure 4. The definitions make use of the semantic function in Figure 5,

$$[\![\star : \tau]\!]^{obs} = \lambda m \in GS. \lambda y \in \mathcal{P}. [\![\tau]\!]^{obs}(m)([\star \mapsto y])$$
$$[\![p : \tau ; \pi]\!]^{obs} = \lambda m \in GS. \lambda x \in \mathcal{P}. \; \text{if} \; (x = p) \; \text{then} \; [\![\tau]\!]^{obs}(m)([\star \mapsto p])$$
$$\text{else} \; [\![\pi]\!]^{obs}(m)(x)$$

Fig. 4. Semantics of the policy language: syntactic category π

$$[\![Y?_{loc}Z]\!]^{obs}(m)(env) = obs \; (env^\dagger \; Y) \; (env^\dagger \; Z) \qquad (\text{where } Y, Z \in \mathcal{P} \cup \{\star\})$$
$$[\![Y?Z]\!]^{obs}(m)(env) = m \; (env^\dagger \; Y) \; (env^\dagger \; Z) \qquad (\text{where } Y, Z \in \mathcal{P} \cup \{\star\})$$
$$[\![\tau_1 \; \text{binop} \; \tau_2]\!]^{obs}(m)(env) = \left([\![\tau_1]\!]^{obs}(m)(env)\right) \; [\![\text{binop}]\!] \; \left([\![\tau_2]\!]^{obs}(m)(env)\right)$$

Fig. 5. Semantics of the policy language: syntactic category τ

essentially mapping the syntactic category τ to an element of $\mathcal{C}_{ES} \to \widehat{\mathbb{N}^3}$. This is again interpreted relative to observations obs and the current trust state $m : GS$, but also relative to an environment, $env : \{\star\} \to \mathcal{P}$, which interprets $\star$ as a name in $\mathcal{P}$. The env function extends trivially to a function of type $\{\star\} \cup \mathcal{P} \to \mathcal{P}$ (the identity on non-$\star$ elements). The semantics of a binop is the corresponding $\sqsubseteq$ or $\preceq$ lub/glb[4], which is $\sqsubseteq$-monotone by Theorem 1.

We can now view a collection of mutually referring policies,

$$\Pi^{obs} = \{[\![\pi_P]\!]^{obs} \mid P \in \mathcal{P}\}$$

as defining a "web of trust", and define a unique monotone function Π_λ^{obs}

$$\Pi_\lambda^{obs} = \langle [\![\pi_P]\!]^{obs} : P \in \mathcal{P} \rangle : GS \to GS$$

with the property that

$$\mathrm{Proj}_Q \circ \Pi_\lambda^{obs} = [\![\pi_Q]\!]^{obs}$$

for all $Q \in \mathcal{P}$. This function essentially takes a piece of global trust information $m : GS$ and gives a piece of global trust information $\Pi_\lambda^{obs}(m) : GS$ which, when applied to p and then to q, returns p's trust in q under π_p, given trust as specified in m. Now, since the trust values $\mathcal{C}_{ES} \to \widehat{\mathbb{N}^3}$ form a complete lattice with the information ordering, and since Π_λ^{obs} is a monotone endo-function on this structure, it has a unique least fixed point. We define the trust information in a web of trust, $\Pi = \{\pi_p \mid p \in \mathcal{P}\}$ with local observations given by obs, as the least fixed point of the induced function, $[\![\Pi]\!]^{obs} \overset{(\mathrm{def})}{=} \mathsf{lfp}\ \Pi_\lambda^{obs}$.

The interested reader is referred to [8, 9] for examples of policies.

4   Transferring Information

The example policy language in the previous section allows principals to share trust information by means of the reference constructs. However, we were implicitly assuming that all principals agree on the event structure used. One event structure describes a particular context, i.e. there is one event structure for each possible way of interacting. It is useful to be able to map trust values between contexts that are somehow related, e.g. if one has only very little information about context ES_1 but much information about a related context ES_2, it is often useful to somehow apply the knowledge of ES_2 to give an estimate in ES_1. We are aiming at formalising the kind of evidential transfer we all employ in every day life, where e.g. observations of an individual A's behaviour with respect to timely payments of bills affects also our trust in A with respect to the question of whether to lend him money. We propose a definition of a morphism of event structures enabling such an information transfer.

Definition 7 (Morphism of event structures). *Let $ES = (E, \leq, \#)$ and $ES' = (E', \leq', \#')$ be event structures. A morphism of event structure, $\eta : ES \to ES'$ is a function $\eta : E' \to 2^E$ which has the following two properties:*

---

[4] We use a strict version of the $\preceq$-lub/glb which is the $\top$-strict extension of $\vee, \wedge :$
$\mathbb{N}^3 \times \mathbb{N}^3 \to \mathbb{N}^3$ to a function $\widehat{\mathbb{N}^3} \times \widehat{\mathbb{N}^3} \to \widehat{\mathbb{N}^3}$ which is also monotone.

1. *Monotonic: For any $e', e'' \in E'$ if $e' \leq' e''$ then we have*

$$\forall e_2 \in \eta(e'') \exists e_1 \in \eta(e') : e_1 \leq e_2$$

2. *Preserves conflict: For any $e', e'' \in E'$ if $e' \#' e''$ then*

$$\forall e_1 \in \eta(e') \forall e_2 \in \eta(e'') : e_1 \# e_2$$

A morphism $\eta : ES \to ES'$ can be thought of as a transfer of evidence from ES to ES'. The idea is that $e \in \eta(e')$ means that an occurrence of e in ES is an indication of the event e' occurring in ES'. We will think of the set $\eta(e')$ as a disjunction of conditions in the sense that e' occurs if there is some $e \in \eta(e')$ which has occurred in ES. If $\eta(e') = \emptyset$ then we say that e' has no enabling condition under η.

Definition 8 (Category of Event Structures, E). *Consider the following categorical data, which we will call the category of event structures, and denote* **E**.

- *Objects are event structures $ES = (E, \leq, \#)$*
- *Morphisms $\eta : ES \to ES'$, are the morphisms of Definition 7.*
- *Identities $1_{ES} : ES \to ES$ are the functions $1_{ES} : E \to 2^E$ given by*

$$1_{ES}(e) = \{e\}$$

- *For $\eta : ES \to ES'$ and $\epsilon : ES' \to ES''$ composition, $\epsilon \circ \eta : ES \to ES''$ is given by the following function $\epsilon \circ \eta : E'' \to_\star 2^E$*

$$\epsilon \circ \eta(e'') = \bigcup_{e' \in \epsilon(e'')} \eta(x')$$

Proposition 3 (E is a category). *The definition of* **E** *yields a category.*

A morphism, $\eta : ES \to ES'$ can then be used to map configurations of ES to configurations of ES' by the mapping, $\bar{\eta} : \mathcal{C}_{ES} \to \mathcal{C}_{ES'}$

$$\bar{\eta}(x) = \{e' \in E' \mid \exists e \in \eta(e') : e \in x\}$$

The axioms of morphisms imply that $\eta(x)$ is a configuration, and the fact that **E** constitutes a category means that we can compose the information transfer functions to obtain information transfer functions.

5   Conclusion

We have proposed a mathematical framework for trust, and a way of deriving trust values from interaction histories. In this framework trust is identified with evidential information, arising from observed behaviour, allowing the estimation of likely future behaviour. The framework is deployed in the SECURE project, and has been used in concrete SECURE prototype applications for e.g. spam filtering [14]. The trust model fits well with the bi-ordered trust structures of [8, 9] and uses ideas from the framework of Weeks [5]. The way that trust values are derived from interaction histories is similar to the way that belief and plausibility functions are derived in [18], and the way in which Jøsang derives his "opinions" from belief-mass assignments in the subjective logic [17]. Event structures can be seen as a generalisation of the traditional frames of discernment from the Dempster-Shafer theory of evidence. If one allows a generalised version of event structures in which the conflict relation is allowed to be a subset of $E \times \mathbf{2}^E$ (where $e\#X$ means that e cannot occur if X has occurred), it is not hard to see that each frame of discernment θ corresponds to an event structure with events $\{\bar{p} \mid p \in \theta\}$, where any event $\bar{p}$ is in conflict only with the set $\{\bar{q} \mid q \in \theta, q \neq p\}$. The understanding of a $\bar{p}$ is the exclusion of the state p. The configurations of the event structure is isomorphic to the poset $(\mathbf{2}^\theta \setminus \emptyset, \subseteq)^{op}$. Furthermore, $x \cap y = \emptyset$ in $\mathbf{2}^\theta \setminus \emptyset$ iff the corresponding configurations are in conflict.

While the problem of transferring trust between related contexts has been discussed, we still need to investigate the usefulness of our formalisation in terms of event structure morphisms in concrete application scenarios. The concept of morphisms seems to be appropriate for evidence transfer, but the exact definition needs some further investigation. As an example, we have considered a generalisation of the event structure morphisms presented, in which we allow $\eta : E' \to \mathbf{2}^{\mathcal{C}^0_{ES}}$, i.e. η is maps events to a disjunction of arbitrary finite configurations instead of only prime configurations (i.e. configurations of the form $\{e \in E \mid e \leq e_0\}$ for some e_0). This generalised definition gives rise to a category containing $\mathbf{E}$ as a subcategory.

Acknowledgements

We would like to thank Marco Carbone, Vladimiro Sassone and the SECURE consortium for their contribution to this work.

References

1. Blaze, M., Feigenbaum, J., Ioannidis, J., Keromytis, A.: The role of trust management in distributed systems security. In Vitek, J., Jensen, C.D., eds.: Secure Internet Programming. Volume 1603 of Lecture Notes in Computer Science., Springer (1999)
2. Blaze, M., Feigenbaum, J., Lacy, J.: Decentralized trust management. In: Proc. IEEE Conference on Security and Privacy, Oakland. (1996)
3. Blaze, M., Feigenbaum, J., Strauss, M.: Compliance checking in the policymaker trust management system. In: Financial Cryptography. (1998) 254–274
4. Blaze, M., Feigenbaum, J., Lacy, J.: KeyNote: Trust management for public-key infrastructure. Springer LNCS **1550** (1999) 59–63
5. Weeks, S.: Understanding trust management systems. In: Proc. IEEE Symposium on Security and Privacy, Oakland. (2001)
6. Ellison, C.M., Frantz, B., Lampson, B., Rivest, R., Thomsa, B., Ylonen, T.: SPKI certificate theory. RFC 2693 (1999)

7. Grandison, T., Sloman, M.: A survey of trust in internet application. IEEE Communications Surveys, Fourth Quarter (2000)
8. Carbone, M., Nielsen, M., Sassone, V.: A formal model for trust in dynamic networks. In: Proceedings from Software Engineering and Formal Methods, SEFM'03, IEEE Computer Society Press. (2003)
9. Krukow, K., Nielsen, M.: Towards a formal notion of trust. In: Proceedings of the 5th ACM SIGPLAN international conference on Principles and Practice of Declarative Programming. (2003)
10. Shmatikov, V., Talcott, C.: Reputation-based trust management. Journal of Computer Security (selected papers of WITS '03) (2004)
11. Mui, L., Mohtashemi, M.: Notions of reputation in multi-agent systems: A review. In: Trust, Reputation, and Security: Theories and Practice, AAMAS 2002 International Workshop, Bologna, Italy, July 15, 2002, Selected and Invited Papers. (2002)
12. Jøsang, A., Ismail, R.: The beta reputation system. In: Proceedings of the 15th Bled Conference on Electronic Commerce, Bled. (2002)
13. Cahill et al., V.: Using trust for secure collaboration in uncertain environments. IEEE Pervasive Computing **2** (2003) 52–61
14. Cahill, V., Signeur, J.M.: Secure Environments for Collaboration among Ubiquitous Roaming Entities. Website: http://secure.dsg.cs.tcd.ie (2003)
15. Nielsen, M., Plotkin, G., Winskel, G.: Petri nets, event structures and domains. Theoretical Computer Science **13** (1981) 85–108
16. Winskel, G., Nielsen, M.: Models for concurrency. Handbook of Logic in Computer Science **4** (1995) 1–148
17. Jøsang, A.: A logic for uncertain probabilities. Fuzziness and Knowledge-Based Systems **9(3)** (2001)
18. Shafer, G.: A Mathematical Theory of Evidence. Princeton University Press (1976)

Issues with Applying Cryptography in Wireless Systems

Valtteri Niemi

Nokia Research Center
P.O. Box 407, FIN-00045
NOKIA GROUP, Finland
valtteri.niemi@nokia.com

Abstract. We survey main cryptographic features in several major wireless technologies. Cellular systems GSM/GPRS and UMTS (3G) are covered, and also shorter range systems Wireless LAN and Bluetooth. Then we continue by presenting problematic areas with applying cryptography in these wireless systems. Several examples are given in each problem area.

1 Introduction

In the first part of this paper we do a brief survey on cryptographic mechanisms in some of the most important wireless technologies. On cellular systems, we first describe security solutions in the GSM technology, the dominant global cellular standard. We continue by showing how the security model and security mechanisms were extended and enhanced in the successor of the GSM system, i.e. in the Universal Mobile Telecommunications System (UMTS), specified in the global 3rd Generation Partnership Project (3GPP). On shorter range wireless technologies we discuss Wireless LAN security, as standardized by IEEE, and also Bluetooth security that is specified by an industry consortium called the Bluetooth SIG (Special Interest Group). A typical use case of WLAN is access to Internet through a WLAN access point from distances up to several hundred meters while a typical Bluetooth use case is communication between two devices, e.g. a mobile phone and an accessory, with a distance in the order of ten meters.

Cryptographic algorithms provide a major tool for wireless security but defining how the algorithms are used as a part of a communication system architecture is a demanding task by itself. The second part of the paper contains issues with applying cryptography in wireless systems of global scale. Several examples of problems are presented. For most of these examples solutions exist also but typically these solutions are not fully satisfactory. We study issues in the following problematic areas: composition of several mechanisms, continuity from legacy systems and equipment to more secure solutions of the future, key management and constraints imposed by other functionalities.

J. Karhumäki et al. (Eds.): Theory Is Forever (Salomaa Festschrift), LNCS 3113, pp. 205–215, 2004.

2   GSM Cryptography

The essential cryptographic algorithms of GSM are explained in this section. The $A3$ algorithm has a one-way property and it is the core of a challenge-response protocol that is needed for authentication of users. Key generation is tightly linked into authentication and another algorithm with one-way property, called $A8$, is used for this purpose. The generated 64-bit key, K_c is subsequently used in the encryption algorithm $A5$ that is embedded in the physical layer of the GSM radio interface. To be more precise, $A3$, $A5$ and $A8$ are names of algorithm families. The GSM security architecture allows each algorithm to be replaced by another one that has the same input-output structure. For encryption, three different stream ciphers A5/1, A5/2 and A5/3 have been standardized so far in European Telecommunication Standards Institute (ETSI).

The situation is even more fragmented for A3 and A8. This is a consequence of the fact that these algorithms need not be standardized at all. The algorithms are executed only in two locations, in the SIM card inside the user terminal, and in the Authentication Centre that is a database in the user's home network. Therefore, each mobile network operator may in principle use its own proprietary algorithms.

In the packet switched domain of the GSM system, i.e. in GPRS (General Packet Radio Service), the radio interface encryption is replaced by encryption on layer three of the radio network. This change has minor effects on the crypto-graphic algorithm input-output structure but there are more substantial effect in the sense that the protection is extended further in the network, i.e. the packet data traffic is encrypted from the terminal all the way to the core network. At the time of writing, there are three different standardized stream ciphers GEA1, GEA2 and GEA3 for GPRS.

The GSM security architecture, like any wide scale security architecture, is a result of a trade-off between cost and security. The parts of the system that were seen as most vulnerable have been protected well while some less vulnerable parts have no protection. As already mentioned above, in circuit-switched part of GSM the encryption covers only the radio interface between the terminal and the base station. It is also worth noticing that the terminal does not execute any explicit authentication of the network, thus leaving the terminal vulnerable against certain types of active attacks. In particular, this refers to a malicious party who has the required equipment to masquerade as a legitimate network element and/or legitimate user terminal.

The GSM security architecture has also been criticized for keeping some essential parts secret, e.g. specifications of the cryptographic algorithms. This secrecy does not create trust on algorithms in the long run because they are not publicly available for analysis with most recent methods. Also, protection based on global secrets is not efficient because these secrets tend to be revealed sooner or later.

3   UMTS Cryptography

The UMTS technology can be seen as a successor of GSM. Indeed, the core
network part of UMTS is an enhanced version of the GSM core network. On the
other hand, there has been more revolutionary development in the radio network
part. The evolution vs. revolution aspect is reflected in the security features. The
UMTS authentication and key agreement mechanisms are executed between the
terminal and the core network; these mechanisms were created by enhancing
GSM-type challenge-response user authentication protocol with a network authentication based on sequence numbers.

In the radio network there is more revolution in security features. Encryption
is provided by $f8$ algorithm on the radio layer two, and as a completely new
feature, integrity protection algorithm $f9$ is applied to signaling messages on
the radio layer three. Also in UMTS, the symbols $f8$ and $f9$ refer to algorithm
families; only the input-output structure is fixed, not the internal structure of
the algorithms. Currently, one version of both algorithms has been standardized,
both based on the publicly specified KASUMI block cipher. At the time of
writing, the 3GPP has begun the process of specifying another pair of algorithms.
While it is not seen probable that KASUMI algorithm would be broken in the
near future, an alternative would clearly increase the overall security level of
UMTS.

In the following subsections we briefly go through all the essential UMTS
security feastures. See [5] for further reading.

3.1   Mutual Authentication

Three entities are involved in the authentication mechanism of the UMTS system: home network, serving network (SN) and terminal, or more specifically
Universal Subscriber Identity Module (USIM, typically in a smart card). The
SN checks subscribers identity (as in GSM) by a challenge-response technique
while, on the other hand, the terminal checks that SN has been authorised by
the home network to do so. As explained earlier, the latter feature is new in
UMTS (when compared to GSM).

The basis for the authentication mechanism is a master key K that is shared
between the USIM and the home network. This is a permanent secret with the
length of 128 bits. The key K is never transferred out from the two locations.
In particular, the user has no way of getting to know her/his master key. A
key agreement procedure is inseparably linked to the mutual authentication. It
provides keys for encryption and integrity protection. These are temporary keys
with the same length of 128 bits. Every time the USIM is authenticated, new
keys are derived from the permanent key K.

3.2   Radio Encryption

Once the user and the network have authenticated each other they may begin
secure communication. As described earlier, a cipher key CK is shared between

the core network and the terminal after a successful authentication. Before encryption can begin, the communicating parties have to agree on the encryption algorithm also. On user side, encryption/decryption takes place in the terminal (not in USIM), and on network side, the Radio Network Controller (RNC) handles encryption/decryption. This means that the cipher key CK has to be transferred from core network to the radio access network, and inside terminal the CK is transferred over the USIM-terminal interface.

The UMTS encryption mechanism is based on a stream cipher concept because it has an inherent advantage that the mask data can be generated before the actual plaintext is known. Then the final encryption is a very fast bit operation.

3.3   Integrity Protection

The purpose of the integrity protection is to authenticate individual control messages. The integrity protection is also used between the terminal and RNC, just like encryption. The integrity key IK is generated during the authentication and key agreement procedure, again similarly as the cipher key.

Output of the integrity protection mechanism is a message authentication code that consists of 32 bits.

3.4   Network Domain Security

The term "network domain security" refers to protection of communication between different 3GPP networks (and network elements). A basic notion on the IPsec-based security in 3GPP systems is the security gateway. All control plane IP communication towards external networks should go via security gateways. These gateways use the Internet Key Exchange (IKE) protocol to exchange IPsec Security Associations between themselves. The protection method for the data traffic is IPsec Encapsulated Security Payload (ESP).

3.5   SIP Security

The 3GPP has specified an IP Multimedia Subsystem (IMS) that is a core network subsystem using Session Initiation Protocol (SIP) for session management and call control. The actual user data traffic (voice, video etc.) is carried over IP and in principle IMS may be built on top of any access technology that supports IP connectivity.

When a User Agent (UA) in the terminal wants to get access to IMS, it typically first connects to the 3GPP radio network. In this process, UMTS security features (described earlier) are utilized: mutual authentication, integrity protection and encryption on the hop between the terminal and RNC. Through the core network, the UA is able to contact IMS nodes using SIP signaling. The first contact in IMS is SIP Proxy, called P-CSCF (Proxy Call Session Control Function). Through it the UA is able to register itself to home IMS. At the same time UA and home IMS authenticate each other, based on permanent shared master secret. They also agree on temporary keys.

The SIP traffic between visited IMS and home IMS is protected by network domain security mechanisms. The security associations used for this purpose are not specific to the UA in question.

Next UA and the P-CSCF negotiate in a secure manner all parameters of the security mechanisms to be used to protect further SIP signaling, e.g. cryptographic algorithms. Finally integrity protection of first hop SIP signaling between UA and P-CSCF is started, using IPsec ESP.

3.6   Recent Developments

At the time of writing, 3GPP is finalizing security mechanisms on following areas:

- -Wireless LAN interworking with 3GPP systems;
- - Multimedia Broadcast/Multicast Service;
- - Generic Authentication architecture;
- - Presence, Messaging, Conferencing services;
- - Authentication framework for network domain security: this adds support of a Public Key Infrastructure (PKI) for key management.

4   Bluetooth Cryptography

The Bluetooth technology includes authentication and key agreement between two peer devices where the cryptoalgorithm SAFER++ is in use in an appropriate mode. In Bluetooth,unlike in cellular systems, the authentication algorithm has to be standardized because it is executed in terminal devices. The keys used in authentication are agreed first by a pairing procedure in which a link key is generated. For radio interface confidentiality, Bluetooth uses a stream cipher tailor-made for this purpose.

It allows fast and simple hardware implementation because it uses linear feedback shift registers (LFSR) as building blocks. The algorithm has strong correlation properties and LFSRs are initialized for each data packet to be encrypted. It is possible to encrypt both point-to-point and point-to-multipoint connections.

The Bluetooth link layer security is based on the strength of the link key. That is dependent of the Bluetooth PIN or it is fetched directly from the application layer. Link layer security mechanisms, i.e authentication and encryption, can be activated directly or from the application. Anonymity protection has to be provided by separate means if needed. Positioning attack is possible because identification of the Bluetooth devices is based on permanent addresses. On the other hand, this hardly constitutes a severe threat in most practical situations, since the coverage of Bluetooth radio is small.

5 WLAN Cryptography

The IEEE 802.11 group has specified the Wireless Local Area Network (WLAN) technology. During that process, also security mechanisms for WLAN were developed, called Wired Equivalent Privacy (WEP). The naming already indicates that the goal was the same as in GSM, i.e. to provide security level comparable to that of wired networks. Unfortunately, the original design of WEP has several weaknesses. For example, the RC4 cipher is used with short initialization values, key management is weak in many implementations and the system lacks integrity protection and replay protection. At the time of writing, the IEEE 802.11 is finalizing a completely new set of security mechanisms, including new cryptoalgorithms.

An industry consortium Wi-Fi Alliance has already endorsed an intermediate set of enhanced set of security mechanisms. This set is called WPA (Wi-Fi Protected Access) and it is also implemented in many products.

Both WPA and the complete security specification (IEEE 802.11i) are based on an authentication and key management framework (IEEE 802.1X). This introduces possibility to use EAP (Extensible Authentication Protocol) [4]. For instance, password-based or public key based authentication can be used in EAP and, thus, also in WPA and in 802.11i. WPA includes also improved usage of RC4 by TKIP (Temporal Key Integrity Protocol). Furthermore, WPA adds message integrity and replay protection.

The complete IEEE 802.11i adds adequate replacements for all features of WEP. For instance, RC4 is replaced by AES. Wider range of network configurations is supported, as well as concepts of personal area networks, roaming and handovers.

6 Composition of Mechanisms

In this section and in the remainder of the paper, we study several issues that have arisen when cryptographic techniques are introduced to wireless systems. These examples of issues try to give an overall picture about the diversity of issues.

The first category consists of issues with composition of security mechanisms. Indeed, it is well known phenomenon in security area that two secure mechanisms can be combined in a way that looks reasonable at first sight but yet it turns out that the composition provides no security at all.

6.1 Tunneled Authentication

As already briefly mentioned in previous section, Extensible Authentication Protocol (EAP) is a general protocol framework that supports multiple authentication mechanisms. It allows a back-end server to implement the actual mechanism while the authenticator element simply passes authentication signaling through.

EAP consists of several Request/Response pairs where Requests are always sent by the network.

We now provide some analysis of the problem. We have an inner protocol that is, for instance, a legacy GSM authentication protocol based on SIM card built into EAP. Then there is an outer protocol, typically in the form of a TLS tunnel. Note here that the inner legacy protocol is usually also in use without any tunnelling, as is the case for GSM. If we continue now with this GSM example, a man-in-the-middle can set up a false cellular base station to ask terminal for responses to challenges.

Even in the case where EAP protocol would be used exclusively in tunnelled mode, authentication of the TLS tunnel relies solely upon terminal actions. There is a weak point here because the terminal user may easily accept an unknown certificate. This kind of dependency on user's actions is typically not acceptable to network operators. Now the session keys are derived from TLS Master Key generated using tunnel protocol (this is the same key as used to create the tunnel before the inner authentication can begin). The result is that the keys potentially derived in the EAP protocol (e.g., the K_c in the case of GSM authentication) are not used for the tunnel anyhow. Therefore, in the best case, where the inner protocol cannot be used without the tunnel, the security depends on user's judgment on certificates, and on the worse case, where the inner protocol can be used also without the tunnel (e.g. GSM case), there is no way of knowing whether the end-point of the tunnel is actually man-in-the middle instead of the legitimate end-point authenticated in the inner protocol.

The main lesson to be learnt from this is not only that there is another case where composing two secure protocols may result in an insecure protocol. It is important to note that using tunnelling to "improve" a remote authentication protocol is very common approach. Known vulnerable combinations include at least HTTP Digest authentication and TLS, PEAP and any EAP subtype, PIC and any EAP subtype.

There are solutions that can be used to fix the problem but usually the exact fix needs to be tailored to the specific protocols. A typical solution could be to create a cryptographic binding between tunneling protocol and the authentication protocol by, for instance, using a one-way function to compute session keys from tunnel secrets (e.g. TLS master key) and EAP secrets (e.g. IK, CK). See [1] for further details on this issue.

6.2   Using IPsec for IMS Message Authentication

As explained in an earlier section, IPsec ESP is used in 3GPP system for SIP message authentication. This implies that the identity to be authenticated is the IP address. On the other hand, charging is based on user's identity at SIP level. This constitutes the first problem in this example. Another problem stems from the fact that SIP level identity is authenticated for registrations and keys are derived at the same time for the IPsec ESP. These keys should be taken into use immediately but how does the IP layer know that its old Security Association is not valid anymore ?

These problems are not at all unsolvable. But the most straight-forward solution that has been created for the first problem, i.e. binding of different layer identities, could also be claimed to be a layer violation and these typically cause unnecessary restrictions for the system architecture. The second problem was solved for 3GPP systems by introducing special handling of port numbers which somebody could call as a misuse of port number semantics.

6.3   Barkan-Biham A5/2 Attack

This recent attack, see [2], exploits weaknesses in GSM cryptographic algorithms. In particular, A5/2 can be broken fast in "ciphertext only" model. A further attack exploits also other legacy features in the GSM security system: A5/2 is a mandatory feature in terminals, call integrity is only based on encryption and the same K_c can in principle be used in different algorithms.

An example attack goes as follows. This allows the attacker to decrypt a strongly encrypted call without using a brute-force method. First the attacker passively catches the challenge $RAND$ from the radio interface, does not care about the response $SRES$, and records the corresponding call encrypted with K_c and A5/3. Then the attack turns active. The attacker replays the stored $RAND$ towards the victim and tells the victim to use the weaker algorithm A5/2. Now the attacker is able to find K_c based on the received encrypted uplink signal. Consequently, the earlier recorded call can also be decrypted by the attacker.

A proposed countermeasure in 3GPP (that is not yet accepted at the time of writing) would create an amendment to the GSM security architecture. It uses the fact that the random challenge $RAND$ is the only variable information sent from home network to the terminal in the authentication. Now we may divide the space of all 128-bit $RAND$ values into different classes with respect to which encryption algorithm is allowed to be used with the K_c derived from this particular $RAND$. A 32-bit flag could indicate to the terminal that this kind of special $RAND$ is in use, and 16 bits would further be used to indicate which algorithms out of total 8 GSM and 8 GPRS encryption algorithms are allowed to be used with the key derived from this special $RAND$. These parameter lengths would imply that the effective length of $RAND$ is reduced from 128 bits to 80 bits. Fortunately, this reduction would not cause any decrease in the overall security level.

7   Continuity

In this section we study issues that are related to an important practical requirement of continuity. We have to make sure that new systems are backward compatible with legacy devices and equipment. There is a two-fold effect. From one hand, handling of legacy easily introduces security holes into the new system. On the other hand, new systems are also legacy of the future, and therefore, potential future requirements should be taken into account as well in the design. Of course, finding the right balance between this kind of future-proofing and current needs is a tricky task.

7.1   3G-WLAN Interworking with EAP-SIM

The abbreviation EAP-SIM refers to an Internet draft that describes how GSM authentication and key agreement protocol can be done in EAP, see [3]. In addition, this mechanism enhances GSM Authentication and key agreement protocol with mutual entity authentication based on the derived key K_c. This is done by utilizing a bundle of (at least two) GSM triplets $(RAND, SRES, K_c)$ in one run of the entity authentication. Therefore, the network authentication is based on (at least) 128-bit secret.

WLAN interworking in 3GPP follows the basic idea of connecting WLAN access zone to the cellular core network. There are several levels of interworking. For instance, we may have shared subscriber database, shared charging and authentication, or even shared services.

Cellular access (also for 3GPP radio network) is possible with either SIM or USIM, and therefore, the same should hold for WLAN access as well. This creates the following problem: enhancements (e.g. mutual authentication) in EAP-SIM fall down if an active GSM attack is possible against the terminal in the cellular side. In particular, an attacker may mount a divide-and-conquer attack against the bundle of triplets by breaking each K_c separately.

The problem can easily be avoided if the same physical SIM cannot be used for both cellular and WLAN domains but then we lose part of the interworking benefits.

7.2   Phased Introduction of Security

This example case of a legacy issue is related to the introduction of security gateways in network-to-network communications, see earlier section on network domain security. The problem is, however, not restricted to this context. The starting point is that communication between networks works well without this additional security measure.

The first problem can be illustrated by the following simplified calculation. Assume 10 % of networks have been upgraded to support security gateways. But then only about 1 % of the total communication volume is protected. This problem actually applies to any added feature, not only to security features.

The second problem is, instead, specific to security. Assume now that 99 % of networks have been upgraded to support security gateways. Then about 98 % of total communication volume is protected. But certainly an active attacker masquerades as one of the remaining 1 % of networks.

7.3   Bluetooth Initialization

Our last example of continuity issues deals with future-proofing.

The original motivation of Bluetooth radio technology was to "replace wires". It makes sense to assume that initial introduction between two devices owned by the same person (e.g. a mobile phone and a headset) occurs in a relatively secure environment (e.g. at home or at the office). In such an environment, an active man-in-the-middle attack is not very probable.

Later many use cases were invented for Bluetooth and suddenly there was a new requirement: to establish secure connections with foreign devices as well, e.g. your mobile phone and a Bluetooth device in a ticket booth. It is not reasonable anymore to assume there are no active men-in-the-middle in this kind of environment.

Lesson to be learnt from this case is that it is always good to leave some safety margin. This is true not only in the quantitative dimension (e.g. key lengths) but also qualitatively, i.e. in the plurality of security functionalities.

8   Key Management

It is a well-known fact in practical cryptography that management of keys is usually a tricky issue.

8.1   Change of Keys: IMS First Hop Protection

In SIP registration there is a possibility for entity authentication (in 3GPP system). At the same time new keys are derived (both at terminal side and at network). For optimal use of resources, we would like to minimize the number of simultaneously valid keys per user. Clearly we have to allow two keys during the change process: the current key (set) and the new key (set).

Now we have a problem. Registration message from the user triggers the creation of new keys on the network side. What if an attacker sends a fake registration message while the change of keys is ongoing ? Should we ignore this message ? Both answers have unpleasant consequences: ignoring is bad if the message is not fake after all but, on the other hand, accepting the message necessarily increases the number of concurrent keys to be stored.

8.2   PKI Issues

Public Key Infrastructure is an area where many issues are well documented in the litterature, see e.g. [6] (see also [7] for fundamentals of public key technology). In this paper we just list a few interesting issues with PKI and certificates: How to deliver new root certificates into terminals that are already on the field ? How to introduce client certificates into legacy systems ? Can we utilize existing authentication and authorization infrastructure ? How to define certificates applicable to future services ?

8.3   Digital Rights Management Issues

This is another area that has been under extensive study during recent years. A few points of specific interest in DRM are listed in the following:

- User may act as an attacker against his own device.
- Use of global secrets implies unwanted "break one break all" phenomenon.
- There are difficult backward compatibility issues in DRM.
- With download applications it is OK to get the key (and rights) to the content afterwards but with streaming applications the key is necessarily needed in advance.

9   Constraints from Outside

We briefly mention a couple of problems on this area. End-to-end protection is problematic because of, at least, the following reasons: addressing and routing, middle proxies, lawful interception, key management.

Meeting lately introduced requirements is always difficult (e.g. Bluetooth initialization) but it is still good that security is taken into account already in early phases of system design, hence an iterative process should be used.

10   Conclusions

Some general conclusions can be crystallized:
- Cryptography is a major tool in making wireless systems secure;
- It is nontrivial to apply general-purpose security tools in a new context;
- We cannot ignore massive legacy systems;
- Incremental security enhancements lead to complex solutions;
- Reasonable safety margins can be justified also in deciding which security features to implement.

References

1. Asokan, N., Niemi, V. and Nyberg K., *Man-in-the-middle in tunneled authentication protocols*, Proceedings of 11th Cambridge Workshop on Security Protocols, Springer Lecture Notes in Computer Science (to appear).
2. Barkan, E., Biham, E. and Keller, N., *Instant Ciphertext-Only Cryptoanalysis of GSM Encrypted Communication*, Proceedings of CRYPTO 2003, Springer Lecture Notes in Computer Science (2003).
3. draft-haverinen-pppext-eap-sim-12, October 2003: "EAP SIM Authentication" (work in progress).
4. Blunk, L. and Vollbrecht, J., *PPP Extensible Authentication Protocol (EAP)*, Internet Engineering Task Force, Request for Comments (RFC) 2284.
5. Niemi, V. and Nyberg, K., *UMTS security*, John Wiley & sons (2003).
6. Gutmann, P., *PKI: It's Not Dead, Just Resting*, IEEE Computer 35(8), pp. 41-49, August 2002.
7. Salomaa, A., *Public-Key Cryptography*, Second Edition, Springer (1996).

On a Tomographic Equivalence Between (0,1)-Matrices[*]

Maurice Nivat

Université Denis Diderot–Case 7014, 2, place Jussieu, F-75251 Paris Cedex 05
mnivat@wanadoo.fr

Abstract. The tomographic problems studied here are associated to reconstructing a matrix when only some local information is given. We investigate a problem of discrete tomography via R-null matrices and prove a result similar to Ryser's Theorem.

1 Introduction

A very important basic fact occurring often in the present paper is as follows: every tiling of the plane by translations of a given $m \times n$ rectangle is invariant by one translation. This invariant translation is either the horizontal translation of length m or the vertical translation of length n (or both in the particular case of a regular tiling).

Moreover, the following is true: assume that one point of the tiling rectangle is marked in some way and each tile in the tiling contains a marked point in the same position. If we then look at the tiled plane through a rectangular $m \times n$ window, exactly one marked point is seen regardless of the position of the window, see Figure 1.

This last fact is not characteristic of rectangles, but is valid also for all pieces P with which one can tile the plane by translation as shown in Figure 1. In fact, a theorem can be stated as:

Theorem 1. *Let $\mathcal{U}$ be a mapping from $\mathbb{Z}^2$ to $\{0,1\}$ and P be a mapping from a finite subset F of $\mathbb{Z}^2$ to $\{0,1\}$ such that $card\{f \in F \mid P(f) = 1\} = 1$. Then the two following assertions are equivalent:*

(1) $\forall z \in \mathbb{Z}^2$, $card\{f \in F \mid \mathcal{U}(z+f) = 1\} = 1$
(2) $\mathbb{Z}^2 = \mathcal{U}^{-1}(1) \oplus F$,

where the symbol $\oplus$ denotes the unambiguous Minkowski sum: $C = A \oplus B$ if and only if:

$$\begin{cases} \forall c \in C \; \exists a \in A, b \in B, \; c = a + b, \\ \forall a_1, a_2 \in A; \; b_1, b_2 \in B; \; a_1 + b_1 = a_2 + b_2 \Rightarrow a_1 = a_2 \text{ and } b_1 = b_2. \end{cases}$$

---

[*] Dedicated to Arto Salomaa for his 70th birthday.

J. Karhumäki et al. (Eds.): Theory Is Forever (Salomaa Festschrift), LNCS 3113, pp. 216–234, 2004.

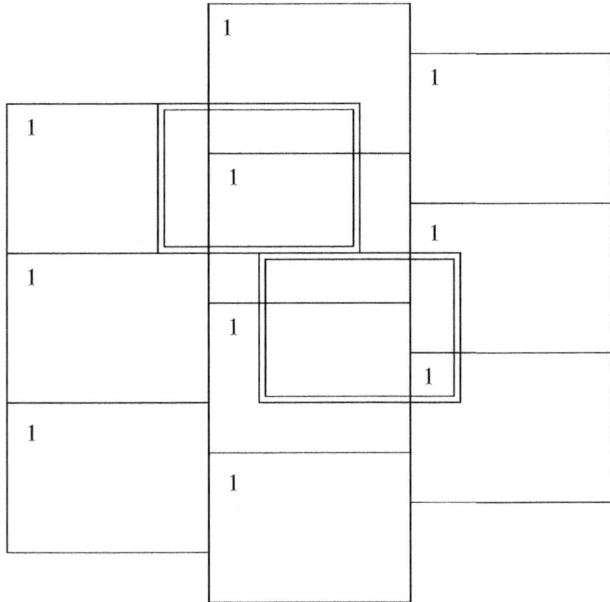

Fig. 1. Part of a tiling which is invariant by a vertical translation showing two positions of the window in which appears exactly one 1

The property (1) says that $\mathcal{U}$ contains exactly one 1 in each position of the window F and property (2) says that $\mathbb{Z}^2$ is tiled by translation of F, even more precisely that, if we surround each 1 in $\mathcal{U}$ by a copy of F such that the symbol 1 is always in the same position in F, we obtain a tiling of $\mathbb{Z}^2$. The notions above have been generalized in [5] to obtain the notion of a homogeneous bidimensional sequence.

Definition 1. *Mapping* $\mathcal{U} : \mathbb{Z}^2 \to \{0,1\}$ *is homogeneous of degree k with respect to a finite window F if and only if*

$$\forall z \in \mathbb{Z}^2 \quad card\{f \in F \mid \mathcal{U}(z+f) = 1\} = k.$$

In [5] we proved a rather surprising result.

Theorem 2. *Mapping* $\mathcal{U} : \mathbb{Z}^2 \to \{0,1\}$ *is homogeneous of degree k with respect to a rectangle R if and only of there exist k disjoint homogeneous sequences of degree 1 (with respect to the same R) such that:*

$$\mathcal{U} = \mathcal{U}_1 + \mathcal{U}_2 + \ldots + \mathcal{U}_k$$

This last result can be nicely rephrased:

If $\mathcal{U} : \mathbb{Z}^2 \to \{0,1\}$ is homogeneous of degree k with respect to F then one can color the 1's with k colors in such a way that in each position of the window there appears one and only one 1 of each color.

Example 1. In Figure 2, the three sequences corresponding to a, c and d are invariant by the translation $(4, 0)$ and the fourth one corresponding to b is invariant by the translation $(0, 3)$.

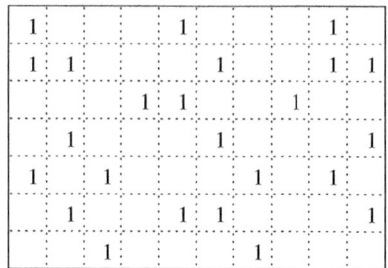 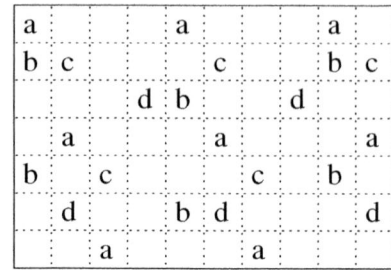

Fig. 2.

We suspect that Theorem 2 is valid for all *exact* windows F. By an exact window we mean a window F such that one can tile the plane by translation of F. Anyway there exists a sequence $\mathcal{U}$ which is homogeneous of degree 1 with respect to F if and only if F is exact (by Theorem 1) and thus Theorem 2 can hold only for exact windows. For the time being we are unable to prove Theorem 2 in this more general case.

2   A Decomposition Theorem for Homogeneous Matrices with Integer Coefficients

We shall deal with matrices rather than sequences. A matrix M of size $p \times q$ is a mapping from

$$\{0, 1, \ldots, p - 1\} \times \{0, 1, \ldots, q - 1\}$$

into a set of coefficients which will be either $\{0, 1\}$ or $\{-1, 0, 1\}$ or $\mathbb{N}$ or $\mathbb{Z}$. We use the notation $[p]$ to denote the set $[p] = \{0, \ldots, p - 1\}$. The definition of homogeneity is now slightly changed.

Definition 2. *Let R be the rectangle $[m] \times [n]$. The matrix $M : [p] \times [q] \to \mathbb{Z}$ is homogeneous of degree k with respect to R if and only if:*

$$\forall (x, y) \in [p - m + 1] \times [q - n + 1]$$

$$\sum \{M(x + i, y + j) \mid (i, j) \in [m] \times [n]\} = k$$

Remark that if the set of coefficients is $\{0, 1\}$ this definition coincides with the previous definition 1.

The matrix $R(M)$ of size $(p-m+1)\times(q-n+1)$ with coefficients $R(M)(x,y) = \sum\{(x+i,y+j) \mid (i,j) \in [m] \times [n]\}$ is called the *R-projection* of a matrix M with coefficients in $\mathbb{Z}$.

A matrix is homogeneous with respect to R if and only if its R-projection is constant. We call an matrix *R-null* if and only if its R-projection is the matrix 0, whose coefficients are all 0.

Theorem 3. *A matrix M with coefficients in $\mathbb{N}$, the set of non negative integers, is homogeneous of degree k with respect to R if and only if it is the sum of k matrices M_1, M_2, ..., M_k with coefficients $\{0,1\}$ which are homogeneous of degree 1.*

The proof is very similar to the proof of theorem 2 given in [5] but slightly more difficult.

Let M and M' be two matrices of the same size $p \times q$ with coefficients in $\mathbb{N}$. We say that M' is smaller than M if and only if for all $(i,j) \in [p] \times [q]$:

$$M'(i,j) \le M(i,j).$$

In order to prove theorem 3 we show that if M is homogeneous of degree k with respect to R, there exists a matrix M' with coefficients in $\{0,1\}$ which is homogeneous of degree 1 with respect to R and smaller than M. Then we can subtract M' from M to obtain $M - M'$ whose coefficients are in $\mathbb{N}$, and obviously $M - M'$ is homogeneous of degree $k - 1$.

Clearly we can repeat this process and eventually write M as a sum of $(0,1)$-matrices which are homogeneous of degree 1.

Now we need a crucial lemma.

Lemma 1. *If M with coefficients in $\mathbb{Z}$ is homogeneous with respect to R, then for all $(x,y) \in [p] \times [q]$ satisfying $(x+m,y+n) \in [p] \times [q]$ one has*

$$M(x,y) + M(x+m,y+n) = M(x+m,y) + M(x,y+n).$$

Proof. Let $s = \sum\{M(x+i,y) \mid 1 \le i \le m-1\}$ and $s' = \sum\{M(x+i,y+n) \mid 1 \le i \le m-1\}$. We clearly have

$$M(x,y) + s = M(x,y+n) + s'$$

and

$$s + M(x+m,y) = s' + M(x+m,y+n),$$

which imply the equality

$$s - s' = M(x,y+n) - M(x,y) = M(x+m,y+n) - M(x+m,y).$$

$\square$

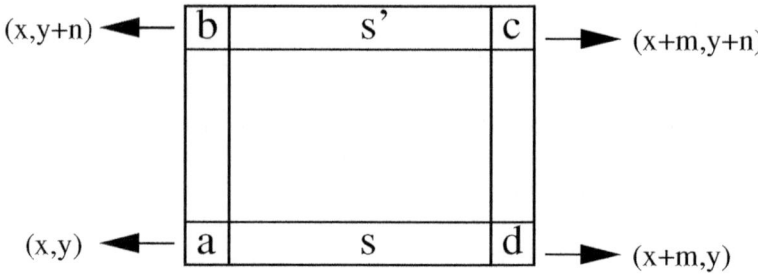

Fig. 3.

Figure 3 helps in visualizing the property of Lemma 1.

We just expressed the fact that the sum of the coefficients in the rectangle whose left inferior corner is (x, y) is equal to the sum of the coefficients in the rectangle whose left superior corner is $(x, y + n)$.

In a similar way, the equality of the sums of the coefficients in the 2 rectangles whose right inferior corner is $(x + m, y)$ and the right superior corner is $(x + m, y + n)$.

Lemma 1 can be easily extended to get:

Lemma 2. *For all $(x, y) \in [p] \times [q]$ and all $\alpha, \beta \in \mathbb{Z}$ the following holds: if M satisfies the conditions of Lemma 1 and $(x + \alpha m, y + \beta n)$ belongs to $[p] \times [q]$, one has*

$$M(x, y) + M(x + \alpha m, y + \beta n) = M(x, y + \beta n) + M(x + \alpha m, y)$$

Proof. The proof is immediate by symmetry and induction.   □

Proof (The proof of Theorem 3). Assume first that M is invariant by the translation $(m, 0)$.

Due to the invariance, for all $(x, y) \in [p] \times [q]$, $\beta \in \mathbb{Z}$ we have that if $x + \beta m \in [p]$, then $M(x + \alpha m, y) = M(x, y)$. Then one can find easily a $(0, 1)$-matrix which is homogeneous of degree 1 with respect to R and smaller than M in the following way:

Take any non-regular coefficients $M(x, y)$; $(x, y) \in [m] \times [n]$ and set $x_0 = x$. Now for all β such that $y + \beta n \in [q]$ then exists a strictly positive coefficient $M(x\beta, y + \beta n)$ with $x\beta \in [m]$. This is obvious since

$$\sum \{M(x, y + \beta n) \mid x \in [m]\} = \sum \{M(x, y) \mid x \in [m]\}$$

and $\sum \{M(x, y) \mid x \in [m]\} \geq M(x, y) > 0$.

Now for each α, β such that $(x\beta + \alpha m, y + \beta n) \in [p] \times [q]$, all the coefficients $M(x\beta + \alpha m, y + \beta n)$ are strictly positive and the $(0, 1)$-matrix M' given by

$$M'(u, v) = \begin{cases} 1, & \text{if } (u, v) = (x\beta + \alpha m, y + \beta n), \\ 0, & \text{otherwise.} \end{cases}$$

is smaller than M and can be subtracted from M. The matrix $M - M'$ is homogeneous of degree $k - 1$ and is also invariant by the translation $(m, 0)$. As mentioned in the beginning of the proof, it follows that a homogeneous matrix having coefficients in $\mathbb{N}$ which is invariant by $(m, 0)$, can be expressed as a sum of $(0, 1)$-matrices which are homogeneous of degree 1 and invariant by $(m, 0)$.

Assume now M is not invariant by the translation $(m, 0)$. Then there exists (x, y) such that $M(x, y)$ and $M(x + m, y)$ are different. We can assume that $M(x, y) > M(x + m, y)$ (the argument in case $M(x, y) < M(x + m, y)$ is exactly the same).

We can show that for all β such that $y + \beta n \in [q]$, $M(x, y + \beta n)$ is strictly positive. This is an immediate consequence of Lemma 2, since

$$M(x, y + \beta n) + M(x + m, y) = M(x, y) + M(x + m, y + \beta n)$$

implies that

$$M(x, y + \beta n) - M(x + m, y + \beta n) = M(x, y) + M(x + m, y)$$

is strictly positive.

1	2	1	1	1		1	2		1	1
			1							
			1		2			2		1
	1	3			2		1	2		
				1						
			1		2			2		1
	2	2		1	1		2	1		1
				1						

Fig. 4.

Let us set $y_0 = y$. Consider any column $x + \alpha m$, $x + \alpha m \in [p]$ and compare the sum

$$\sum \{M(x, y - h + j) \mid j \in [n]\}$$

of n consecutive coefficients in the column x containing $M(x, y)$ for some $h < n$ with the sum

$$\sum \{M(x + \alpha m, y - h + j) \mid j \in [n]\}$$

The sums are equal. Then two cases are possible:

- $\forall j \in [n]$ $M(x + \alpha m, y - h + j)) = M(x, y - h + j)$. This implies, by Lemma 2, that $M(x + \alpha m, j) = M(x, j)$ for all $j \in [q]$.

2	1	1		2		1	
		1					
	1		2		2	1	
1	2		1	1	1		
		(1)					
	1		2		2	1	
1	2		1	1	1		
		1					

Fig. 5.

1	1	1		1			
	1		2		2		
	2		1		1		
	1		2		2		
	2		1		1		

Fig. 6.

We can then take $y_\alpha = y$ and be sure that

$$M(x + \alpha m, y_\alpha + \beta n)$$

is strictly positive for all β such that $y_\alpha + \beta n \in [q]$.
- There exists j such that $M(x, \alpha m, y - h + j) > M(x, y - h + j)$.
  Then if we set $y_\alpha = y - h + j$ we are sure by an argument used above that $M(x + \alpha m, y_\alpha + \beta n)$ is strictly positive for all β such that $y_\alpha + \beta n \in [q]$.

Now the matrix M' defined as

$$M'(u, v) = \begin{cases} 1, & \text{if } (u, v) = (x + \alpha m, y + \beta n), \\ 0, & \text{otherwise.} \end{cases}$$

is a $(0, 1)$-matrix which is homogeneous of degree 1, invariant by the translation $(0, n)$ and smaller than M.

This completes the proof.   □

Example 2. Let $m = n = 3$. The matrix in Figure 4 homogeneous of degree 5. The circled element can only belong to a homogeneous matrix of degree 1 invariant by the horizontal translation. We easily find one (we have the choice between 2) and subtract it from M to obtain matrix in Figure 5. The matrix in Figure 5 is homogeneous of degree 4.

The circled element can only belong to a "vertical" homogeneous submatrix. We find one and delete it to obtain matrix of Figure 6. It is homogeneous of degree 3. This last matrix can be decomposed in only one sum of 2 "horizontal" and one "vertical" homogeneous of degree 1.

We can now easily prove a theorem of decomposition for homogeneous matrices with coefficients in $\mathbb{Z}$ and R-null matrices.

Theorem 4. *A matrix with coefficients in $\mathbb{Z}$ is homogeneous of degree k with respect to R if and only if it is a difference of two sums of homogeneous $(0, 1)$-matrices which are homogeneous of degree 1 with respect to R.*

The number of elements in these two sums can be bounded by:

$$k + \sum \{M(x,y) \mid M(x,y) < 0, (x,y) \in [p] \times [q]\}$$

and

$$k + \sum \{M(x,y) \mid M(x,y) > 0, (x,y) \in [p] \times [q]\}$$

Proof. Let M be a homogeneous matrix with coefficients in $\mathbb{Z}$. Consider a negative coefficient $M(x,y) = -a$, $a > 0$.

Take any $(0,1)$-matrix M' which is homogeneous of degree 1 and satisfies $M'(x,y) = 1$. $M + aM'$ is a matrix with coefficients in $\mathbb{Z}$ which is homogeneous of degree $k + a$ and satisfies:

$$(M + aM')(i,j) \geq M(i,j) \text{ for all } (i,j) \in [p] \times [q].$$

Moreover, $M + aM'$ has at least one negative coefficient less than M.

Repeating this process until all the negative coefficients disappear we can write M as the difference $M_1 - M_2$ where M_1 and M_2 have non-negative coefficients, are homogeneous, M_1 of degree $k + \sum \{-M(x,y) \mid M(x,y) < 0 \quad (x,y) \in [p] \times [q]\}$.

It may be more economical to have all the positive coefficient disappear if the sum of the positive coefficients is less than the absolute value of the sum of the negative coefficients.

When M has been written as the difference $M_1 - M_2$ we obtain the theorem by decomposing M_1 and M_2 into sums of homogeneous $(0, 1)$-matrices of degree 1. □

3   *R*-null Matrices

Clearly if M and M' have the same R projection then $M-M'$ is R-null.

Studying R-null matrices is a natural way to study the equivalence between matrices defined by the equality of their R-projection.

We can see the problem of constructing a $(0,1)$-matrix with a given R-projection as a problem of discrete tomography: we are given a family of local pieces of information on a set of pixels distributed in a rectangle and the problem is to retrieve this information.

The first problem of discrete tomography appearing in the literature is the problem of constructing a $(0,1)$-matrix with given row sums and column sums. Solutions to this problem were given by Ryser and, independently by Gale.

Let $r_0, \ldots, r_{p-1}$ and $c_0, \ldots, c_{q-1}$ be two sequences of non negative integers such that

$$\sum \{r_i \mid i \in [p]\} = \sum \{c_j \mid j \in [q]\}.$$

Can one find a $(0,1)$-matrix of size $p \times q$ such that

$$\forall i \in [p] \text{ we have } \sum \{M(i,j) \mid j \in [q]\} = r_i$$

and

$$\forall j \in [q] \text{ we have } \sum \{M(i,j) \mid i \in [p]\} = c_j.$$

The r_i's are called row sums (in more recent literature the vector $\langle r_0, \ldots, r_{p-1} \rangle$ is called the horizontal projection) and the c_j's are called columns sums (the vector $\langle c_0, \ldots, c_{q-1} \rangle$ is the vertical projection). What interests us here is the study of the equivalence defined by:

M is equivalent to M' if and only if M and M' have the same horizontal and vertical projection (Ryser).

An elementary Ryser transformation amounts to exchange in a matrix two 1's in position (x, y), $(x+h, y+l)$ with two 0's in position $(x+h,y)$ and $(x,y+l)$, see Figure 7.

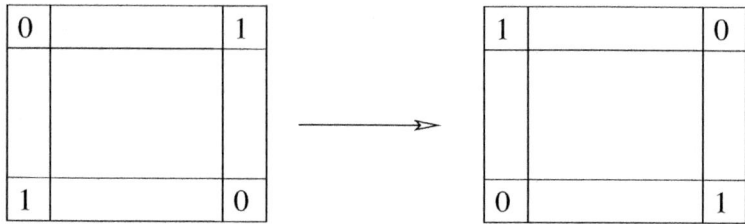

Fig. 7.

Clearly such a transformation leaves the two projections invariant. The nice result of Ryser is that if M and M' are equivalent then one can transform M into M' by a sequence of elementary transformations.

The matrix in Figure 8 is obtained by performing two sequences Ryser elementary transformations.

Introducing matrices with coefficients in $\{-1, 0, 1\}$ we can state Ryser's result as follows.

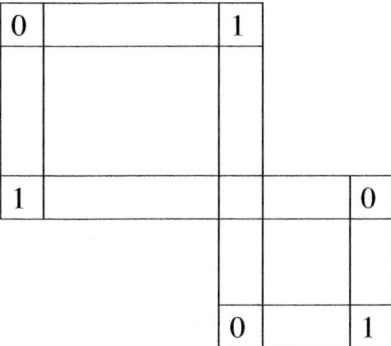

Fig. 8.

Theorem 5 (Ryser's theorem). *Every matrix with coefficients in $\{-1, 0, 1\}$ whose horizontal and vertical projections are the constant vector equal to 0 is a sum of matrices of the form:*

$M'(x, y) = 0$ *but for* $(x, y) \in \{(x_0, y_0), (x_1, y_1), (x_0, y_1), (x_1, y_0)\}$ *for some x_0, x_1, y_0, y_1 such that $x_0 \neq x_1$ and $y_0 \neq y_1$ for which one has*

$$M'(x_0, y_0) = M'(x_1, y_1) = 1$$

and

$$M'(x_0, y_1) = M'(x_1, y_0) = -1.$$

Here we can prove a very similar theorem. Let us say that a row $\{M(x, y) \mid y \in [q]\}$ of matrix M is m-null if and only if the sum of m consecutive entries of that row is always 0. We define n-null columns in the same way.

Note that a m-null row is invariant by the translation $(m, 0)$. Obviously $a_1 + a_2 + \ldots + a_m = a_2 + a_3 + \ldots + a_m + a_{m+1}$ imply $a_1 = a_{m+1}$.

Adding a m-null row or a n-null column to a given matrix M obviously does not change the R-projection. Moreover, if M and M' have the same R-projections then one can obtain M' from M by adding to M a number of m-null rows and n-nulls columns.

We can now state the following theorem.

Theorem 6. *The set of all matrices whose all entries are 0's except for a m-null row or a n-null column, is a generating subset of the vector space of R-null matrices.*

Remark 1. Note that the set of m-null rows and n-null columns is not a basis of the vector space for they are not linearly independent as proved by the example in Figure 9.

Proof. We first give a very easy one.

	1	−1		1	−1		1
	−1	1		−1	1		−1
	1	−1		1	−1		1
	−1	1		−1	1		−1
	1	−1		1	−1		1

Fig. 9.

In the decomposition of a R-null matrix in a sum of homogeneous $(0,1)$ or $(0,-1)$ -matrices of degree 1 or -1 it is clear that the number of $(0,1)$-matrices will be the same as the number of $(-1,0)$-matrices. Thus M is a sum

$$M = M_1 + \ldots + M_k$$

of matrices of the form $H - H'$, where both H and H' are $(0,1)$-matrices homogeneous of degree 1.

We then prove that each matrix $H - H'$ is a sum of m-null rows and n-null columns. Consider first the case H and H' are both invariant by the translation $(m, 0)$. The figures illustrate the proof.

	1			1			1			1		
		−1			−1			−1			−1	
1			1			1			1			1
	−1			−1			−1			−1		
		1			1			1			1	
−1			−1			−1			−1			−1
	1			1			1			1		

Fig. 10.

We can make all the (nonzero) rows start with a 1 or a -1 in the first column by adding m-null rows. To the matrix of Figure 10 we add the set of m-null rows shown in Figure 11 to obtain the sum in Figure 12 which is obviously composed of n-null columns.

If the rows containing the 1's and -1's are the same, then obviously all the rows are m-null.

Now consider $H - H'$ where H is invariant by $(m, 0)$ and H' is invariant by $(0, n)$, see Figure 13. Note that the two 0's which appear come from a 1 and a -1 occurring in the same position.

1	-1		1	-1		1	-1		1	-1		
-1		1	-1		1	-1		1	-1		1	-1
-1	1			-1	1		-1	1		-1	1	
1			-1	1		-1	1		-1	1		-1	1
1	-1		1	-1		1	-1		1	-1		1

Fig. 11.

By adding m-null rows we can have all the rows containing 1's start in the first column (Figure 14). As a result, we get Figure 15.

Now it suffices to have all the columns containing -1's contain -1's in the rows containing 1's, we can achieve that by adding n-null columns to obtain a matrix which has only m-nulls rows. $\qquad\square$

Since addition is commutative and a matrix whose rows (resp. columns) are all m-null (resp. n-null) is invariant by the translation $(m, 0)$ (resp $(0, n)$) we can state:

Corollary 1. *Every R-null matrix M is the sum of M_1 and M_2 where M_1 is $(m \times 1)$-null and M_2 is $(1 \times n)$-null.*

4   An Alternate Proof of Theorem 6

Consider an $m \times n$-null matrix M of size $p \times q$. We can add to M a matrix M_1 with m-null rows in order that $M + M_1$ has only 0's in its leftmost column.

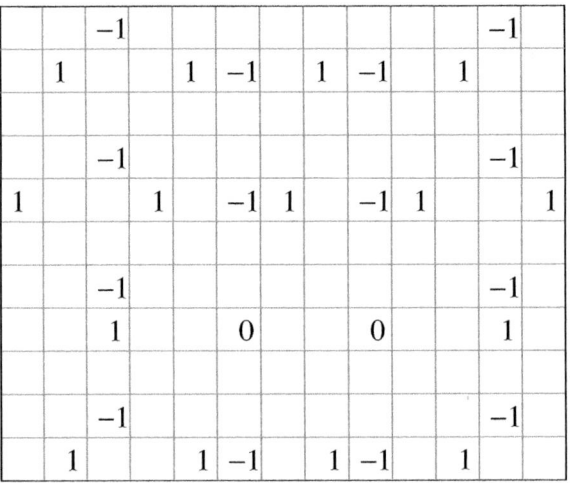

Fig. 12.

Fig. 13.

If $M(i,0) = a_i$ is different from 0 we add one row

$$a_i\, 0 \,\ldots\, -\, a_i\, 0 \,\ldots\, a_i\, 0 \,\ldots\, -\, a_i\, 0 \,\ldots$$

where a_i appears in the positions βm, $\beta \in \mathbb{N}$, and $-a_i$ in the positions $\beta m + h$ for some h between 1 and $m - 1$. Clearly such a row is m-null.

$M_1(i, *)$ is a row of this type for all i such that $M(i,0) \neq 0$, and is a line full of 0's if $M(i,0) = 0$.

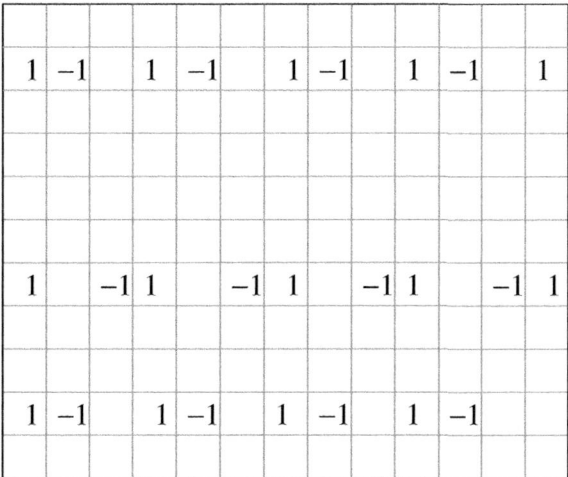

Fig. 14.

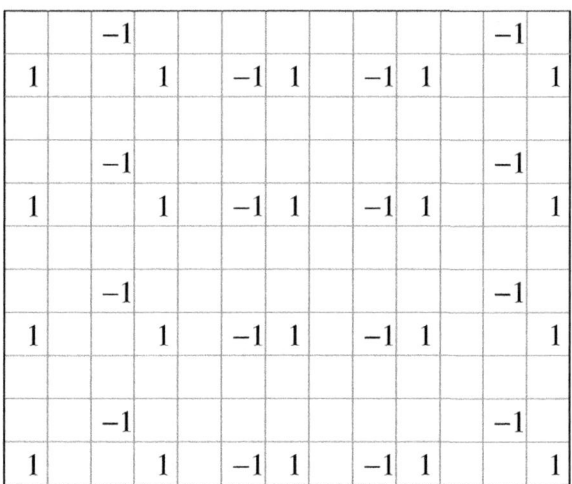

Fig. 15.

The first column of $M + M_1$ is full of 0's, and this implies that all the columns of rank βm, $\beta \in \mathbb{N}$, of $M + M_1$ are m-null columns. Since $(M + M_1)(i, 0) = 0$, for all i,

$$\sum \{(M + M_1)(i + h, j + k) \mid h \in [n], 1 \leq k \leq m - 1\} = 0$$

and, since $\sum \{(M + M_1)(i + h, j + 1 + k) \mid h \in [n], k \in [m]\} = 0$, we have for all i

$$\sum \{(M + M_1)(i + h, m) \mid h \in [n]\} = 0.$$

The sum of n consecutive coefficients in the column of rank m is equal to 0 and obviously this is also true for all the columns of rank βm, $\beta \in \mathbb{N}$.

Let $\overline{M}_1$ be the matrix whose columns are full of 0's but for the column of rank βm which is the opposite of the n-column of rank βm of $M + M_1$. Then in $M + M_1 + \overline{M}_1$ all the columns of rank βm are full of 0's.

We can repeat the process and find M_2 whose rows are m-null such that in $M + M_1 + \overline{M}_1 + M_2$ the column 1 is full of 0's, and we can keep the columns of rank βm full of 0's.

Then the columns of rank $\beta m + 1$, $\beta \geq 1$, are, by the same argument as above n-null columns. Whence we can find $\overline{M}_2$ whose columns are n-null and such that in $M + M_1 + \overline{M}_1 + M_2 + \overline{M}_2$ the columns of rank βm and $\beta m + 1$ are full of 0's.

Eventually we can write M as a sum of a matrix with m-null rows and a matrix with n-null columns.

Example 3. Let M be 4×3-null as in Figure 16. One can see that the columns 4 and 8 of $M + M_1$ are 3-null. In Figure 17, the column 5 of the last matrix is 3-null. In the last matrix of Figure 18, the two remaining columns are 3-null.

M:

-2				2	1			-3
1						-1		4
2	2	-3		-1	1	-2		
1	-2		-1	5	-1		-1	
		4	-3	-1		3	-3	3
4	-1	-2		1	-2	-1		2
-2				2	1			-3

M_i:

2	-2			2	-2			2
-1		1		-1		1		-1
-2		2		-2		2		-2
-1			1	-1			1	-1
-4		4		-4		4		-4
2		-2		2		-2		2

$M + M_i$:

	-2			4	-1			-1
	1		-1					3
	2	-1		-3	1			-2
	-2			4	-1			-1
		4	-3	-1		3	-3	3
	-1	2		-3	-2	3		-2
		-2		4	1	-2		-1

Fig. 16.

Eventually we have

$$M = (-M_1 - M_2 - M_3) + (-\overline{M}_1 - \overline{M}_2 - \overline{M}_3)$$

where the first matrix has only 4-null rows and the second has only 3-null columns.

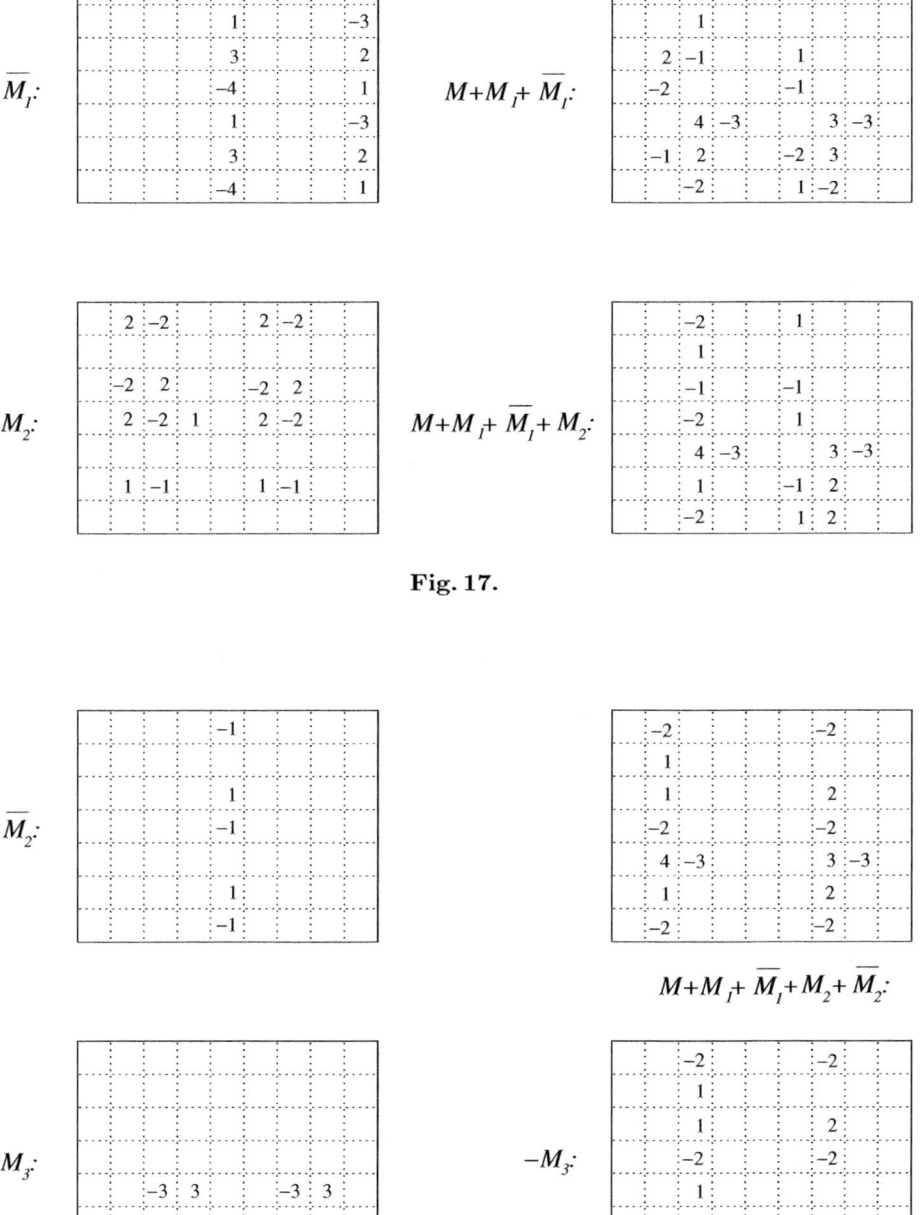

Fig. 17.

Fig. 18.

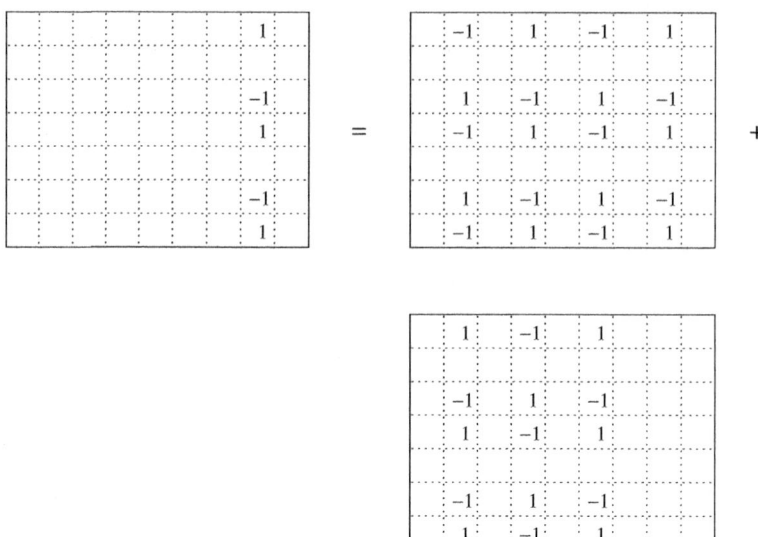

Fig. 19.

Remark 2. We can describe a basis of the vector space of $p \times q$-null matrices.

We take all the matrices that have only one row which is not full of 0's and this row is of the form

$$(1\,0\,\ldots\,1\,0\,\ldots)(1\,0\,\ldots\,1\,0\ldots)\ldots$$

with 1's in position βm and $\beta m + h$ for some h between 1 and $m - 1$.

There are $(m - 1)q$ such matrices, $m - 1$ for each of the q rows. We take all the matrices which have only one column which is not full of 0's and this column is of the form

$$(\ldots)^T(0\,1\,\ldots\,0\,1)^T(\ldots\,0\,1\,\ldots\,0\,1)^T$$

with 1's in positions αn and $\alpha n + k$ for some k between 1 and $n - 1$.

There are $(n - 1)p$ such matrices, $n - 1$ for each of the p columns. The set of these matrices certainly generates the whole vector space.

But we can express the last $p - 1$ columns as linear combination of the other columns and of the rows, see Figure 19, and we have a decomposition as in Figure 20.

The dimension of the vector space is then

$$(n - 1)p + (m - 1)q - (p - 1)(q - 1)$$

and this is compatible with the fact that if we know the elements in the $(p - 1)$ first columns and the $(q - 1)$ first rows of a $p \times q$-null matrix, then we know the matrix.

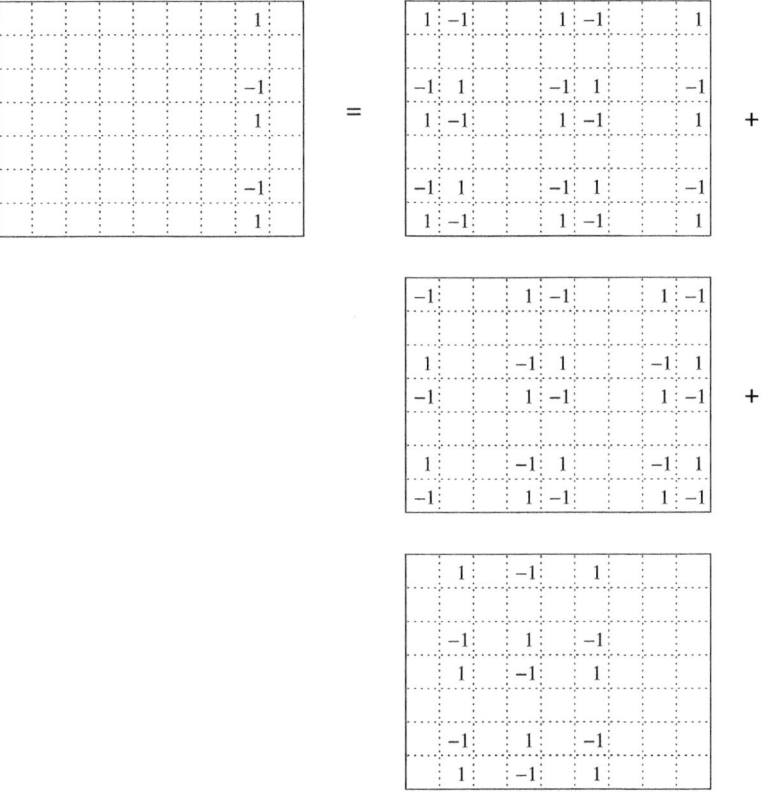

Fig. 20.

Thanks

The author did part of this work in the Centro di Modelisacion Matematica of the University of Chile. He wishes to thank Eric Goles, the president of the CONICYT and Rafael Correa, the director of the CMM who made his stay there possible. The author had very fruitful discussions in the CMM with Eric Goles and Ivan Rapaport. The work was completed in the University of Siena and the author thanks Simone Rinaldi and Andrea Frosini who have made his satay possible Indeed the origin of the paper lies in discussions with the late Alberto del Lungo, professor in the university of Siena who died suddenly on June 1st 2003.

References

1. H.Ryser , Combinatorial properties of matrices of zeros and ones, Canad. J. Math. 9, 371-377 (1957).

2. D. Gale, A theorem on flows in networks, *Pacif. J. Math.*, 7 (1957) 1073-1082.
3. G. T. Herman and A. Kuba (eds.), *Discrete Tomography: Foundations, Algorithms and Applications*, Birkhauser Boston, Cambridge, MA (1999).
4. R. Tijdeman and L. Hajdu, An algorithm for discrete tomography, *Linear Algebra and Its Applications*, 339 (2001) 147-169.
5. M. Nivat, Sous-ensembles homogènes de $\mathbb{Z}^2$ et pavages du plan, *C. R. Acad. Sci. Paris*, Ser. I 335 (2002) 83-86.

P Systems with Tables of Rules

Gheorghe Păun[1,2], Mario J. Pérez-Jiménez[2], and Agustín Riscos-Núñez[2]

[1] Institute of Mathematics of the Romanian Academy
PO Box 1-764, 014700 Bucureşti, Romania
[2] Research Group on Natural Computing
Department of Computer Science and Artificial Intelligence
University of Sevilla
Avda. Reina Mercedes s/n, 41012 Sevilla, Spain
gpaun,marper,ariscosn@us.es

Abstract. In the last time, several efforts were made in order to remove the polarization of membranes from P systems with active membranes; the present paper is a contribution in this respect. In order to compensate the loss of power represented by avoiding polarizations, we introduce *tables* of rules: each membrane has associated several sets of rules, one of which is non-deterministically chosen in each computation step. Three universality results for tabled P systems are given, trying to use rules of as few as possible types. Then, we consider tables with *obligatory rules* – rules which must be applied at least once when the table is applied. Systems which use tables with at most one obligatory rule are proven to be able to solve SAT problem in linear time. Several open problems are also formulated.

1 Introduction

In membrane computing, the P systems with active membranes have a special place, because of the fact that they provide biologically inspired means to solve computationally hard problems: by using the possibility to divide membranes, one can create an exponential working space in a linear time, which can then be used in a parallel computation for solving, e.g., **NP**-complete problems in polynomial or even linear time. Details can be found in [7], [8], as well as in the comprehensive page from the web address http://psystems.disco.unimib.it.

One of the important ingredients of P systems with active membranes is the polarization of membranes: besides a label, each membrane also has an "electrical charge", one of + (positive), − (negative), 0 (neutral). These electrical charges correspond only remotely to biological facts; by sending ions outside, cells and cell compartments can get polarizations, but this is not a very common phenomenon. Starting from this observation and also as a mathematical challenge, in the last time several efforts were made to avoid using polarizations.

However, the question seems not to be a simple one, and the best result obtained so far was to reduce the number of "electrical charges" to two; this is achieved in [1], where both the universality and the possibility of solving SAT

J. Karhumäki et al. (Eds.): Theory Is Forever (Salomaa Festschrift), LNCS 3113, pp. 235–249, 2004.
© Springer-Verlag Berlin Heidelberg 2004

in linear time are proven for P systems with active membranes and only two polarizations. When completely removing the polarizations, similar results are obtained (see [2], [3]) only by compensating the loss of power (of "programming" possibilities) by using additional ingredients, such as the possibility of changing the labels of membranes, the division of non-elementary membranes, etc.

The present paper goes into the same direction of research: we get rid of polarizations and we "pay" this by structuring the sets of rules associated with each membrane by considering *tables* of rules, like in Lindenmayer systems. Specifically, several sets of rules are associated with each membrane, and in each step of a computation we non-deterministically choose one of these sets, and its rules are used in the maximally parallel manner. The use of tables can have a biological motivation, in the same way as the tables from L systems theory have a biological origin: the change of environmental conditions (for instance, of seasons) can select specific evolution rules for different times (different seasons).

The use of tables proves to be helpful in what concerns the computing power: we get universality for systems of a rather reduced forms, with only a few types of rules used, and without polarizations.

An important *problem* remains unsolved: can tables compensate polarizations also in what concerns the possibility to solve hard problems in polynomial time? A possible negative answer to this problem would be a very nice finding: in view of the result from [1], it would follow that passing from one polarization (all membranes neutral) to two polarizations makes possible the step from the complexity class **P** to **NP**.

If, however, we add a further ingredient – at the first sight not very powerful – to tabled P systems, namely designating in each table *obligatory* rules, which should be used at least once when applying the table, then we can solve SAT in linear time. The construction uses at most one obligatory rule in each table.

2 P Systems with Active Membranes

We assume the reader to be familiar with basic elements of membrane computing, e.g., from [7], but, for the sake of completeness, we recall here the definition of the class of P systems we work with, those with active membranes (and electrical charges).

Such a system is a construct

$$\Pi = (O, \mu, w_1, \ldots, w_m, R),$$

where:

1. $m \geq 1$ (the initial degree of the system);
2. O is the alphabet of *objects*;
3. μ is a *membrane structure*, consisting of m membranes, labeled in a one-to-one manner with elements of $H = \{1, 2, \ldots, m\}$;
4. $w_1, \ldots, w_m$ are strings over O, describing the *multisets of objects* placed in the m regions of μ;

5. R is a finite set of *developmental rules*, of the following forms:

(a) $[\, a \rightarrow v \,]_h^e$,

for $h \in H, e \in \{+, -, 0\}, a \in O, v \in O^*$

(object evolution rules, associated with membranes and depending on the label and the charge of the membranes, but not directly involving the membranes, in the sense that the membranes are neither taking part in the application of these rules nor are they modified by them);

(b) $a[\ ]_h^{e_1} \rightarrow [\, b \,]_h^{e_2}$,

for $h \in H, e_1, e_2 \in \{+, -, 0\}, a, b \in O$

(*in* communication rules; an object is introduced in the membrane, possibly modified during this process; also the polarization of the membrane can be modified, but not its label);

(c) $[\, a \,]_h^{e_1} \rightarrow [\ ]_h^{e_2} b$,

for $h \in H, e_1, e_2 \in \{+, -, 0\}, a, b \in O$

(*out* communication rules; an object is sent out of the membrane, possibly modified during this process; also the polarization of the membrane can be modified, but not its label);

(d) $[\, a \,]_h^e \rightarrow b$,

for $h \in H, e \in \{+, -, 0\}, a, b \in O$

(dissolving rules; in reaction with an object, a membrane can be dissolved, while the object specified in the rule can be modified);

(e) $[\, a \,]_h^{e_1} \rightarrow [\, b \,]_h^{e_2} [\, c \,]_h^{e_3}$,

for $h \in H, e_1, e_2, e_3 \in \{+, -, 0\}, a, b, c \in O$

(division rules for elementary membranes; in reaction with an object, the membrane is divided into two membranes with the same label, possibly of different polarizations; the object specified in the rule is replaced in the two new membranes by possibly new objects).

We have omitted the rules for dividing non-elementary membranes, usually identified as being "of type (f)".

It is worth noting that the rules of all types are non-cooperative (and that there are no further ingredients involved, such as a priority relation, for controlling the spplication of rules). In the customary definition of P systems with active membranes, the initial membranes of μ are not necessarily labeled in a one-to-one manner, but there is no loss of generality in the assumption that the labels are unique: we can relabel the membranes with the same label and then duplicate the necessary rules. Moreover, because in what follows we only consider that by membrane division we obtain membranes with the same label, the labels present in the system are always from the set $\{1, 2, \ldots, m\}$ present at the beginning (maybe some of them used several times, because of the division of membranes). Therefore, the set H of labels is specified by μ, it can be omitted when specifying the system.

The rules of type (a) are applied in the parallel way (all objects which can evolve by such a rule should do it), while the rules of types $(b), (c), (d), (e)$ are used sequentially, in the sense that one membrane can be used by at most one rule of these types at a time. In total, the rules are used in the non-deterministic maximally parallel manner: all objects and all membranes which can evolve, should evolve. Only halting computations give a result, and the result is the number of objects expelled into the environment during the computation; the set of numbers computed in this way by the various halting computations in Π is denoted by $N(\Pi)$.

By $NOP_{m,n,p}(pol_3, a, b, c, d, e)$ we denote the family of sets $N(\Pi)$ computed as sketched above by systems starting with at most m membranes, using membranes of at most n types, at most p membranes being simultaneously present, and using all types of rules; when rules of a certain type are not used the corresponding letter a, b, c, d, e will be missing. Also, when membrane division rules are not used, we will specify only the number of membranes in the initial configuration (hence, only m) as a subscript of NOP. The parameter pol_3 indicates the fact that one uses three polarizations.

Further details can be found in [7] – including the proof of the following result. (We denote by REG, CF, CS, RE the families of regular, context-free, context-sensitive, and of recursively enumerable languages. In general, for a family FL of languages, NFL denotes the family of length sets of languages in FL. Therefore, NRE is the family of Turing computable sets of natural numbers.)

Theorem 1. $NOP_3(pol_3, a, b, c) = NRE$.

The number of polarizations were decreased to two in [1]; with the previous notations, the result can be written as:

Theorem 2. $NOP_2(pol_2, a, c) = NRE$.

Note that the result from Theorem 1 was improved at the same time in the number of polarizations, the number of membranes, and the number of types of rules used.

In [3] and [2] rules of types $(a) - (e)$ without polarizations were considered. Because "no polarization" means "neutral polarization", we add the subscript 0 to the previous letters identifying the five types $(a_0) - (e_0)$ of rules.

The power of polarizationless P systems with active membranes is not precisely known, but it was shown in [2] that they are able to compute at least the Parikh images of languages generated by matrix grammars without appearance checking.

Because the notion of a matrix grammar will be also used below, we introduce it here in its general form.

A *matrix grammar* (with appearance checking) is a construct $G = (N, T, S, M, F)$, where N and T are disjoint alphabets, $S \in N$, M is a finite set of sequences of the form $(A_1 \to x_1, \ldots, A_n \to x_n)$, $n \geq 1$, of context-free rules over $N \cup T$ (with $A_i \in N, x_i \in (N \cup T)^*$, in all cases), and F is a set of occurrences of rules in M (N is the nonterminal alphabet, T is the terminal alphabet, S is the axiom, while the elements of M are called matrices).

For $w, z \in (N \cup T)^*$ we write $w \implies z$ if there is a matrix $(A_1 \to x_1, \ldots, A_n \to x_n)$ in M and the strings $w_i \in (N \cup T)^*, 1 \leq i \leq n+1$, such that

$w = w_1, z = w_{n+1}$, and, for all $1 \leq i \leq n$, either $w_i = w_i' A_i w_i''$, $w_{i+1} = w_i' x_i w_i''$, for some $w_i', w_i'' \in (N \cup T)^*$, or $w_i = w_{i+1}$, A_i does not appear in w_i, and the rule $A_i \rightarrow x_i$ appears in F. (The rules of a matrix are applied in order, possibly skipping the rules in F if they cannot be applied – therefore we say that these rules are applied in the *appearance checking* mode.)

The language generated by G is defined by $L(G) = \{w \in T^* \mid S \Longrightarrow^* w\}$. The family of languages of this form is denoted by MAT_{ac}. If the set F is empty, then the grammar is said to be without appearance checking.

It is known that $CF \subset MAT \subset MAT_{ac} = RE$ and $NREG = NCF = NMAT \subset NCS$ (for instance, the one-letter languages in MAT are known to be regular, [6]).

A matrix grammar $G = (N, T, S, M, F)$ is said to be in the *binary normal form* if $N = N_1 \cup N_2 \cup \{S, \#\}$, with these three sets mutually disjoint, and the matrices in M are in one of the following forms:

1. $(S \rightarrow XA)$, with $X \in N_1, A \in N_2$,
2. $(X \rightarrow Y, A \rightarrow x)$, with $X, Y \in N_1, A \in N_2, x \in (N_2 \cup T)^*, |x| \leq 2$,
3. $(X \rightarrow Y, A \rightarrow \#)$, with $X, Y \in N_1, A \in N_2$,
4. $(X \rightarrow \lambda, A \rightarrow x)$, with $X \in N_1, A \in N_2$, and $x \in T^*, |x| \leq 2$.

Moreover, there is only one matrix of type 1 (that is why one uses to write it in the form $(S \rightarrow X_0 A_0)$, in order to fix the symbols X, A present in it), and F consists exactly of all rules $A \rightarrow \#$ appearing in matrices of type 3; $\#$ is a trap-symbol, because once introduced, it is never removed. A matrix of type 4 is used only once, in the last step of a derivation.

For each matrix grammar there is an equivalent matrix grammar in the binary normal form. Details can be found in [4] and in [11].

3   Tables of Rules

In the "standard" P systems with active membranes there is specified only one set of rules; because the membranes are present in the rules, we precisely know where each rule is to be applied. A possible generalization is to consider several sets of rules – for uniformity with L systems, we call them *tables* – such that in each step of a computation a table is used, non-deterministically chosen (the rules of the selected table are applied in the maximally parallel manner, as mentioned in the previous section).

This case corresponds to having *global tables*; a more relaxed variant is to consider *local tables*, sets of rules associated with each membrane.

Specifically, for each membrane i we can consider sets $R_{i,1}, \ldots, R_{i,k_i}$ of rules, for some $k_i \geq 1$, all of the rules from sets $R_{i,j}, 1 \leq j \leq k_i$, involving membrane i. In a step of a computation, we apply the rules from one of the tables associated with each membrane, as usual, in the maximally parallel non-deterministic manner with respect to the chosen table.

If we are allowed to "evolve" a region by means of a table for which no rule is actually applied, then the local tables can be combined in global tables, hence in this case the local version is weaker than the global one. However, there is no difference from the computational point of view (at least in the cases investigated in the next section): systems with local tables (and restricted types of rules) are equivalent with Turing machines; moreover, the proofs are based on systems with one or two membranes, with the "main work" of two-membranes systems done in the inner membrane, hence choosing tables which change nothing in one of the regions do not change the generated set of numbers.

In what follows we will consider only local tables, that is why we choose a more restricted – also, more natural – definition of a transition step: if there are tables by which a region can effectively evolve (at least a rule of these tables can be effectively applied), then one of these tables must be chosen. Otherwise stated, we cannot choose a table with no applicable rule if there are tables with applicable rules. This restriction both corresponds to the notions of parallelism and synchronization, basic in membrane computing, and it is also useful in the proofs below.

In systems with tables (either local or global) we have two levels of non-determinism: in each step we first non-deterministically choose one table (in the local case, associated with each membrane), and then we use the rules of the chosen table in a non-deterministic manner (observing the restriction of maximal parallelism for the chosen table). The standard definition of P systems corresponds to the case where we have only one table (at the level of the whole system).

The fact that we use (local) tables is indicated by adding tab to the notations from the previous section.

We do not know whether the number of tables associated with membranes matters (that is, whether it induces an infinite hierarchy of the computed sets of numbers) or normal form theorems like that known for ET0L systems (two tables are enough, see [9]) are true also in our case. In view of this *open problem* it could be better to indicate also the maximal number of tables used, writing tab_s for using at most s tables, but we do not deal with this aspect here.

The usefulness of using tables is intuitively obvious, because by clustering the rules in "teams of rules" we can control in a more precise way the work of the system. This is illustrated also by the following simple **example**: consider the system

$$\Pi = (\{a,b\}, [\ ]_1, a, R_{1,1}, R_{1,2}, R_{1,3}),$$
$$R_{1,1} = \{[\ a \to aa\ ]_1\},$$
$$R_{1,2} = \{[\ a \to b\ ]_1\},$$
$$R_{1,3} = \{[\ b\ ]_1 \to a\}.$$

After using $n \geq 0$ times the first table (thus producing 2^n copies of a), we can end the computation by using once the second table, and then 2^n times the third one. Consequently, $N(\Pi) = \{2^n \mid n \geq 1\} \in NOP_1(tab, a_0, c_0)$, a set of numbers which is not in $NMAT$.

4   Universality Results

The usefulness of tables is illustrated also by the results below: the computational universality is obtained without polarizations for various reduced combinations of types of rules.

The first result uses rules of the first three types (hence neither membrane dissolution nor membrane division operations).

Theorem 3. $NOP_2(tab, a_0, b_0, c_0) = NRE$.

Proof. We only (have to) prove the inclusion $\supseteq$, and to this aim we use the equality $NRE = MAT_{ac}$. Let us consider a matrix grammar with appearance checking $G = (N, \{a\}, S, M, F)$ in the binary normal form, hence with $N = N_1 \cup N_2 \cup \{S, \#\}$ and with matrices of the four types mentioned in Section 2. All matrices of M are supposed to be labeled in an injective manner with $m_i, 1 \leq i \leq n$ (hence i uniquely identifies the matrix). Each terminal matrix $(X \to \lambda, A \to x)$ is replaced with $(X \to f, A \to x)$, where f is a new symbol (the label of the matrix remains unchanged).

We construct the tabled P system with active membranes, Π, with the components:

$$O = N_1 \cup N_2 \cup \{Z_i, Z'_i, \langle i \rangle \mid 1 \leq i \leq n\} \cup \{a, a', e, f, \#\},$$
$$\mu = [\,[\ ]_2\,]_1,$$
$$w_1 = \lambda,$$
$$w_2 = X_0 A_0 e, \text{ where } (S \to X_0 A_0) \text{ is the initial matrix of } G,$$

and the following tables (by U we denote the set $N_1 \cup \{Z_i, Z'_i, \langle i \rangle \mid 1 \leq i \leq n\}$).

1. For each matrix $m_i : (X \to Y, A \to x)$ in M of types 2 or 4, we consider the tables

$$R_{2,i} = \{[\ X \to Z_i\ ]_2,\ [\ A\ ]_2 \to [\ ]_2\langle i \rangle,\ [\ e\ ]_2 \to \#\}$$
$$\cup\ \{[\ \alpha \to \#\ ]_2 \mid \alpha \in U\},$$
$$R'_{2,i} = \{[\ Z_i \to Z'_i\ ]_2,\ \langle i \rangle[\ ]_2 \to [\ \langle i \rangle\ ]_2\}$$
$$\cup\ \{[\ \alpha \to \#\ ]_2 \mid \alpha \in U\},$$
$$R''_{2,i} = \{[\ Z'_i \to \lambda\ ]_2,\ [\ \langle i \rangle \to xY\ ]_2\}$$
$$\cup\ \{[\ \alpha \to \#\ ]_2 \mid \alpha \in U\}.$$

2. For each matrix $m_i : (X \to Y, A \to \#)$ in M of type 3, we consider the table

$$R_{2,i} = \{[\ X \to Y\ ]_2,\ [\ A \to \#\ ]_2\}$$
$$\cup\ \{[\ \alpha \to \#\ ]_2 \mid \alpha \in U\}.$$

3. We also consider the following tables:

$$R_{2,f} = \{[\ f \to \lambda\ ]_2\}$$
$$\cup\ \{[\ \alpha \to \#\ ]_2 \mid \alpha \in U \cup N_2\},$$
$$R_{2,a} = \{[\ a\ ]_2 \to [\ ]_2 a',\ [\ \# \to \#\ ]_2\},$$
$$R_1 = \{[\ a'\ ]_1 \to [\ ]_1 a,\ [\ \# \to \#\ ]_1\}.$$

We have the equality $N(\Pi) = \{n \mid a^n \in L(G)\}$. Indeed, we start with the multiset $X_0 A_0 e$ in the central membrane; assume that we have here a multiset $X w e$ for some $X \in N_1$ and $w \in (N_2 \cup \{a\})^*$. There is only one table for membrane 1, sending out a copy of a (provided that there are copies of a' in the skin region), and using the trap-rule $\# \to \#$ provided that the object $\#$ is present; in this latter case, the computation will never stop. If applied in membrane 2 when $X w e$ is here, the table $R_{2,f}$ will introduce the trap-object $\#$, and this happens also if we use any table of the forms $R'_{2,i}, R''_{2,i}$. Thus, we can apply only a table of type $R_{2,i}$ for m_i a matrix of M. That matrix should be either of the form $m_i : (X \to Y, A \to x)$ (of type 2 or of type 4), or of the form $m_i : (X \to Y, A \to \#)$ (of type 3): if the first rule of the matrix is $\alpha \to \beta$ with $\alpha \neq X$, then the trap-object is introduced.

The case of a matrix of type 3 is simpler: if A is present, then the trap-object is introduced, and the computation will never stop (because of the table $R_{2,a}$, which can be used forever). If A is not present, then we just change X into Y. Thus, the simulation of the matrix m_i of type 3 is correct.

If we choose to simulate a matrix of types 2 or 4, then it must have the second rule of the form $A \to x$, for A as specified by the table $R_{2,i}$: if the rule $[\,A\,]_2 \to [\ ]_2 \langle i \rangle$ is not used, thus "keeping busy" the membrane, then the rule $[\,e\,]_2 \to \#$ must be used, and the computation will never stop (table R_1 can be applied forever).

In the next step we have to continue the simulation of the matrix m_i by using the corresponding table $R'_{2,i}$. This is the only table which can be applied without introducing the trap-object $\#$. In this way, $\langle i \rangle$ comes back to membrane 2, and Z_i is replaced by Z'_i. In the next step, again only one table can be used without introducing the trap-object, namely $R''_{2,i}$. It erases the object Z'_i and replaces $\langle i \rangle$ with xY, thus completing the simulation of the matrix.

At any moment, if any object a is present in membrane 2, then table $R_{2,a}$ can be used and a is sent out (first transformed into a' in the skin region).

The system is returned to a configuration with the contents of membrane 2 as in the beginning, hence the process can be iterated. When the object f is introduced, no table $R_{2,i}, R'_{2,i}, R''_{2,i}$ can be used. By means of $R_{2,f}$ we check whether any symbol from N_2 is present, hence whether the derivation in G is terminal. The computation in Π ends by sending out all copies of a, hence $N(\Pi)$ equals the length set of the language $L(G)$.   □

In the previous proof, the role of rules of type $(b_0), (c_0)$ (besides sending the result outside the system) was to ensure that only one object A is replaced by x, thus correctly simulating the second rule of a matrix $(X \to Y, A \to x)$ of types 2 or 4. This can be done also by using rules of type (e_0).

Theorem 4. $NOP_{2,2,3}(tab, a_0, c_0, e_0) = NRE$.

Proof. As above, we consider a matrix grammar with appearance checking $G = (N, \{a\}, S, M, F)$ in the binary normal form, with the matrices of M labeled in an injective manner with $m_i, 1 \le i \le n$, and each terminal matrix $(X \to \lambda, A \to x)$ replaced with $(X \to f, A \to x)$, where f is a new symbol.

We now construct the tabled P system with active membranes Π, with the components:

$$O = N_1 \cup N_2 \cup \{Z_i, \langle i \rangle \mid 1 \leq i \leq n\} \cup \{a, a', d, e, f, \#\},$$
$$\mu = [\, [\ \ ]_2 \,]_1,$$
$$w_1 = \lambda,$$
$$w_2 = X_0 A_0 e, \text{ where } (S \rightarrow X_0 A_0) \text{ is the initial matrix of } G,$$

and the following tables (by U we denote the set $N_1 \cup \{Z_i, \langle i \rangle \mid 1 \leq i \leq n\}$).

1. For each matrix $m_i : (X \rightarrow Y, A \rightarrow x)$ in M of types 2 or 4, we consider the tables

$$R_{2,i} = \{[\ X \rightarrow Z_i\ ]_2,\ [\ A\ ]_2 \rightarrow [\ \langle i \rangle\ ]_2[\ d\ ]_2,$$
$$[\ d \rightarrow \#\ ]_2,\ [\ e\ ]_2 \rightarrow [\ \#\ ]_2[\ \#\ ]_2\}$$
$$\cup \{[\ \alpha \rightarrow \#\ ]_2 \mid \alpha \in U \cup \{a\}\},$$
$$R'_{2,i} = \{[\ Z_i \rightarrow \lambda\ ]_2,\ [\ \langle i \rangle \rightarrow xY\ ]_2,\ [\ d \rightarrow \#\ ]_2\}$$
$$\cup \{[\ \alpha \rightarrow \#\ ]_2 \mid \alpha \in U \cup \{a\}\}.$$

2. For each matrix $m_i : (X \rightarrow Y, A \rightarrow \#)$ in M of type 3, we consider the table

$$R_{2,i} = \{[\ X \rightarrow Y\ ]_2,\ [\ A \rightarrow \#\ ]_2,\ [\ d \rightarrow \#\ ]_2\}$$
$$\cup \{[\ \alpha \rightarrow \#\ ]_2 \mid \alpha \in U \cup \{a\}\}.$$

3. We also consider the following tables:

$$R_{2,f} = \{[\ f \rightarrow \lambda\ ]_2,\ [\ d \rightarrow \#\ ]_2\}$$
$$\cup \{[\ \alpha \rightarrow \#\ ]_2 \mid \alpha \in U \cup N_2\},$$
$$R_{2,d} = \{[\ d\ ]_2 \rightarrow d,\ [\ a\ ]_2 \rightarrow [\ \ ]_2 a',\ [\ \# \rightarrow \#\ ]_2\},$$
$$R_1 = \{[\ a'\ ]_1 \rightarrow [\ \ ]_1 a,\ [\ \# \rightarrow \#\ ]_1\}.$$

The equality $N(\Pi) = \{n \mid a^n \in L(G)\}$ follows in a similar way as in the previous proof, this time with the interplay of rules $[\ A\ ]_2 \rightarrow [\ \langle i \rangle\ ]_2[\ d\ ]_2$ and $[\ e\ ]_2 \rightarrow [\ \#\ ]_2[\ \#\ ]_2$ ensuring that the second rule of each matrix of type 2 or 4 is correctly simulated (used exactly once): if the second rule is used, then the computation never stops, hence $[\ A\ ]_2 \rightarrow [\ \langle i \rangle\ ]_2[\ d\ ]_2$ must be used. In this way, membrane 2 is divided. In the first copy of the membrane we have the object $\langle i \rangle$, which will complete the simulation of the matrix. In the second copy of the membrane, the one containing the object d, we cannot use any table which contains the rule $d \rightarrow \#$, hence the only continuation is by using the table $R_{2,d}$. This dissolves the membrane, and its objects, left free in the skin region, will no longer evolve. The matrices of type 3 are again simulated in only one step of a computation in Π. All copies of object a are immediately sent out of membrane 2 (to prevent their duplication when dividing the membrane), and from the skin region are sent out of the system. We leave the details to the reader and conclude that the system correctly simulates the matrix grammar G. $\qquad\square$

One of the difficulties in the previous proofs was to inhibit the parallelism of using the rules of type (a_0). In membrane computing, the usual way to do this is by using *catalysts*, distinguished objects which never evolve, but can enter rules of the form $ca \to cv$, where a is a single object, which evolves under the control of the catalyst c. This idea can be considered also for P systems with active membranes, allowing rules of type (a_0) of the form $[\ ca \to cv\ ]_i$, where c is a catalyst, a is an object and v a multiset of objects. (When specifying a system with catalysts, the set C of catalysts is explicitly given after the set of objects.) We indicate the use of catalysts by writing cat_r in the notation for families of numbers computed by systems of a given type as above; r indicates the fact that at most r catalysts are used.

The previous results have the following counterpart for the catalytic case – with only two types of rules being used, and with only one membrane (note that one catalyst suffices).

Theorem 5. $NOP_1(tab, cat_1, a_0, c_0) = NRE$.

Proof. We consider again a matrix grammar with appearance checking $G = (N, \{a\}, S, M, F)$ in the binary normal form, with each terminal matrix $(X \to \lambda, A \to x)$ replaced with $(X \to f, A \to x)$, where f is a new symbol, and we construct the tabled P system with catalysts Π, with the components:

$$O = N_1 \cup N_2 \cup \{a, c, d, f, \#\},$$
$$C = \{c\},$$
$$\mu = [\ \ ]_1,$$
$$w_1 = X_0 A_0 d, \text{ where } (S \to X_0 A_0) \text{ is the initial matrix of } G,$$

and the following tables.

1. For each matrix $m_i : (X \to Y, A \to x)$ in M of types 2 or 4, we consider the table

$$R_{1,i} = \{[\ X \to Y\ ]_1,\ [\ cA \to cx\ ]_1,\ [\ cd \to c\#\ ]_1\}$$
$$\cup \{[\ Z \to \#\ ]_1 \mid Z \in N_1 \cup \{f\}\}.$$

2. For each matrix $m_i : (X \to Y, A \to \#)$ in M of type 3, we consider the table

$$R_{1,i} = \{[\ X \to Y\ ]_1,\ [\ A \to \#\ ]_1\}$$
$$\cup \{[\ Z \to \#\ ]_1 \mid Z \in N_1 \cup \{f\}\}.$$

3. We also consider the following tables:

$$R_{1,f} = \{[\ f \to \lambda\ ]_1,\ [\ d \to \lambda\ ]_1\}$$
$$\cup \{[\ \alpha \to \#\ ]_1 \mid \alpha \in N_1 \cup N_2\},$$
$$R_{1,a} = \{[\ a\ ]_1 \to [\ \ ]_1 a,\ [\ \# \to \#\ ]_1\}.$$

This time, the matrices m_i of types 2 and 4 are simulated by a single table, of type $R_{1,i}$: the first rule must be used (that is, the symbol X must be present), otherwise the trap-object is introduced; similarly, the rule $cA \to cx$ must be used, otherwise the catalyst will evolve together with the available object d and again the trap-object is introduced. Each matrix m_i of type 3 is simulated by the corresponding table $R_{1,i}$. After introducing the object f, no table as above can be used (without introducing the trap-object), hence we have to use $R_{1,f}$, which checks whether the derivation in G is terminal. At any time, the copies of object a are sent out by means of the table $R_{1,a}$. Consequently, $N(\Pi) = \{n \mid a^n \in L(G)\}$, and this completes the proof. □

The previous result is relevant in view of the fact that transition P systems with only one catalyst (not with active membranes) are not known to be universal, while the universality was proved for the case of using two catalysts [5].

5   Tables with Obligatory Rules

The idea of distinguishing some rules of each table and imposing that these rules are applied at least once when the tables are selected has at least two motivations. First, this is a way to also ensure the fact that a selected table does not leave unchanged the objects from the region where it is applied. Then, it reminds the matrices from matrix grammars, whose rules are all applied when applying a matrix. However, having several obligatory rules in the same table is a way to make the system cooperative: if both $a \to u$ and $b \to v$ must be simultaneously used at least once, then $ab \to uv$ must be used at least once (but the two cases are not equivalent, because besides evolving one a and one b, by rules $a \to u$ or $b \to v$ we can separately evolve further copies of a or of b, respectively).

That is why in what follows we allow at most one obligatory rule in each table. Such a rule is marked with a dot; when the table is used, its obligatory rule must be used at least once, otherwise the table is not allowed to be chosen.

This apparently small change in the definition of tabled P systems is powerful enough in order to lead to fast solutions (making use of membrane division) to computationally hard problems.

Theorem 6. *Tabled P systems with active membranes using obligatory rules (at most one in each table) can solve* SAT *in linear time; the construction is uniform, and the system is deterministic.*

Proof. Let us consider a propositional formula $\gamma = C_1 \wedge \cdots \wedge C_m$, consisting of m clauses $C_j = y_{j,1} \vee \cdots \vee y_{j,k_j}$, $1 \leq j \leq m$, where $y_{j,i} \in \{x_l, \neg x_l \mid 1 \leq l \leq n\}$, $1 \leq i \leq k_j$ (there are used n variables). Without loss of generality, we may assume that no clause contains two occurrences of some x_l or two occurrences of some $\neg x_l$ (the formula is not redundant at the level of clauses), or both x_l and $\neg x_l$ (otherwise such a clause is trivially satisfiable, hence can be removed). Therefore, in each clause there are at most n literals.

We codify γ, which is an instance of SAT with size parameters n and m, by the multiset

$$w(\gamma) = \{s_{j,i} \mid y_{j,r} = x_i, \text{ for some } 1 \leq i \leq n, 1 \leq j \leq m, 1 \leq r \leq k_j\}$$
$$\cup \{s'_{j,i} \mid y_{j,r} = \neg x_i, \text{ for some } 1 \leq i \leq n, 1 \leq j \leq m, 1 \leq r \leq k_j\}.$$

(We replace each variable x_i from each clause C_j with $s_{j,i}$ and each negated variable $\neg x_i$ from each clause C_j with $s'_{j,i}$, then we remove all parentheses and connectives. In this way we pass from γ to $w(\gamma)$ in a number of steps which is linear with respect to $n \cdot m$.)

We construct the P system Π with the following components:

$$O = \{a_i \mid 1 \leq i \leq n+1\} \cup \{t_i, f_i \mid 1 \leq i \leq n\} \cup \{r_i, r'_i \mid 1 \leq i \leq m\}$$
$$\cup \{d_i \mid 1 \leq i \leq m+1\} \cup \{c_i \mid 0 \leq i \leq 2n+m+3\}$$
$$\cup \{s_{j,i}, s'_{j,i} \mid 1 \leq j \leq m, 1 \leq i \leq n\} \cup \{\text{yes}, \text{no}\},$$
$$\mu = [\,[\,[\ ]_3\,]_2\,]_1,$$
$$w_1 = \lambda, \ w_2 = c_0, \ w_3 = a_1,$$
$$R_{1,1} = \{[\ \text{yes}\ ]_1 \to [\ ]_1\text{yes}, [\ \text{no}\ ]_1 \to [\ ]_1\text{no}\},$$
$$R_{2,1} = \{[\ d_{m+1}\ ]_2 \to \text{yes}, [\ c_{2n+m+3}\ ]_2 \to \text{no}\}$$
$$\cup \{[\ c_i \to c_{i+1}\ ]_2 \mid 0 \leq i \leq 2n+m+2\},$$
$$R_{3,d} = \{[\ a_i\ ]_3 \to [\ t_i\ ]_3[\ f_i\ ]_3 \mid 1 \leq i \leq n\},$$
$$R_{3,i,t} = \{[\ t_i \overset{\bullet}{\to} a_{i+1}\ ]_3\}$$
$$\cup \{[\ s_{j,i} \to r_j\ ]_3, [\ s'_{j,i} \to \lambda\ ]_3 \mid 1 \leq j \leq m\}, \text{ for each } i = 1, 2, \ldots, n,$$
$$R_{3,i,f} = \{[\ f_i \overset{\bullet}{\to} a_{i+1}\}$$
$$\cup \{[\ s'_{j,i} \to r_j\ ]_3, [\ s_{j,i} \to \lambda\ ]_3 \mid 1 \leq j \leq m\}, \text{ for each } i = 1, 2, \ldots, n,$$
$$R_{3,0} = \{[\ a_{n+1} \overset{\bullet}{\to} d_1\ ]_3\}$$
$$\cup \{[\ r_j \to r'_j\ ]_3 \mid 1 \leq i \leq m\},$$
$$R_{3,j} = \{[\ r'_j \overset{\bullet}{\to} \lambda\ ]_3\}$$
$$\cup \{[\ d_i \to d_{i+1}\ ]_3 \mid 1 \leq i \leq m\}, \text{ for each } j = 1, 2, \ldots, m,$$
$$R_{3,m+1} = \{[\ d_{m+1}\ ]_3 \to [\ ]_3 d_{m+1}\}.$$

There is no object in the skin membrane, while region 2 contains only the counter c_0, which will continuously increase its subscript, by means of table $R_{2,1}$. The "main work" is done in membrane 3. In the beginning, we have here the object a_1, hence the only applicable table is $R_{3,d}$, which divides the membrane, at the same time expanding the object a_1 to the truth values $t_1 = true$ and $f_1 = false$ of variable x_1. In the next step, the only tables which can be applied in the two membranes with label 3 are $R_{3,1,t}$ and $R_{3,1,f}$: the obligatory rules select the tables in a precise way. At the same time with the passage from t_1, f_1 to copies of a_2, we also introduce all clauses which are satisfied by t_1 and f_1, respectively (encoded by the variable r_1). The process continues now with a_2, then with a_3, and so on, until expanding all variables and introducing all clauses satisfied by these truth-assignments.

Therefore, after $2n$ steps we get 2^n membranes 3, containing the clauses satisfied by the 2^n possible truth-assigments for the n variables.

In step $2n + 1$ the only table which can be applied for membranes 3 is $R_{3,0}$: a_{n+1} is replaced with d_1 (which will check whether there is any membrane where all clauses are satisfied), and all r_j are primed.

From now on, for at most m steps, we use the tables $R_{3,j}$, $1 \leq j \leq m$. (Because these tables use primed versions of objects r_j, they were not applicable before using table $R_{3,0}$ – and this was the reason of priming.) Each of these tables removes the occurrences of one r'_j; because this operation is done by an obligatory rule, this is a way to check that the respective r'_j is present. At the same time, the subscript of the object d from each membrane 3 increases by one. If in a given membrane 3 there are copies of r'_j for all $j = 1, 2, \ldots, m$, then the respective object d reaches the subscript $m + 1$, which indicates the fact that the corresponding truth-assignment has satisfied all clauses of γ. If a given membrane 3 does not contain copies of all r'_j, $1 \leq j \leq m$, then that membrane cannot evolve m steps, hence the local object d remains of the form d_j with $j \leq m$.

Simultaneously, the object from region 2 arrives at the form c_{2n+m+1}.

If at least one membrane 3 contains the object d_{m+1} (hence the formula is satisfiable), then in step $2n + m + 2$ we use the table $R_{3,m+1}$ and the object d_{m+1} is sent to membrane 2 (at the same time, in region 2 we get c_{2n+m+2}). If no membrane 3 sends out the object d_{m+1}, hence the formula is not satisfiable, then the objects d_j with $j \leq m$ remain inside these membranes – but c_{2n+m+1} evolves to c_{2n+m+2} in region 2.

Now, in step $2n + m + 3$, if any object d_{m+1} is present in region 2, then one of them will dissolve membrane 2, and will produce the object yes, which is left free in the skin region; in the next step, this object will leave the system, thus signaling that the formula is satisfiable. Because membrane 2 is dissolved, the object c_{2n+m+3} (obtained in step $2n+m+3$) also remains free in the skin region, where it cannot evolve any more. If no object d_{m+1} is present in region 2, then this membrane is not dissolved, c will get the subscript $2n + m + 3$ and then in step $2n + m + 4$ will exit membrane 2 transformed in no; in the next step, this object exits the systems, signaling that the formula is not satisfiable.

Thus, either we get yes outside the system in step $2n + m + 4$, or no in step $2n + m + 5$, and these objects correctly indicate whether or not γ is satisfiable.

The system Π can be constructed in polynomial time by a Turing machine, starting from n and m, and it works in a deterministic manner (after each reachable configuration there is at most one next configuration which can be correctly reached).   □

If we are more interested in the time our system works than in the time of constructing it or in its deterministic behavior, then the answer to a given instance of SAT can be obtained in $n + m + 4$ steps, by considering a system constructed in a *semi-uniform manner* (starting directly from an instance of the problem) in the following way.

For a given formula γ as above, for $l = 1, 2, \ldots, n$, we denote

$$sat(t_l) = \{r_j \mid \text{there is } 1 \leq i \leq k_j \text{ such that } y_{j,i} = x_l\},$$
$$sat(f_l) = \{r_j \mid \text{there is } 1 \leq i \leq k_j \text{ such that } y_{j,i} = \neg x_l\}.$$

Then, we construct the system Π with:

$$O = \{a_i, t_i, f_i \mid 1 \leq i \leq n\} \cup \{r_i \mid 1 \leq i \leq m\}$$
$$\cup \{b_i \mid 0 \leq i \leq n + m + 2\} \cup \{c_i \mid 0 \leq i \leq n + m + 3\}$$
$$\cup \{\text{yes}, \text{no}\},$$
$$\mu = [\,[\,[\ ]_3\,]_2\,]_1,$$
$$w_1 = \lambda, \ w_2 = c_0, \ w_3 = b_0 a_1 a_2 \ldots a_n,$$
$$R_{1,1} = \{[\,\text{yes}\,]_1 \to [\ ]_1\text{yes}, \ [\,\text{no}\,]_1 \to [\ ]_1\text{no}\},$$
$$R_{2,1} = \{[\,b_{n+m+2}\,]_2 \to \text{yes}, \ [\,c_{n+m+3}\,]_2 \to [\ ]_2\text{no}\}$$
$$\cup \{[\,c_i \to c_{i+1}\,]_2 \mid 0 \leq i \leq n + m + 2\},$$
$$R_{3,i} = \{[\,a_i\,]_3 \xrightarrow{\bullet} [\,t_i\,]_3[\,f_i\,]_3\}$$
$$\cup \{[\,b_j \to b_{j+1}\,]_3 \mid 0 \leq j \leq n - 1\}, \ \text{for each } i = 1, 2, \ldots, n,$$
$$R_{3,n+1} = \{[\,b_n \xrightarrow{\bullet} b_{n+1}\,]_3\}$$
$$\cup \{[\,t_i \to sat(t_i)\,]_3, \ [\,f_i \to sat(f_i)\,]_3 \mid 1 \leq i \leq n\},$$
$$R_{3,n+1+i} = \{[\,r_i \xrightarrow{\bullet} \lambda\,]_3\}$$
$$\cup \{[\,b_{n+j} \to b_{n+j+1}\,]_3 \mid 1 \leq j \leq m\}, \ \text{for each } i = 1, 2, \ldots, m,$$
$$R_{3,n+m+2} = \{[\,b_{n+m+1}\,]_3 \to [\ ]_3 b_{n+m+2}\}.$$

This time, in the first n steps we divide membrane 3 again and again, by means of the obligatory rules of tables $R_{3,i}$, $1 \leq i \leq n$, which expand the objects a_i to the truth values $t_i = true$ and $f_i = false$ of variable x_i. The order of using tables $R_{3,i}$ is arbitrary, but after n steps we get the same configuration irrespective of this order: 2^n membranes 3, containing the 2^n truth-assignments of the n variables, as well as the object b_n (at the same time, in membrane 2 we have obtained c_n).

In step $n + 1$, in region 3 we can only apply $R_{3,n+1}$, which replaces b_n with b_{n+1} and each t_i, f_i by the clauses satisfied by these truth values (specifically, t_i is replaced by $sat(t_i)$ and f_i by $sat(f_i)$).

From now on, for at most m steps, we use the objects $b_{n+j+1}, 1 \leq j \leq m$, in the same way as objects d_j were used in the previous proof, in order to check whether or not at least one truth-assignment has satisfied all clauses. If this is the case, then at least one membrane 3 will contain the object b_{n+m+1}, which will exit to membrane 2, will dissolve it in step $n + m + 3$, and will produce the object yes, which then leave the system. If not, c_{n+m+3} will exit membrane 2 (in step $n + m + 4$) transformed in no, which will exit the system in one further step.

The system Π can be constructed in polynomial time by a Turing machine, starting from γ (only the tables $R_{3,i}$ directly depend on the formula), and the system is clearly confluent.

6 Final Remarks

Contributing to the "campaign" of removing polarizations from P systems with active membranes, we have obtained several universality results for systems without polarizations, but having the rules structured in *tables*. When tables with (at most one) obligatory rules are used, **NP**-complete problems can be solved in linear time – this is illustrated with SAT problem.

Two important problems have remained *open*: (i) are systems without polarizations and without tables (maybe with catalysts) universal? (ii) can **NP**-complete problems be solved in polynomial time by means of tabled P systems with active membranes without polarizations (and without obligatory rules)?

Acknowledgements. The support of this research through the project TIC2002-04220-C03-01 of the Ministerio de Ciencia y Tecnología of Spain, cofinanced by FEDER funds, is gratefully acknowledged.

References

1. A. Alhazov, R. Freund, Gh. Păun, P systems with active membranes and two polarizations, in *Proc. Second Brainstorming Week on Membrane Computing*, Sevilla, February 2004, TR 01/04 of Research Group on Natural Computing, Sevilla University, 2004, 20–36.
2. A. Alhazov, L. Pan, Polarizationless P systems with active membranes, *Grammars*, 7, 1 (2004).
3. A. Alhazov, L. Pan, Gh. Păun, Trading polarizations for labels in P systems with active membranes, submitted 2003.
4. J. Dassow, Gh. Păun, *Regulated Rewriting in Formal Language Theory*, Springer, Berlin, 1989.
5. R. Freund, L. Kari, M. Oswald, P. Sosik, Computationally universal P systems without priorities: two catalysts are sufficient, *Theoretical Computer Science*, in press.
6. D. Hauschild, M. Jantzen, Petri nets algorithms in the theory of matrix grammars, *Acta Informatica*, 31 (1994), 719–728.
7. Gh. Păun, *Computing with Membranes: An Introduction*, Springer, Berlin, 2002.
8. M. Pérez-Jiménez, A. Roméro-Jimenez, F. Sancho-Caparrini, *Teoría de la complejidad en modelos de computación celular con membranas*, Kronos Editorial, Sevilla, 2002.
9. G. Rozenberg, A. Salomaa, *The Mathematical Theory of L Systems*, Academic Press, New York, 1980.
10. G. Rozenberg, A. Salomaa, eds., *Handbook of Formal Languages* (3 volumes), Springer, Berlin, 1997.
11. A. Salomaa, *Formal Languages*, Academic Press, New York, 1973.

Some Properties of Multistage Interconnection Networks

Azaria Paz

Computer Science Department
Technion—IIT
Haifa, 32000 Israel
and The Netanya Academic College—Israel

Abstract. In a previous paper [Pa2002] the author developed a theory of decomposition into prime factors of multistage interconnection network. Based on that theory some new properties of such networks are investigated and proved.

1 Preliminary

Multistage Interconnection Networks (MIN's) play an important role in the design of the hardware and operating systems of computers, in particular of parallel computers, and they enable efficient communication algorithms between processors and memories, see e.g. [L92]. We shall be concerned in this paper mainly with one such MIN, the Butterfly network, some of its isomorphic networks and some of its extensions. In general, MIN's can be described by an n-layered graph defined as below

Definition 1. *An n-layered graph is a graph* $G = (X_1, X_2, ..., X_n, E_2, E_3 \cdots E_n)$ *where the* X_i*'s are sets of vertices, and the* X_i *vertices are connected only to the* X_{i+1} *vertices by the edges* E_{i+1}.

Clearly every layer of an n-layered graph can be depicted as a bipartite graph with X_i, X_{i+1} as vertices and E_{i+1} as edges. Thus an n-layered graph is a concatenation of $n-1$ bipartite graphs and every such bipartite graph will be called a stage. Even and Litman introduced in [EL97] a technique, the "Cross Product" technique, and showed that this technique enables the representation of several well known n-layered networks as a cross product of simple such networks. This author extended their technique in [Pa2002] into a full decomposition theory. The basic definitions and properties and some main results from [Pa2002] are reproduced below.

Definition 2. *Let* $B_1 = (X_1, Y_1, E_1)$ *and* $B_2 = (X_2, Y_2, E_2)$ *be two bipartite graphs. Their* cross-product *is the bipartite graph* $G_3 = (X_3, Y_3, E_3)$ *such that* $X_3 = X_1 \times X_2$, $Y_3 = Y_1 \times Y_2$ *('* $\times$ *' represents the Cartesian product operation) and* $((x_{1i}, x_{2j}), (y_{1k}, y_{2l})) \in E_3$ *if and only if* $(x_{1i}, y_{1k}) \in E_1$ *and* $(x_{2j}, y_{2l}) \in E_2$.
We shall use the notation ' $\times$ *' for the cross-product operation of bipartite graphs as defined above. For a given bipartite graph* $B = (X, Y, E)$ *we shall refer to* X *and* Y *as the floor and the ceiling, respectively, of* B.

J. Karhumäki et al. (Eds.): Theory Is Forever (Salomaa Festschrift), LNCS 3113, pp. 250–258, 2004.

Definition 3. *Let B_1 and B_2 be two bipartite graphs. We shall say that B_1 is isomorphic to B_2 (notation: $B_1 \sim B_2$) if there are $1-1$ and onto mappings ψ and ϕ from X_1 to X_2 and from Y_1 and Y_2 such that $(x_i, y_j) \in E_1$, iff $(\psi(x_i), \phi(y_j)) \in E_2$.*

It is easy to see that the cross product operation is associative and communicative up to isomorphism of the resulting graphs.

Definition 4. *Let $G_1 = (X_1, ..., X_n, E_2, ..., E_n)$ and $G_2 = (X'_1, ..., X'_n, E'_2, ..., E'_n)$ be two n-layered graphs. $G_1 \times G_2 = G_3 = (X''_1, ..., X''_n, E''_2, ..., E''_n)$ is the n-layered graph such that*

$$(X_i, X_{i+1}, E_{i+1}) \times (X'_i, X'_{i+1}, E'_{i+1}) = (X''_i, X''_{i+1}, E''_{i+1}) \ .$$

Definition 5. *Two n-layered graphs $G_1 = (X_1, \ldots, X_n, E_2, \ldots, E_n)$ and $G_2 = (X'_1, \ldots, X'_n, E'_2, \ldots, E'_n)$ are isomorphic if for all $i, 1 \le i \le n - 1$, the bipartite graphs $B_i = (X_1, X_{i+1}, E_{i+1})$ and $B'_i = (X'_i, X'_{i+1}, E'_{i+1})$ are isomorphic and the isomorphisms ψ_i and ϕ_i can be defined in a way such that for all $2 \le i \le n - 1$ $\psi_i(X_i) = \phi_{i-1}(X_{i-1})$ (see Definition 3).*

Remark. The notion of isomorphism introduced here differs from the standard definition of graph isomorphism in that its is sensitive to the identities of the vertices of the graph. The notion introduced here would have deserved a different name. Nevertheless, we choose to use this name in order to comply with its name common in the literature investigating multilayered graphs.

It follows from the definitions that the graph $G = G_1 X G_2 X, \ldots, X G_k$ is isomorphic to all graphs of the form $G_{\pi(1)} X \cdots X G_{\pi(k)}$ where π is any permutation of $(1, \ldots, k)$.

2   The BCP Family of Graphs

Consider the primitive simple bi-partite graphs labeled a, b, c and 1 shown below:

$$
\begin{array}{cccc}
0 & 0 \ \ 1 & 0 \ 1 & 0 \\
a: \ \bigwedge \ ; & b: \ \bigvee \ ; & c: | \ \ | \ ; & 1: \ | \\
0 \ \ 1 & 0 & 0 \ 1 & 0
\end{array}
$$

Denote $\Sigma = \{a, b, c\}$, let $\sigma \in \Sigma$ and let B_σ be the graph whose label is σ. Then for any word $w \in \Sigma^*$, $w = \sigma_1 \cdots \sigma_k$ w represents the graph $B_w = B_{\sigma_1} \times \cdots \times B_{\sigma_k}$. Let $B_\sigma = (X_\sigma, Y_\sigma, E_\sigma)$ and $B_w = (X_w, Y_w, E_w)$.

Then, by our definitions $X_w = X_{\sigma_1} \times \cdots \times X_{\sigma_k}$ and $Y_w = Y_{\sigma_1} \times \cdots \times Y_{\sigma_k}$. We shall label the vertices of X_w and Y_w according to the following procedure: The vertices in X_{σ_i} and Y_{σ_i} are labeled by either 0 or 1 as per the definition of B_{σ_i}. The vertices in X_w and Y_w correspond to k-tuples of zeroes and ones. Order the vertices in X_w and Y_w according to the lexicographic order of the corresponding k-tuples and then label those vertices with consecutive integers, starting with

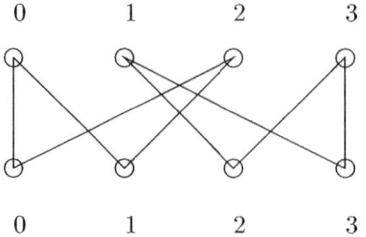 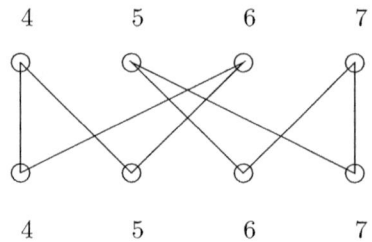

Fig. 1. The graph B

0, according to the order of their n-tuples, the notation (X_w) and (Y_w) will be used to denote the sets X_w and Y_w when ordered as above.

The BCP family of graphs is defined as $L_1 = \{B : B = B_w, w \in \Sigma^*\}$.

An Example. Consider the graph B given above.

The reader can verify that $B = B_{cbca} = B_c B_b B_c B_a$.

For the sake of simplicity we shall represent graphs in the BCP family by their label, e.g., the graph $B_{a^2bc^2} = B_a \times B_a \times B_b \times B_c \times B_c$ will be denoted by $w = a^2bc^2 \in \Sigma^*$.

3   The n-LCP Family of Graphs

Let $w_1, w_2, ..., w_n$ be a sequence of graphs in the BCP family such that the size of the ceiling of w_i is equal to the size of the floor of w_{i+1} for $1 \leq i \leq n - 1$. We can construct an n-layered graph from the above graphs by identifying the vertices in the floor of w_{i+1} with the vertices in the ceiling of w_i in their given order.

Denote this graph by the "page"

$$\begin{bmatrix} w_1 \\ w_2 \\ \vdots \\ w_n \end{bmatrix}$$

Definition 6. *The n-LCP (n-layered cross-product) family of graphs is the set of all graphs of the form*

$$\begin{bmatrix} w_1 \\ \vdots \\ w_n \end{bmatrix}$$

such that $w_i \in BCP$ for $1 \leq i \leq n$ and the vertices in the floor of w_{i+1} (whose number is equal to the number of vertices in the ceiling of w_i) are identified with the vertices in the ceiling of w_i, in their given order.

An Example. The n-layered Ω network [La75] can be described as follows:

1. The number of vertices in the floor and ceiling of every layer is equal to 2^n.
2. All layers have the i-th vertex in the floor, $i = 1 \dots 2^n$, connected to the vertices $2i - 1$ and $2i$ modulo 2^n in the ceiling. It is easy to see that the omega network is represented by the page, containing n words, below.

$$\left.\begin{bmatrix} a \; c^{n-1} \; b \\ \vdots \\ a \; c^{n-1} \; b \end{bmatrix}\right\} n \text{ rows}$$

4   Prime n-Factors

We define below a subset of $O(n^2)$ graphs in the n-LCP family which have the "primality" property, i.e., all the graphs in the family can be represented as cross-products of those factors and the primes themselves cannot be factorized into simpler graphs.

Definition 7. *A prime graph in the n-LCP family is a graph which can be represented by a page as below:*

a. *Choose a row $0 \leq i \leq n$ and write a in row i. Choosing the row 0 means that a is omitted altogether from the page.*
b. *Assume we choose position i for a. Choose a row $i < j \leq n + 1$ and write b in row j. Choosing the $n + 1$ row means that b is omitted altogether from the page.*
c. *Write 1 (1 represents the graph $|$) in rows $k, k < i$ or $k > j$, if such rows exist (i.e., if $i > 1$ or $j < n$) and write c in rows $t, i < t < j$ if such rows exist (i.e., if $j > i + 1$).*

Several primes in n-LCP are shown below.

$$\begin{bmatrix} c \\ c \\ c \\ c \end{bmatrix} \quad \begin{bmatrix} 1 \\ a \\ b \\ 1 \end{bmatrix} \quad \begin{bmatrix} 1 \\ 1 \\ 1 \\ a \end{bmatrix} \quad \begin{bmatrix} a \\ c \\ c \\ b \end{bmatrix}$$

Remark. As the number of possible locations for a is $n + 1$ and for the i-th location of a, there are $n - i + 1$ possible locations for b, we have that the total number of n-primes is $\sum_{j=1}^{n+1} j = \frac{(n+1)(n+2)}{2}$. It follows from the definition that the primes do not factor into simpler factors, if we disregard the trivial prime consisting of a page with a single 1 in every row, which is not included in the above definition.

The Graphs corresponding to the primes described above are shown below.

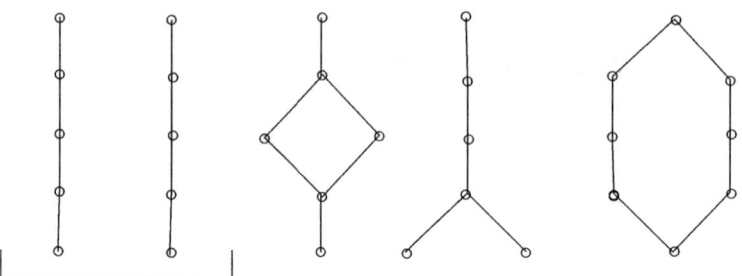

5   Previous Results

The following results have been obtained in [Pa2002].

- A polynomial algorithm is provided (and its correctness proved) that enables the factorization of any graph in n-LCP into prime factors. The complexity of the algorithm is shown to be $O(n^2 \log(E))$ where n is the number of layers of the graph and $\overline{E}$ is the maximal number of edges in a layer of the graph.
- The following theorem, showing that the factorization provided by the algorithm is unique, up to isomorphism, is also proved in [Pa2002].

Theorem 1. *Let N_1 and N_2 be isomorphic networks in n-LCP. Let $N_1 = f_1 \times \cdots \times f_{k_1}$, $N_2 = g_1 \times \cdots \times g_{k_2}$ be two factorizations of N_1 and N_2, respectively, into prime factors, not necessarily distinct. Then $k_1 = k_2$, and the two factorizations contain the same factors with the same multiplicity.*

In the sequel we shall use the notation below.
Notation. Consider the prime graphs in the n-LCP family as described in Definition 7. We shall denote those graphs as below:
$X_{i,j}, 1 \le i \le j < n$, denotes the diamond shaped-graph with an a at layer i, a b at layer $j+1$, c's between the a and the b and ones below the b and above the a.
$X_{i,n}, 1 \le i \le n$, denotes the fork-shapes graph with an a at layer i, c's below it and ones above it $Y_{1,j}, 1 \le i \le j \le n$, denotes the Y-shaped graph with a b at layer j, c's above it and ones below it. S_n is the prime with c in all its layers.

Examples: The Baseline Network

The Baseline network ([WF80]) can be described recursively as follows. The one layered Baseline has two vertices in its floor and in its ceiling and both floor vertices are connected to both ceiling vertices. The $(i + 1)$-layered Baseline is constructed from two identical i-layered baselines set on layers 2 to $i+1$. The first layer set on top of the two i-layered baselines is equal to the (constant) layer of the Ω network. Thus the 3-layered baseline and n-layered baseline representation is shown below.

$$
\begin{array}{cccc}
a & c^{n-1} & \cdots & b \\
c & a & c^{n-2} & b \\
\vdots & & \vdots & \\
c^{n-1} & \cdots & & a\ b \\
\end{array}
$$

$$
\begin{array}{cccc}
a & c & c & b \\
c & a & c & b \\
c & c & a & b \\
\end{array}
$$

3-layered n-layered

Applying the factorization algorithm we get for the n-layered Baseline $BL(n)$ the factorization below:

$$BL(n) = X_{1,n} \times X_{2,n} \times \cdots \times X_{n,n} \times Y_{1,n} \times Y_{1,n-1} \times \cdots \times Y_{1,1}$$

The Butterfly network [L92] can be described by

$$
\begin{array}{cccc}
a & b & c^{n-1} & \\
c & a & b & c^{n-2} \\
\vdots & & & \\
c^{n-1} & & a & b \\
\end{array}
$$

and decomposes as:

$$BY(n) = X_{1,n} \times \cdots \times X_{n,n} \times Y_{1,1} \times \cdots \times Y_{1,n}$$

In a similar way one can show that the Ω network decomposes as:

$$\Omega(n) = X_{n,n} \times \cdots \times X_{1,n} \times Y_{1,n} \times \cdots \times Y_{1,1}$$

Thus, the three networks, the Omega, the Butterfly and the Baseline decompose into the same factors and are therefore isomorphic.

6   Some New Results

As shown in the previous Sections 3 and 5 the Omega network, the Baseline network and the Butterfly network have the same factorization consisting of all the Y-shaped factors (the $Y_{1,i}$ factors) and all the fork-shaped factors (The $X_{i,n}$ factors), and therefore are isomorphic. In fact any network factorizing into a permutation of the above factors is isomorphic to the Butterfly network. We shall denote all the networks that are isomorphic to the Butterfly by B-networks. While, as mentioned above, the order of the factors in any B-network is immaterial, thus order becomes relevant when B-networks are combined. We shall consider two types of combinations which have been studied in the literature.

(a) Two identical networks connected in tandem i.e. the networks are combined in a way such that the ceiling of the top layer of one network is identified with the floor of the bottom layer of the second. If H is a B-network, then we shall denote this operation as H/H.

(b) Two identical networks are connected in a way such that one network is connected to the mirror image of the second network where the floor of the bottom layer of one network is identified with the ceiling of the top layer of the mirror image (i.e. top-bottom reversal) of the second network. We shall denote this connection as $H/\tilde{H}$ ($\tilde{H}$ denotes the mirror, top down reversal, of H).

A well know combined network is the Beneś network [L92]. The Beneś network can be described as $H/\tilde{H}$ where H is the Baseline network (see description in Sect. 5). An important property of the Beneś network is that it is rearrangeable, i.e. for any mapping π of the inputs to the outputs it is possible to construct edge-disjoint paths in the network linking the i-th vertex in the floor of the network to the $\pi(i)$-th vertex in the ceiling of the network (see [L92, Theorem 3.10, p.452]). It is well known that the Beneś network is isomorphic to $BY/\tilde{BY}$, where BY is the Butterfly network (of the same order). See e.g. [EL97]. Based on the theory of decomposition into prime factors of networks described in Sect. 5, we can now generalize the isomorphism range and prove the theorem below.

Theorem 2. *Let $H_1(n)$ and $H_2(n)$ be any two n-layered B-networks. Then $H_1(n)/\tilde{H}_1(n)$ is isomorphic to $H_2(n)/\tilde{H}_2(n)$.*

Proof (of theorem). As mentioned before the set of factors of any n-layered B-network $H(n)$ contains the $Y_{1,i}$ and the $X_{i,n}, 1 \leq i \leq n$, factors and no other factors. In the mirror image of $H(n)$ the $Y_{1,i}$ factors will transform into $X_{n-i+1,n}$ factors, the $Y_{1,j}$ factors will transform into X_{n-j+1} factors, and the order of the corresponding factors in $\tilde{H}(n)$ is the same as the order of the source factors in $H(n)$. The factors generated in $H(n)/\tilde{H}(n)$ are depicted below where, for the sake of saving space, we represent the prime factors in transposition, as rows instead of columns, $()_m$ representing a prime with m layers.

(i) A factor X_{in} in $H(n)$ represented as $(1, \ldots, 1, \overset{i}{a}, c, \ldots, c)_n^T$ of $H(n)$ generates the factor $(c, \ldots, c, \overset{n-i+1}{b}, 1, \ldots, 1)_n^T$ in $\tilde{H}(n)$. Those two factors combine in $H(n)/\tilde{H}(n)$ into the factor $(1, \ldots, 1, \overset{i}{a}, c, \ldots, c, \overset{2n-i+1}{b}, 1, \ldots, 1)_{2n}^T$ whose label is $X_{i,2n-i}$ (see definitions in Sect. 5).

(ii) A factor $Y_{1,j}$ of $H(n)$ represented as $(c, \ldots, c, \overset{j}{b}, 1, \ldots, 1)_n^T$ generates the factor $(1, \ldots, 1, \overset{n-j+1}{a}, c, \ldots, c)_n^T$ of $\tilde{H}(n)$. Those factors do not combine in $H(n)/\tilde{H}(n)$ but rather "stretch out" into the factors $(c, \ldots, cjb, 1, \ldots, 1)_{2n}^T$ and $(1, \ldots, 1, \overset{2n-j+1}{a}, c, \ldots,)_{2n}^T$ whose labels are Y_{1j} and $X_{2n-jh,2n}$ respectively. All in all, the set of all factors of $H(n)/\tilde{H}(n)$, for **any** n-layered B-network H consists of the following factors:
(i) the diamond shaped factors $X_{i,2n-i}$ $1 \leq i \leq n$
(ii) The Y-shaped factors $Y_{1,j}$, $1 \leq j \leq n$
(iii) The fork shaped factors $X_{2n-j+1,2n}$, $1 \leq j \leq n$.

Thus all the $2n$-layered networks $H(n)/\tilde{H}(n)$ such that $H(n)$ is an n-layered B-networks are isomorphic. The theorem is thus proved. $\qquad\square$

It was also mentioned in the literature that the $H/\tilde{H}$ combination is not isomorphic to the H/H combination for some B-networks (e.g. the Butterfly) while for some other networks the above combinations are isomorphic (e.g. the Baseline network—see [EL97]). We provide now a characterization of the B-networks H such that $H/\tilde{H}$ is isomorphic to H/H.

Definition 8. *Let H be an n-layered B-network. Assume that the order of the fork-shaped factors in the factorization of H is $X_{i_1,n}, X_{i_2,n}, \ldots, X_{i_n,n}$ and the order of the Y shaped factors is $Y_{1,j_1}, Y_{1,j_2}, \ldots, Y_{1,jn}$. H will be called a* **matching** *network if $(i_1, i_2, \ldots, i_n) = (j_n, j_{n-1}, \ldots, j_1)$. H will be called* **touching** *if $(i_1, i_2, \ldots, i_n) = j_1, j_2, \ldots, j_n)$.*

Theorem 3. *Let H be an n-layered B-network. Then H/H is isomorphic to $H/\tilde{H}$ if an only is H is matching.*

Remark 1. The reader can verify that the Baseline network and the Butterfly network described in Sect. 5 are matching and touching correspondingly.

The proof of Theorem 3 is similar to the proof of Theorem 2. The reader can verify that, if and only if H is matching, then the order of the Y-shaped factors in $\tilde{H}$ generated by the fork shaped factors in H is the same as the order of the Y-shaped factors in H and that the set of factors of H/H is equal to the set of factors of $H/\tilde{H}$ which was shown in the proof of Theorem 2, given that H is matching.

Corollary 1. *Let H_1 and H_2 be two n-layered B-networks. Then $H_1/\tilde{H}_1$ is isomorphic to $H_2/\tilde{H}_2$ and, if and only if H_2 is matching, then $H_1/\tilde{H}_1 \sim H_2/\tilde{H}_2 \sim H_2/H_2$. The power of the Theorems 2,3 and their corollary stems from the fact that all the combined B-networks of type $H/\tilde{H}$ are rearrangeable, a very important property for communication purpose.*

Finally we can prove the following.

Theorem 4. *Let H_1 and H_2 be any two n-layered B-networks. If both H_1 and H_2 are touching then H_1/H_1 is isomorphic to H_2/H_2.*

The proof which is similar to the proof of Theorem 2 is left to the reader. It can easily be verified that the Butterfly network is touching and it is known that $BY(n)/B\tilde{Y}(n)$ is not isomorphic to $BY(n)/BY(n)$. Thus follows directly from Theorem 3. The question whether $BY(n)/BY(n)$ is rearrangeable is an open problem (see [EL97]). It may be easier to approach this problem using an isomorphic network which is homogeneous. E.g. as the Ω network is also touching we may try to analyze the network $\Omega(n)/\Omega(n)$ which is isomorphic to $BY(n)/BY(n)$. As all the layers of $\Omega(n)/\Omega(n)$ have the same form, a $c^{n-1} b$, we can rephrase the problem in a more general form, i.e.:

1. Is it possible to construct a rearrangeable network by concatenating layers of the form $a\ c^{n-1}\ b$?
2. If the answer to (1) is "yes" then find the minimal k such that the network consisting of k layers of the form $a\ c_{n-1}\ b$ is rearrangeable.

References

[EL97] Even, S., Litman, A.: Layered Cross Product—A Technique to Construct Interconnection Networks, Networks **29** (1997) 219–223.
[La75] Lawrie, D.H.: Access and Alignment of Data in an Array Processor", *IEEE Trans. Computers* **C24** (1975) 1145–1155.
[L92] Leighton, F.T.: Introduction to Parallel Algorithms and Architectures: Arrays, Trees and Hypercubes, Morgan Kaufman Publ., San Mateo, CA, (1992).
[Pa2002] Paz, A.: A Theory of Decomposition into Prime Factor of Layered Interconnection Networks—a New Version, technical report #892, Computer Science Dept., Technion—IIT, April 2002.
[WF80] Wu, C., Feng, T.: On a Class of Multistage Interconnection networks, *IEEE Trans. Computers* **C29** (1980) 694–702.

Structural Equivalence of Regularly Extended E0L Grammars:
An Automata Theoretic Proof[*]

Kai Salomaa[1] and Derick Wood[2]

[1] School of Computing, Queen's University, Kingston, Ontario K7L 3N6, Canada
ksalomaa@cs.queensu.ca
[2] Department of Computer Science, Hong Kong University of Science & Technology,
Clear Water Bay, Kowloon, Hong Kong
dwood@cs.ust.hk

Abstract. Regularly extended E0L grammars allow an infinite number of rules for a given nonterminal provided that the set of right sides of the rules for each nonterminal is a regular language. We show that structural equivalence remains decidable for regularly extended E0L grammars.

1 Introduction

The decidability of structural equivalence of context-free grammars is a classical result [9, 13] and a simplified automata theoretic decidability proof is known from [19]. The question remains decidable for E0L grammars [10–12, 17] which are parallel context-free grammars. In the case where the parallel derivation is controlled by a finite set of tables, that is, we have ET0L grammars, structural equivalence is already undecidable [16]. On the other hand, if the tables are restricted to be homomorphisms [18], or if we consider the strong equivalence that compares also the sequences of tables used [8], this question again becomes decidable.

All of the above results are obtained for context-free type grammars with a finite set of rules, and the corresponding syntax trees have nodes of bounded arity. Due to many applications in document grammars, recently there has been much interest in regularly extended context-free type grammars [1–3, 5], as well as, in tree automata operating on unranked trees [3, 4].

It has been shown that structural equivalence remains decidable also for regularly extended (sequential) context-free grammars [5]. In this paper we establish the decidability of structural equivalence for regularly extended E0L grammars. Our proof uses tree automata, following the approach from [19, 17]. Naturally the syntax trees of (regularly extended) E0L grammars cannot be recognized by

---

[*] The research of the first author was supported under the Natural Sciences and Engineering Research Council of Canada grant OGP0147224 and that of the second was supported under the grant HKUST6166/00E from the Research Grants Council of the Hong Kong SAR.

J. Karhumäki et al. (Eds.): Theory Is Forever (Salomaa Festschrift), LNCS 3113, pp. 259–267, 2004.
© Springer-Verlag Berlin Heidelberg 2004

finite tree automata and the automata used in [17] relied on an explicit height counting mechanism. To simplify the constructions, here instead of an explicit height counting capability we restrict the form of the input trees and show that equivalence of the tree automata on the restricted set of input trees can be decided effectively.

The decidability proof for regularly extended context-free grammars in [5] uses a grammatical approach where the given grammars are transformed into a normal form and arbitrary grammars are shown to be structurally equivalent if and only if the normal forms are identical. Similarly a grammatical approach is used in [10–12] for deciding the structural equivalence of E0L grammars with finitely many rules. Already for sequential and parallel grammars with finitely many rules, the proof establishing the uniqueness of the normal form grammar is more involved than the decidability proof based on tree automata, but it has the advantage of explicitly giving a grammatical normal form. We do not know whether a normal form grammar can be constructed for regularly extended E0L grammars such that it is unique for structurally equivalent grammars.

2   Regularly Extended E0L Grammars and Tree Automata

We briefly recall some definitions concerning regularly extended E0L grammars and tree automata. For all unexplained notions on formal languages we refer the reader to [20]. A general reference for L systems is [14] and for information on tree automata we refer the reader to [6, 7].

The cardinality of a finite set A is $\#A$ and the power set of A is $\wp(A)$. When there is no danger of confusion, a singleton set $\{b\}$ is denoted simply as b. All alphabets we consider are finite. The set of words (respectively, nonempty words) over an alphabet A is A^* (respectively, A^+) and λ denotes the empty word.

A *regularly extended E0L grammar*, or reE0L grammar, is a tuple

$$G = (V, \Sigma, S, P),\tag{1}$$

where V is an alphabet of nonterminals, Σ is an alphabet of terminals ($V \cap \Sigma = \emptyset$), $S \in V$ is the start nonterminal and $P \subseteq V \times (V \cup \Sigma)^*$ is a set of productions such that for any $B \in V$ the language

$$\{w \in (V \cup \Sigma)^* \mid (B, w) \in P\}$$

is regular.

Note that we allow that the set of productions can be infinite. As usual we denote productions $(B, w) \in P$ as $B \rightarrow w$.

The productions of P define in the well-known way the parallel one-step rewrite relation $\Rightarrow \subseteq (V \cup \Sigma)^* \times (V \cup \Sigma)^*$ defined by setting $w_1 \Rightarrow w_2$ if and only if

$$w_1 = B_1 \cdots B_n, \ w_2 = u_1 \cdots u_n, \ B_i \rightarrow u_i \in P, \ i = 1, \ldots, n.$$

The language generated by the grammar G is

$$L(G) = \{w \in \Sigma^* \mid S \Rightarrow^* w\}.$$

Note that our definition does not allow the rewriting of terminals, that is, the reE0L grammar is synchronized [14]. Thus the one-step rewrite relation is in fact a subset of $V^* \times (V \cup \Sigma)^*$. It is well-known that this restriction does not cause any restriction in terms of the family of languages generated.

We assume that notions such as the root, a leaf and the height of a tree are known. The height of a tree with a single node is defined to be zero. If C is a (finite) set, by a C-tree we mean a rooted ordered tree the nodes of which are labeled by elements of C. Note that the same symbol of C may be used to label nodes having different numbers of children, that is, we consider trees where the node labels have variable arity. The set of all C-trees is denoted F_C.

Next we define the syntax-trees of an reE0L grammar. In the following G is always as in (1).

Denote $\Xi = V \cup \Sigma \cup \{\hat{\lambda}\}$. Here $\hat{\lambda}$ is a new symbol that will be used to label nodes corresponding to the empty word. A leaf of $t \in F_{\Xi}$ is said to be a non-λ leaf if it is labeled by some element other than $\hat{\lambda}$.

The set of syntax trees of G, $S(G)$, is a subset of F_{Ξ} defined inductively as follows. A tree with one node labeled by a symbol of Ξ is in $S(G)$. Assume that $t \in S(G)$ and the non-λ leaves of t are $u_1, \ldots, u_m$, $m \geq 1$, where u_i is labeled by $B_i \in V$, and $B_i \to a_1^i \cdots a_{k_i}^i \in P$, $k_i \geq 0$, $a_j^i \in V \cup \Sigma$, $j = 1, \ldots, k_i$, $i = 1, \ldots m$. Then the tree t' is in $S(G)$ if t' is obtained from t by attaching, for each node u_i, k_i children that are labeled, respectively, by the symbols $a_1^i, \ldots a_{k_i}^i$. If $k_i = 0$, the node u_i has one child labeled by $\hat{\lambda}$.

According to the above definition the root of a syntax tree can be labeled by any symbol of Ξ. This is useful in inductive constructions where we consider derivations beginning with any grammar symbol. Below the terminal syntax trees are required to represent derivations that begin with the start nonterminal and produce a terminal word.

A syntax tree $t \in S(G)$ is said to be *terminal* if the root of t is labeled by the start nonterminal S and all leaves of t are labeled by elements of $\Sigma \cup \{\hat{\lambda}\}$. The set of terminal syntax trees of G is denoted $TS(G)$. The terminal syntax trees correspond in the obvious way to parallel derivations of the grammar G yielding terminal words.

For $t \in F_{\Xi}$ we denote by yield(t) the word over $V \cup \Sigma$ obtained by concatenating from left to right the labels of all non-λ leaves of t. Then

$$L(G) = \{\text{yield}(t) \mid t \in TS(G)\}.$$

The structure of a syntax tree $t \in S(G)$, str(t) is the tree obtained from t by replacing the label of each internal node of t by ϖ. Here ϖ is a new symbol not appearing in $V \cup \Sigma$. Essentially, the structure trees can be considered to have labels only for the leaves. The special symbol ϖ is used only to make it possible to define transitions of a tree automaton without a separate assumption concerning unlabeled nodes.

The set of terminal structure trees of G is

$$STS(G) = \{\text{str}(t) \mid t \in STS(G)\}.$$

The following definition generalizes the notion of E0L structural equivalence [10–12, 17] for regularly extended grammars.

Definition 1. *Let G_1 and G_2 be regularly extended E0L grammars. The grammars G_1 and G_2 are* structurarly equivalent *if*

$$STS(G_1) = STS(G_2).$$

To conclude this section we recall some basic definitions concerning tree automata that will be needed for the proof of our main result. Differing from the usual model of tree automata operating on trees over ranked alphabets, we consider trees over unranked alphabets, see e.g. [4].

Let Ω be a finite alphabet that is used to label the trees. A *regularly extended bottom-up tree automaton* is a tuple

$$M = (\Omega, Q, Q_F, \delta), \tag{2}$$

where Q is a finite set of states, $Q_F \subseteq Q$ is a set of accepting states and δ associates to each $\tau \in \Omega$ a relation $\tau_\delta \subseteq Q^* \times Q$ such that for every $q \in Q$, $\tau \in \Omega$ the set

$$L(q, \tau) = \{w \in Q^* \mid (w, q) \in \tau_\delta\} \tag{3}$$

is regular. In the following, unless otherwise mentioned, by a tree automaton we always mean a regularly extended bottom-up tree automaton.

By an M-configuration we mean an $(\Omega \cup Q)$-tree where elements of Q occur only as labels of leaves and the set of M-configurations is $\text{conf}(M)$. The computation relation of M, $\vdash_M \subseteq \text{conf}(M) \times \text{conf}(M)$, is defined as follows. Let $t, t' \in \text{conf}(M)$. Then $t \vdash_M t'$ if t' is obtained from t by replacing a subtree $\tau(q_1, \ldots, q_m)$, $\tau \in \Omega$, $q_1, \ldots, q_m \in Q$ by a single node labeled by $p \in Q$ such that

$$(q_1 \cdots q_m, p) \in \tau_\delta.$$

Note that if above τ labels a leaf of t (that is, $m = 0$), then τ is replaced by a state p such that $(\lambda, p) \in \tau_\delta$.

The forest, or tree language, $(\subseteq F_\Omega)$ recognized by M is

$$L(M) = \{t \in F_\Omega \mid (\exists q \in Q_F)\, t \vdash^*_M \bar{q}\}.$$

Above $\bar{q}$ denotes the tree with a single node labeled by q.

A tree automaton (2) is *deterministic* if for each $\tau \in \Omega$ and $w \in Q^*$ there is at most one $p \in Q$ such that $(w, p) \in \tau_\delta$. It is well known that for an arbitrary nondeterministic bottom-up tree automaton we can construct an equivalent deterministic automaton. A forest is regular if it accepted by a (deterministic or nondeterministic) bottom-up tree automaton.

The following lemma is proved using the standard direct product construction. First we just need to add a "dead state" for the second automaton M_2 which ensures that the computation of M_2 always reaches the root. The set of accepting states of the direct product automaton consists of all pairs where the first component is an accepting state of M_1 and the second component is not an accepting state of M_2. It is easy to verify that the state transition relation obtained from the direct product construction satisfies condition (3), that is, the constructed automaton is also a regularly extended tree automaton.

Lemma 1. *Let $M_i = (\Omega, Q_i, Q_{F,i}, \delta_i)$, $i = 1, 2$, be deterministic tree automata. Then we can effectively construct a deterministic tree automaton M such that*

$$L(M) = L(M_1) - L(M_2).$$

3 The Main Result

Clearly the set of syntax trees (or structures of syntax trees) of an reE0L grammar is not regular since a finite tree automaton cannot check that the tree represents a parallel derivation. However, we can use tree automata to decide structural equivalence by restricting consideration to balanced input trees or, strictly speaking, to input trees where each path from the root to a non-λ node is of equal length. Such trees correspond to parallel E0L derivations and then it is sufficient for the tree automaton to check that the derivation is correct locally. For the decidability result it is essential that the automaton can be constructed to be deterministic.

The following notation is introduced to deal with "almost balanced" trees corresponding to derivations of an reE0L grammar, where leaves labeled by $\hat{\lambda}$ can occur at any depth. Let Ω be an alphabet and $\tau \in \Omega$. The set

$$\mathrm{BAL}_\Omega(\tau) \subseteq F_\Omega \tag{4}$$

is defined to consist of all Ω-trees t such that

(i) The symbol τ occurs only as a label of leaves in t.
(ii) If u_1, u_2 are any leaves of t not labeled by τ then the distance of u_1 and u_2 from the root of t is the same.

The above conditions mean that t is a balanced tree with the exception that leaves labeled by the special symbol τ can occur at any level.

Let G be an reE0L grammar as in (1) and $\Omega = \varpi \cup \Sigma \cup \hat{\lambda}$. Then we can note that all structures of terminal syntax trees of G belong to $\mathrm{BAL}_\Omega(\hat{\lambda})$.

The following lemma is an extension of the well known corresponding result for (extended) context-free grammars with the additional requirement that we consider only "almost parallel" input trees.

Lemma 2. *Let $G = (V, \Sigma, S, P)$ be an reE0L grammar and $\Omega = \varpi \cup \Sigma \cup \hat{\lambda}$. Then we can effectively construct a deterministic tree automaton M such that*

$$L(M) \cap \mathrm{BAL}_\Omega(\hat{\lambda}) = STS(G). \tag{5}$$

Proof. Choose

$$M = ((\varpi \cup \Sigma \cup \hat{\lambda}), (\wp(V) \cup \bar{\Sigma} \cup q_\lambda), \mathcal{V}, \delta),$$

where $\bar{\Sigma} = \{\bar{\sigma} \mid \sigma \in \Sigma\}$, $\mathcal{V} = \{U \subseteq V \mid S \in U\}$ and δ is defined as follows.

(i) For $\sigma \in \Sigma$, we set $\sigma_\delta = \{\lambda\} \times \{\bar{\sigma}\}$.
(ii) $\hat{\lambda}_\delta = \{\lambda\} \times \{q_\lambda\}$.
(iii) Let $U_1, \ldots, U_m \in \wp(V)$, $m \geq 1$. Then $(U_1 \cdots U_m, U) \in \varpi_\delta$ for

$$U = \{B \in V \mid (\exists B_i \in U_i, i = 1, \ldots, m)\ B \to B_1 \cdots B_m \in P\}.$$

The set U is the only set X such that $(U_1 \cdots U_m, X) \in \varpi_\delta$.

Since P is the production set of an reE0L grammar we know that for any set $U \subseteq V$ the set

$$\{U_1 \cdots U_m \mid (\forall B \in U)(\exists B_i \in U_i, i = 1, \ldots, m)\ B \to B_1 \cdots B_m \in P\}\ (\subseteq \wp(V)^*)$$

is an intersection of finitely many regular languages, and hence regular. This means that the rules defined in (iii) can be used in a regularly extended tree automaton. Directly by their definition, all the rules (i), (ii) and (iii) are deterministic.

Note that rules (i) mean that at a leaf labeled by σ the computation begins in state $\bar{\sigma}$. Similarly, rule (ii) says that at leaves labeled by $\hat{\lambda}$ the computation begins in state q_λ (which is a new symbol that we use instead of the rather cumbersome notation $\hat{\bar{\lambda}}$). Let t be any tree of height at least one. Using the definition of the rules (iii) and induction on the height of a tree $t \in \mathrm{BAL}_\Omega(\hat{\lambda})$ we see that M reaches the root of t in a state $U \in \wp(V)$ where

$U = \{B \in V \mid$
$\qquad (\exists t' \in S(G))$ such that the root of t' is labeled by B and $str(t') = t\}$.

Note that if $t \in \mathrm{BAL}_\Omega(\hat{\lambda})$ has some internal nodes labeled by symbols other than ϖ then $U = \emptyset$ and the computation of M becomes blocked before reaching the root.

By the choice of the set of accepting states $\mathcal{V}$ the above means that the equation (5) holds. □

It is well known that emptiness of regularly extended tree automata is decidable. In the below two lemmas we show that also emptiness modulo the set of "almost balanced" trees can be decided effectively.

We say that a tree t is *k-bounded* if any node of t has at most k children.

Lemma 3. *Given a tree automaton* $M = (\Omega, Q, Q_F, \delta)$ *we can effectively compute a constant k such that the following condition holds.*

Let $\tau \in \Omega$ *and suppose that* $t \in L(M) \cap \mathrm{BAL}_\Omega(\tau)$. *Then there exists a k-bounded tree* $t' \in L(M) \cap \mathrm{BAL}_\Omega(\tau)$ *such that the height of t' is less than or equal to the height of t.*

Proof. For $q \in Q$ and $\omega \in \Omega$ let $n_{q,\omega}$ denote the number of states of the minimal deterministic finite automaton for the language $L(q,\omega)$ from (3). Choose

$$k = \max_{q \in Q, \omega \in \Omega} n_{q,\omega}. \tag{6}$$

Consider an arbitrary $t \in L(M) \cap \mathrm{BAL}_\Omega(\tau)$ and let u be a node of t having m children, where $m > k$. Let C be an accepting computation of M on t and let $(q_1 \cdots q_m, p) \in \omega_\delta$ be the computation step used at the node u, where $\omega \in \Omega$ is the label of u and $q_1, \ldots, q_m, p \in Q$. Since $m > k$, the equation (6) implies that there exist $1 \le i < j \le m$ such that $q_1 \cdots q_i q_{j+1} \cdots q_m \in L(p,\omega)$, that is,

$$(q_1 \cdots q_i q_{j+1} \cdots q_m, p) \in \omega_\delta. \tag{7}$$

Let t' be the tree that is obtained from t by deleting the $(i+1)$st, $\ldots$, jth immediate subtrees of the node u. By (7) and the fact that C is an accepting computation, M has an accepting computation on t'. Also we note that if r is any tree in $\mathrm{BAL}_\Omega(\tau)$ and we delete a number of immediate subtrees of a node v of r, the resulting tree is still in $\mathrm{BAL}_\Omega(\tau)$ assuming that we do not delete all the immediate subtrees of v. Thus $t' \in \mathrm{BAL}_\Omega(\tau)$.

By repeating the above process as long as there are nodes having more than k children, we obtain a k-bounded tree that is in the forest $L(M) \cap \mathrm{BAL}_\Omega(\tau)$ and each step in the process either preserves the height of the tree or decreases it. The latter can happen if all the remaining subtrees of the given node have all leaves labeled by the special symbol τ.

Also it is clear that given the transition relation of the tree automaton M we can effectively compute the constant k.   □

For the below result recall that the notation $\mathrm{BAL}_\Omega(\tau)$ is as defined in (4).

Lemma 4. *Given a tree automaton* $M = (\Omega, Q, Q_F, \delta)$ *and* $\tau \in \Omega$ *we can effectively decide whether or not*

$$L(M) \cap \mathrm{BAL}_\Omega(\tau) = \emptyset.$$

Proof. Let $t \in \mathrm{BAL}_\Omega(\tau)$. The ith level of t is defined to consist of nodes that are at distance i from the root. Thus

all subtrees rooted at level i nodes and having

a non-τ leaf have the same height. (8)

By a non-τ leaf we mean a leaf node labeled by a symbol in $\Omega - \tau$.

Assume that M accepts t and let C be the accepting computation of M on t. Let A_i denote the set of states that the computation C reaches at nodes on the ith level, where $0 \le i \le \mathrm{height}(t)$. Now if $\mathrm{height}(t) > 2^{\#Q}$ we can guarantee that there exist $0 \le i < j \le \mathrm{height}(t)$ such that $A_i \subseteq A_j$. Let t' be the tree obtained from t by replacing each subtree r rooted at the ith level with a subtree s rooted at the jth level such that computation C reaches the root of r and the root of s in the same state. Thus M accepts t'. Also, by (8), $t' \in \mathrm{BAL}_\Omega(\tau)$.

By repeating the above process we see that if $L(M) \cap \mathrm{BAL}_\Omega(\tau)$ is not empty then it contains a tree of height at most $2^{\#Q}$. Now the result of Lemma 3 implies that $L(M) \cap \mathrm{BAL}_\Omega(\tau)$ contains a k-bounded tree of height at most $2^{\#Q}$ where k can be effectively computed when the tree automaton M is given. This means that in order to decide emptiness of $L(M) \cap \mathrm{BAL}_\Omega(\tau)$, it is sufficient to check whether $L(M)$ accepts a constant number of candidate trees belonging to $\mathrm{BAL}_\Omega(\tau)$ where the candidate trees can be effectively found.   □

Now we are ready to prove the main result.

Theorem 1. *Structural equivalence of regularly extended E0L grammars is decidable.*

Proof. Assume we are given reE0L grammars G_1 and G_2 with terminal alphabet Σ. Without loss of generality we can assume that both grammars use the same terminal alphabet, otherwise we consider the union of the terminal alphabets. Let $\Omega = \varpi \cup \Sigma \cup \hat{\lambda}$. By Lemma 2, we can construct deterministic regularly extended tree automata M_i such that

$$L(M_i) \cap \mathrm{BAL}_\Omega(\hat{\lambda}) = STS(G_i), \quad i = 1, 2.$$

By Lemma 1 we can construct (deterministic) tree automata $N_{1,2}$ and $N_{2,1}$ such that

$$L(N_{i,j}) \cap \mathrm{BAL}_\Omega(\hat{\lambda}) = (L(M_i) \cap \mathrm{BAL}_\Omega(\hat{\lambda})) - (L(M_j) \cap \mathrm{BAL}_\Omega(\hat{\lambda})), \quad \{i,j\} = \{1,2\}.$$

Now G_1 and G_2 are structurally equivalent if and only if $L(N_{i,j}) \cap \mathrm{BAL}_\Omega(\hat{\lambda}) = \emptyset$, when $(i,j) = (1,2)$ and $(i,j) = (2,1)$. By Lemma 4 these conditions can be tested algorithmically.   □

Lemma 2 can be immediately modified to show that we can construct a deterministic tree automaton M such that $L(M) \cap \mathrm{BAL}_\Omega(\hat{\lambda}) = TS(G)$, where $G = (V, \Sigma, S, P)$ and $\Omega = V \cup \Sigma \cup \hat{\lambda}$. Recognizing the syntax trees of G is in fact easier than recognizing the structures of syntax trees of G (in both cases modulo the set of "almost balanced" trees) since in the case of syntax trees the rules (iii) in the construction of the proof of Lemma 2 do not need to consider all possible nonterminals appearing as left side of the rule, in syntax trees the nonterminal is given as the label of the node. This means that the proof of Theorem 1 gives immediately the following:

Corollary 1. *For given reE0L grammars G_1 and G_2 we can effectively decide whether or not $TS(G_1) = TS(G_2)$, i.e., whether or not the grammars are syntax equivalent.*

To conclude we can note that the algorithm obtained from the proof of Theorem 1 is extremely inefficient and the complexity of the structural equivalence of reE0L grammars remains open. From [15] we get a lower bound for the complexity since the structural equivalence of E0L grammars with finitely many rules is already hard for deterministic exponential time.

References

1. Albert, J., Giammarresi, D., Wood, D.: Normal form algorithms for extended context-free grammars. Theoret. Comput. Sci. **267** (2001) 35–47.
2. Berstel, J., Boasson, L.: Formal properties of XML grammars and languages. Acta Informatica **38** (2002) 649–671.
3. Brüggemann-Klein, A., Wood, D.: Caterpillars: A context-specification technique. Mark-up Languages: Theory & Practice **2** (2000) 81–106.
4. Brüggemann-Klein, A., Wood, D.: The regularity of two-way nondeterministic tree automata languages. Internat. J. of Foundations of Computer Science **13** (2002) 67–81.
5. Cameron, H.A., Wood, D.: Structural equivalence of regularly extended context-free grammars and SGML DTDs. Manuscript, 1996.
6. Gécseg, F., Steinby, M.: Tree Automata. Académiai Kiadó, Budapest, 1984.
7. Gécseg, F., Steinby, M.: Tree languages. In: Rozenberg, G., Salomaa, A. (eds.): Handbook of Formal Languages, Vol. III, Springer-Verlag (1997) 1–68.
8. Istrate, G.: The strong equivalence of ET0L grammars. Inf. Process. Lett. **62** (1997) 171–176.
9. McNaughton, R.: Parenthesis grammars. J. Assoc. Comput. Mach. **14** (1967) 490–500.
10. Niemi, V.: A normal form for structurally equivalent E0L grammars. In: Rozenberg, G., Salomaa, A. (eds): Lindenmayer Systems: Impacts on Theoretical Computer Science, Computer Graphics, and Developmental Biology. Springer-Verlag (1992) 133–148.
11. Ottmann, Th., Wood, D.: Defining families of trees with E0L grammars. Discrete Applied Math. **32** (1991) 195–209.
12. Ottmann, Th., Wood, D.: Simplifications of E0L grammars. In: Rozenberg, G., Salomaa, A. (eds): Lindenmayer Systems: Impacts on Theoretical Computer Science, Computer Graphics, and Developmental Biology. Springer-Verlag (1992) 149–166.
13. Paull, M., Unger, S.: Structural equivalence of context-free grammars. J. Comput. System Sci. **2** (1968) 427–463.
14. Rozenberg, G., Salomaa, A.: The Mathematical Theory of L Systems. Academic Press, New York, 1980.
15. Salomaa, K., Wood, D., Yu, S.: Complexity of E0L structural equivalence. RAIRO Theoretical Informatics **29** (1995) 471–485.
16. Salomaa, K., Wood, D., Yu, S.: Structural equivalence and ET0L grammars. Theoret. Comput. Sci. **164** (1996) 123–140.
17. Salomaa, K., Yu, S.: Decidability of structural equivalence of E0L grammars. Theoret. Comput. Sci. **82** (1991) 131–139.
18. Salomaa, K., Yu, S.: Decidability of EDT0L structural equivalence. Theoret. Comput. Sci. **276** (2002) 245–259.
19. Thatcher, J.W.: Tree automata: an informal survey. In: Aho, A.V. (ed.): Currents in the Theory of Computing. Prentice Hall, Englewood Cliffs, NJ (1973) 143–172.
20. Wood, D.: Theory of Computation. John Wiley & Sons, New York, NY, 1987.

Complexity of Evolving Interactive Systems*

Peter Verbaan[1], Jan van Leeuwen[1], and Jiří Wiedermann[2]

[1] Institute of Information and Computing Sciences, Utrecht University,
Utrecht, the Netherlands
[2] Institute of Computer Science, Academy of Sciences of the Czech Republic,
Prague, Czech Republic

Abstract. We study a versatile model of evolving interactive computing: lineages of automata. A lineage consists of a sequence of interactive
finite automata, with a mechanism of passing information from each automaton to its immediate successor. Lineages enable a definition of a
suitable complexity measure for evolving systems. We show several complexity results, including a hierarchy result.

1 Introduction

It is commonly recognised that the Turing machine model, which has long been
the basis for theoretical computer science, fails to capture all the characteristics
of modern day computing (cf. [5, 11, 13]). When we think of a modern networked
computing system, we see a device that can interact with its environment and
that can be changed over time (by installing new hardware or upgrading the
software). The system is 'always on' and a computation on such a device can
extend arbitrarily, which implies the existence of potentially infinite computations. Networked machines and the programs that run on them are examples of
evolving interactive systems ([4, 6]). Further examples are given in [14, 15].

If we look at a (deterministic) computation (or a transduction) from a traditional point of view, the entire input and the underlying system are fixed the
moment we start the computation. In a more realistic setting, we want a user
to generate the input interactively and allow the user to alter the system's behaviour during the computation. Last but not least, computations can be potentially never-ending. Systems that allow these kinds of computations are called
evolving interactive systems. In this paper, we define lineages of automata, a
simple yet elegant model which captures the evolving aspect of computational
systems in a natural way. It turns out that even this simple model is more powerful than classical Turing machines. This was observed in [4, 6], where lineages of
automata were shown to be equivalent to so-called interactive Turing machines
with advice. The latter machines are known to possess super-Turing computing
power.

A lineage is a sequence of interactive, finite automata with a mechanism of
passing information from each automaton to its immediate successor and the

---
* The research of the third author was partially supported by GA ČR grant No.
201/02/1456.

potential to process infinite input streams. Every automaton in the sequence can be seen as a temporary instantiation of the system, before it changes into a different automaton. We study the properties of lineages through the translations they realize. Lineages of interactive finite automata (or: transducers) have been introduced in [6].

The concept of transducers acting on infinite input streams (ω-transducers) is not new. For example, [9] gives an overview of the theory of finite devices that operate on infinite objects. In the field of non-uniform complexity theory, sequences of computing devices are common-place ([1]). It is the idea of combining these concepts and allowing some form of communication between the devices in the sequence that is new and that allows for a closer modelling of system evolution. The approach leads to several new fundamental questions that are settled in this paper.

We can measure the "speed of growth" (i.e. "growth complexity") of a lineage by a function that relates the index of each automaton to its size. That is, the complexity of a lineage is a function g such that the n-th automaton in the sequence has $g(n)$ states. Using this measure, we can divide the translations computed by evolving systems into classes based on the complexity of the lineages that realize them. Our main result states that this division is non-trivial and leads to strict hierarchies, i.e. for every positive, non-decreasing function g, there is a translation that can be realized by a lineage of complexity g, but not by any lineage of lower complexity.

The structure of the paper is as follows. In section 2, we define lineages. Next, in section 3, we define a novel measure of complexity on translations and establish the hierarchy result.

1.1   Notation

In most literature, the term transducer is used to denote an automaton with output capability. Every time we use the word automaton, the term transducer could be substituted for it without ill effects.

We use Σ and Ω to denote alphabets. We use the notations $\Sigma^\star$, Σ^ω for the sets of finite and infinite strings over Σ respectively and Σ^∞ for the union of $\Sigma^\star$ and Σ^ω. We call a partial function from Σ^∞ to Ω^∞ a *translation*.

For a string $x \in \Sigma^\infty$ of length at least n, we denote the n-th symbol of x by x_n, or $(x)_n$, to improve readability. We write $x_{[i:j]}$ for $x_i x_{i+1} \ldots x_j$. We also use the projection functions for tuples, π_n, which are defined straightforwardly, i.e., π_n 'returns' the n-th component of a tuple that serves as the argument of π_n, for any n.

Let D be a subset of Σ^∞. We define the n-th *prefix domain* $P^n(D)$ as the set of all prefixes of length n of strings in D,

$$P^n(D) = \{ \, x_{[1:n]} \mid x \in D \, \} \, . \tag{1}$$

We define a topology on Σ^∞ as follows. Let u be a finite string over Σ. Then the set of all possible extensions of u,

$$\mathbb{B}(u) := \{ \, x \in \Sigma^\infty \mid u \text{ is a prefix of } x \, \} \tag{2}$$

is a *basis set*. Let S be a subset of Σ^∞. We call S an *open set* if it is a union of basis sets. A set is *closed* if it is the complement of an open set.

2   Modelling Evolving Interactive Systems by Lineages

As we explained in the introduction, the models of classical computability theory are not sufficient to capture all aspects of modern computing systems. This calls for theories that better describe these systems. Various extensions of classical models of computation have been studied that capture non-classical features of modern systems in some way (cf. [4, 12]). In this paper, we develop a new model for computing systems, initially outlined in [6]: lineages, inspired by a similar notion in evolutionary biology.

The building blocks of the model are a generalisation of Mealy automata. These automata process potentially infinite input streams and produce potentially infinite output streams, one symbol at a time. We do not assume that the input is provided on an input tape. Instead, the automaton reads its input from a single input port. One symbol is read from this port at each step. Similarly, the output goes to a single output port, one symbol at a time. In contrast to classical models, the input stream does not have to be known in advance, and can be adjusted at any time by an external agent, based on previous in- and output symbols. This allows the environment to interact with the automaton.

We model the evolutionary aspects by considering sequences of automata. Each automaton in the sequence represents the next evolutionary phase of the system. The way in which this sequence develops need not be described recursively in general. When a transition occurs from one automaton to its successor, the information that the automaton has accumulated over time must be preserved in some way. This is done by requiring that every automaton has a subset of its states in common with its immediate successor.

Definition 1. *An automaton is a 6-tuple* $A = (\Sigma, \Omega, Q, I, O, \delta)$, *where* Σ *and* Ω *are non-empty finite alphabets,* Q *is a set of states,* I *and* O *are subsets of* Q, *and* $\delta : Q \times \Sigma \to Q \times \Omega$ *is a (partial) transition function.* Σ *is the input alphabet, and* Ω *is the output alphabet. We call* I *the set of* entry *states and* O *the set of* exit *states.*

Definition 2. *Let* $\mathcal{A}$ *be a sequence of automata* A_1, A_2, $\ldots$, *with the automata* $A_i = (\Sigma, \Omega, Q_i, I_i, O_i, \delta_i)$ *such that* $O_i \subseteq I_{i+1}$ *for every* i. *We call* $\mathcal{A}$ *a* lineage *of automata, or a* lineage *for short.*

We do not require a recursive recipe for constructing the sequences of automata. The elements in $Q_i - O_i$ are called *local states* (of A_i). The first automaton, A_1, has an initial state $q_{\text{in}} \in I_1$. Usually, I_1 contains only the initial state of A_1, and I_{i+1} equals O_i. See Fig. 1 for an example.

A lineage $\mathcal{A}$ operates on elements of Σ^∞. On an input string x, at any time, only one automaton processes x. The automaton that processes x at a particular time is called *active* (at that time). Initially, A_1 is the active automaton, and it

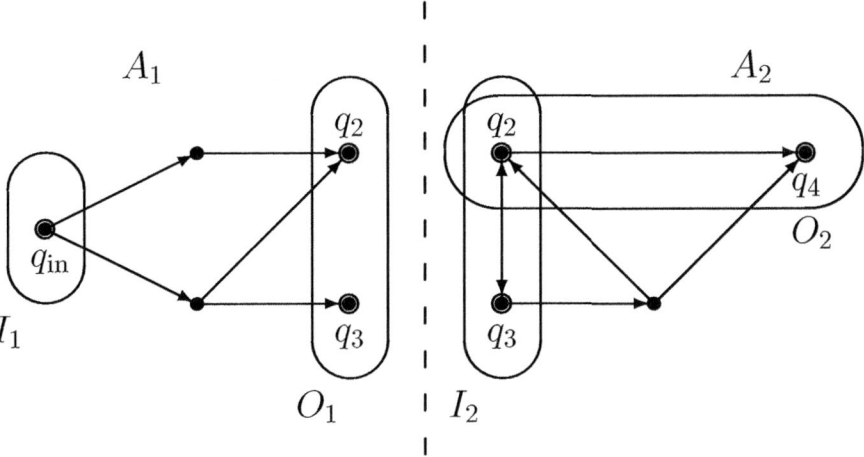

Fig. 1. Part of a lineage $\mathcal{A}$. The set of exit states of A_1 is a subset of the set of entry states of A_2

starts processing x. Whenever an active automaton A_i enters an exit state q, it turns the control over to A_{i+1}, which then becomes the active automaton. This is done by letting A_{i+1} start processing the remainder of x, beginning in state q (which is an entry state of A_{i+1} by definition). This is called *updating*, and A_i is the i-th *update* of $\mathcal{A}$.

Formally, let Q be the union of all Q_i and let $x \in \Sigma^\infty$ be an input to a lineage $\mathcal{A}$. Using simultaneous recursion, we define a sequence of states $(q_j)_{j \geq 1}$ in Q and a sequence of integers $(m_j)_{j \geq 1}$, with m_j representing the index of the active automaton at time j, as follows:

$$q_1 = q_{\text{in}} \ ,$$
$$m_1 = 1 \ ,$$

$$q_{j+1} = \pi_1 \left(\delta_{m_j}(q_j, x_j) \right) \ , \tag{3}$$
$$m_{j+1} = \begin{cases} m_j + 1 & \text{if } q_{j+1} \in O_{m_j} \\ m_j & \text{if } q_{j+1} \in Q_{m_j} - O_{m_j} \\ \text{undefined if } q_{j+1} \text{ is undefined} \end{cases} .$$

Note that q_{j+1} and m_{j+1} depend on $x_{[1:j]}$. Therefore, we also write $q_{j+1}(x_{[1:j]})$ and $m_{j+1}(x_{[1:j]})$ to emphasize the dependence. If q_j is defined for every $j \leq |x| + 1$, then we say that x is a valid input to $\mathcal{A}$. In this case, the *output of $\mathcal{A}$ on x* is the string $y \in \Omega^\infty$ such that $y_j = \pi_2 \left(\delta_{m_j}(q_j, x_j) \right)$, for every $j \geq 1$.

Definition 3. *Let $\mathcal{A}$ be a lineage. For every $n \geq 1$, we define the partial function $\phi^{\mathcal{A},n} : \Sigma^n \to \Omega^n$ by letting $\phi^{\mathcal{A},n}(x)$ be the output of $\mathcal{A}$ on x if x is a valid input and undefined otherwise, for every x of length n. We say that $\phi^{\mathcal{A},n}$ is realized by the lineage $\mathcal{A}$. In general, for a partial function $\psi : \Sigma^n \to \Omega^n$, we say that ψ is realizable, if there is a lineage $\mathcal{A}$ such that ψ equals $\phi^{\mathcal{A},n}$.*

Since there are only finitely many strings of length n, for any lineage $\mathcal{A}$ and integer n, the translation $\phi^{\mathcal{A},n}$ can be realized by a single finite-state automaton, which justifies the definition. There is no need to restrict our attention to finite strings.

Definition 4. *Let $\mathcal{A}$ be a lineage. We define the partial function $\Phi^{\mathcal{A}} : \Sigma^{\infty} \to \Omega^{\infty}$ by letting $\Phi^{\mathcal{A}}(x)$ be the output of $\mathcal{A}$ on x if x is a valid input and undefined otherwise, for every infinite string x. We say that $\Phi^{\mathcal{A}}$ is non-uniformly realized by the lineage $\mathcal{A}$. In general, for a partial function $\Psi : \Sigma^{\infty} \to \Omega^{\infty}$, we say that Ψ is non-uniformly realizable, if there is a lineage $\mathcal{A}$ such that Ψ equals $\Phi^{\mathcal{A}}$.*

For many lineages $\mathcal{A}$, the translation $\Phi^{\mathcal{A}}$ is not realizable by a single finite-state automaton and not even by a Turing transducer, see [10] for details. For the remainder of this paper, we consider translations on infinite strings, unless stated otherwise. See also [6].

Let Φ be a translation and n an integer. We say that $(\Phi)_n$ [3] *depends only on the first n input symbols*, if there is a function $f : \Sigma^n \to \Omega$, such that $(\Phi(x))_n$ equals $f(x_{[1:n]})$ for every x in the domain of Φ. A non-uniformly realizable translation has this property for every n, since

$$(\Phi(x))_n = \pi_2\left(\delta_{m_n}(q_n, x_n)\right) = \pi_2\left(\delta_{m_n\left(x_{[1:n-1]}\right)}\left(q_n\left(x_{[1:n-1]}\right), x_n\right)\right) . \qquad (4)$$

Let $\mathcal{A} = A_1, A_2, A_3, \ldots$ be a lineage of automata and let n and m be integers. In a slight abuse of notation, we say that A_m *is able to process all strings of length n*, if $m_n(x_{[1:n]}) \leq m$ for every string x. In other words, if for any string x, the lineage $\mathcal{A}$ needs less than m updates to process the first n symbols of x.

2.1   Alternative Characterisations of Lineages

In [10], we show that lineages of automata are equivalent to interactive Turing Machines with advice. Among other things, this shows that the theory of lineages has deep connections with non-uniformity theories.

The following two Propositions provide a useful characterisation of non-uniformly realizable translations.

Proposition 1. *Let Φ be a non-uniformly realizable translation. Then the domain of Φ is closed and $(\Phi)_n$ depends only on the first n input symbols, for every n.*

Proof. Let D be the domain of Φ and let $\mathcal{A}$ be a lineage that non-uniformly realizes Φ. Let $x \notin D$ be an infinite string and consider a run of $\mathcal{A}$ on x. Because x is not in the domain of Φ, it must be the case that at some point, after processing a prefix of length $n - 1$ of x, the automaton that is active at that time, say A_k, is in a certain state q, such that there is no transition from q with x_n as input. If this moment would not occur, then $\mathcal{A}$ would never halt during the run, and x would be in the domain.

---
[3] Where $(\Phi)_n$ is shorthand for the function $x \mapsto (\Phi(x))_n$.

Let y be a string in $\mathbb{B}(x_{[1:n]})$ and consider a run of $\mathcal{A}$ on y. Since the first n symbols of x and y are the same, the computation will halt at the same point during the computation, so y cannot be in the domain of Φ. Hence $\mathbb{B}(x_{[1:n]})$ does not intersect D, which implies that D is closed.

It follows from (4), that $(\Phi(x))_n$ depends only on the first n input symbols.

□

Proposition 2. *Let Φ be a translation. Suppose the domain D of Φ is closed and $(\Phi)_n$ depends only on the first n input symbols, for every n. Then Φ is non-uniformly realizable by a lineage of size $|P^n(D)|$.*

Proof. Let D be the domain of Φ. Let f_n be the function with domain $P^n(D)$, such that $(\Phi(x))_n = f_n(x_{[1:n]})$ for every $x \in D$. By the assumptions of the Proposition, f_n is well-defined.

We construct a lineage $\mathcal{A}$ that non-uniformly realizes Φ. Every state of $\mathcal{A}$ will correspond to a prefix u of a string in D. We label the corresponding state by $[\![u]\!]$. Define the set of states of A_n for $n \geq 1$ by

$$I_n = \{ \ [\![u]\!] \ \mid \ u \in P^{n-1}(D) \ \} \ ,$$
$$O_n = \{ \ [\![u]\!] \ \mid \ u \in P^n(D) \ \} \ . \tag{5}$$

The trick is to choose the states such that I_n is a subset of O_n for every n. The initial state of A_1 is $[\![\epsilon]\!]$. The transition function δ_n is defined by

$$\delta_n([\![u]\!], a) = \begin{cases} ([\![ua]\!], f_n(ua)) & \text{if } ua \in P^n(D) \\ \text{undefined} & \text{otherwise} \end{cases} \ . \tag{6}$$

By the definition of f_n, we infer that the lineage $\mathcal{A}$ produces $\Phi(x)$ on input $x \in D$.

Suppose on the other hand that $x \notin D$. Since D is closed, there is a basis set $\mathbb{B}(u)$ that contains x, which does not intersect D. It follows that $u \notin P^{|u|}(D)$. Since the transition functions are not defined on u, we see that x is not a valid input to $\mathcal{A}$. Hence $\mathcal{A}$ non-uniformly realizes Φ. Note that $\mathcal{A}$ is of size $|P^n(D)|$.

□

3   The Complexity of Lineages

In this section, we develop the notion of complexity of lineages and establish several results which allow us to compare many translations, based on the complexity of the lineages that non-uniformly realize them.

The processing power of an automaton is directly related to the number of states. An automaton with more states is able to distinguish among a greater number of different situations. It can apply different actions to each situation it can recognise, thus adding more diversity to a computation.

For a lineage, which is a sequence of automata, the number of states of each of the constituent automata contributes to the computing power of the lineage. Therefore, we use a function to describe the complexity of a lineage.

Definition 5. *The complexity of a lineage $\mathcal{A}$ is a function g such that for every n, the number of states of A_n equals $g(n)$. We say that a translation Φ is of complexity g if there is a lineage $\mathcal{A}$ of complexity g that non-uniformly realizes Φ. We define the complexity class SIZE(g) as the class of non-uniformly realizable translations of complexity g or less.*

First, we give an upper bound on the complexity of any non-uniformly realizable translation.

Proposition 3. *Let Φ be a non-uniformly realizable translation over an alphabet of size c. Then Φ can be non-uniformly realized by a lineage of size at most c^n.*

Proof. By Proposition 1 and Proposition 2, we can obtain a lineage $\mathcal{A}$ of size $|P^n(D)|$ that non-uniformly realizes Φ. There are at most c^n strings of length n, so $|P^n(D)| \leq c^n$.

□

3.1 Complexity Classes and the Functions That Represent Them

We have expressed the complexity of an evolving interactive system by a positive integer-valued function. Conversely, we ask which positive integer-valued functions represent a complexity class. Some functions do not naturally correspond to a complexity class, e.g. the super-exponential functions (Proposition 3). If a function is non-decreasing and has a "growth rate"[4] that is bounded by a constant, then it corresponds to a complexity class. If it is not, then we take a suitable function that is nowhere greater than the original function and consider its corresponding class. This is made precise below.

Let $g : \mathbb{N} \to \mathbb{N}$ be an arbitrary positive non-decreasing function and c a positive integer. Define the function $g_c(n)$ by

$$
\begin{aligned}
g_c(1) &= \min\{\, g(1) \quad, c \qquad \} \\
g_c(n+1) &= \min\{\, g(n+1), c \cdot g_c(n) \,\}
\end{aligned}
\tag{7}
$$

It follows that for every n,

– $g_c(n) \leq g(n)$,
– $g_c(n) \leq c^n$,
– $g_c(n) \leq g_c(n+1)$, and
– $g_c(n+1) \leq c \cdot g_c(n)$.

In other words, g_c is a positive, non-decreasing function that is bounded by g and c^n, with a "growth rate" that is bounded by c.

To show that the class SIZE(g_c) is not empty, we construct a translation $\Phi^{g,c}$ that is in this class. We also show that $\Phi^{g,c}$ is not in SIZE(h), for any function h such that $h(n) < g_c(n)$ for some n. The construction is based on the following idea: suppose $\mathcal{A}$ is a lineage that non-uniformly realizes a translation

---

[4] The growth rate of a function g is defined as $g(n+1)/g(n)$

Φ. Two input-prefixes u and v (not necessarily of the same size) are considered inequivalent when there is an infinite string x and an integer i such that $\Phi(ux)$ and $\Phi(vx)$ both exist and $(\Phi(ux))_{|u|+i} \neq (\Phi(vx))_{|v|+i}$. This is impossible if $\mathcal{A}$ enters the same state after processing either u or v. Thus, to make sure that an automaton of a lineage has at least k states, we need to make sure that there are at least k inequivalent inputs to choose from, once this automaton is active.

First, we establish the domain of $\Phi^{g,c}$. Let Σ be an alphabet of size c. For every n, we choose $g_c(n)$ strings of length n, such that they are prefixes of the $g_c(n+1)$ strings of length $n+1$. The details are given in Construction 1.

Construction 1. Label the letters from Σ as a_1 through a_c, and let C_n be the chosen subset of size $g_c(n)$ of Σ^n. We proceed recursively.

$$C_1 = \{\, a_i \quad | \quad i \leq g_c(1) \,\} \ . \tag{8}$$

Assume C_n is chosen. Using integer division, we write $g_c(n+1) = l \cdot g_c(n) + m$, for unique integers l and m, with $0 \leq m < g_c(n)$. It follows that $1 \leq l \leq c$. Let $u_1, \ldots, u_m$ be m different strings in C_n. Now take

$$C_{n+1} = \{\, ua_i \quad | \quad u \in C_n \wedge i \leq l \,\} \cup \{\, u_j a_{l+1} \quad | \quad j \leq m \,\} \ . \tag{9}$$

It is left to the reader to verify that C_{n+1} contains $g_c(n+1)$ strings of length $n+1$. Note that C_{n+1} is well-defined, as either $l < c$ or $l = c$ and $m = 0$.

We call a string in C_n a *choice*. Note that ua_1 is a choice if u is a choice. Consider an infinite sequence of choices, such that each choice is a prefix of its successor. Such a sequence defines a unique infinite string, such that each of its prefixes is a choice. Let Δ be the set of infinite strings x such that each prefix of x is a choice. The translation $\Phi^{g,c}$ will be defined on the domain Δ.

Construction 2. Consider the family of functions $f_{k,l,m} : \Sigma^\infty \to \Sigma^{1+l}$, defined by

$$f_{k,l,m}(x) = \begin{cases} x_k^{1+l} & \text{if } 0 < k \leq m \\ x_1^{1+l} & \text{otherwise} \end{cases} , \tag{10}$$

for $k, l, m \in \mathbb{N}$. Let $\psi : \mathbb{N} \to \mathbb{N} \times \mathbb{N}$ be a surjective function that attains each value infinitely often. Then the translation Φ is defined by

$$\Phi(x) = f_{\psi(1),1}(x) f_{\psi(2),2}(x) f_{\psi(3),3}(x) \ldots \tag{11}$$

Finally, we define the translation $\Phi^{g,c}$ by

$$\Phi^{g,c}(x) = \begin{cases} \Phi(x) & \text{if } x \in \Delta \\ \text{undefined} & \text{otherwise} \end{cases} . \tag{12}$$

The translation $\Phi^{g,c}$ can be non-uniformly realized by a lineage of automata. To prove this, we need the following two Lemmas.

Lemma 1. Δ *is a closed set.*

Proof. Suppose $x \notin \Delta$. Then there is a prefix u of x such that u is not a choice. Let y be an infinite string in $\mathbb{B}(u)$. Since u is a prefix of y, it follows that $y \notin \Delta$. We conclude that Δ is closed.

$\square$

Lemma 2. *For every n, the function $\pi_n \circ \Phi^{g,c}$ depends only on the first n input symbols.*

Proof. Let $x \in \Delta$ be an infinite string. The output of $\Phi(x)$ consists of infinitely many concatenations of strings of the form $f_{\psi(m),m}(x)$. For every integer $m \geq 1$, the string $f_{\psi(m),m}(x)$ starts at index

$$i_m = \left(\sum_{t=1}^{m-1} |f_{\psi(t),t}(x)| \right) + 1 \geq m . \tag{13}$$

This string consists of multiple copies of x_k, for a certain $k \leq m$. Note that k does not depend on the particular choice of x. The n-th symbol of $\Phi(x)$ belongs to $f_{\psi(m),m}(x)$ for a certain m. Obviously $n \geq i_m$, so $k \leq m \leq i_m \leq n$. It follows that $\pi_n(\Phi^{g,c}(x)) = x_k$ for a certain $k \leq n$.

$\square$

Combining Lemmas 1, 2 and Proposition. 2, we conclude that $\Phi^{g,c}$ can be non-uniformly realized. Next, we will examine the complexity of $\Phi^{g,c}$. Proposition. 4 shows that $\Phi^{g,c}$ is of complexity g_c, while Prop. 5 tells us that any lineage with a complexity less than g_c cannot non-uniformly realize $\Phi^{g,c}$.

Proposition 4. *The translation $\Phi^{g,c}$ can be non-uniformly realized by a lineage $\mathcal{A}$ that updates at every step, such that A_n has $g_c(n)$ states.*

Proof. Let $\mathcal{A}$ be the lineage from the proof of Proposition. 2. We see that $P^n(\Delta)$ equals C_n. It follows that $\mathcal{A}$ is of size g_c.

$\square$

For the proof of Proposition 5, we need the following Lemma.

Lemma 3. *Let k, l and $n \geq 1$ be integers such that $k, l \geq n$. Let x and y be infinite strings such that $x_n \neq y_n$. Then there is an i such that*

$$\pi_{k+i}(\Phi(x)) \neq \pi_{l+i}(\Phi(y)) . \tag{14}$$

Proof. Assume $k \geq l$. Let $t = k - l$. Choose an integer $m \geq l$ such that $\psi(m) = (n, t)$. Then $f_{\psi(m),m}(x) = f_{n,t,m}(x) = x_n^{1+t}$, since $n \leq m$. It follows that $\Phi(x)$ contains a string x_n^{1+t}, starting at an index $i_m \geq m$, namely the string $f_{\psi(m),m}(x)$. Similarly, $\Phi(y)$ contains a string y_n^{1+t}, starting at the same index, see Fig. 2. But then

$$\pi_{i_m+t}(\Phi(x)) \neq \pi_{i_m}(\Phi(y)) , \tag{15}$$

since $x_n \neq y_n$. Now $i_m \geq m \geq l$, so $i_m = l + i$ for a certain i, and $i_m + t = l + i + (k - l) = k + i$. Therefore

$$\pi_{k+i}(\Phi(x)) \neq \pi_{l+i}(\Phi(y)) . \tag{16}$$

$\square$

$$t + 1$$

x_n	x_n	$\cdots$	x_n

$i_m + t$

y_n	y_n	$\cdots$	y_n

i_m

Fig. 2. Part of the outputs of Φ, on the inputs x and y. Starting in position i_m, the outputs contain a sequence of x_n's and y_n's respectively, of size $t + 1$ each. As a result, $\pi_{i_m+t}(\Phi(x)) = x_n \neq y_n = \pi_{i_m}(\Phi(y))$

Let u and v be two different choices of length n. Let u' be a choice that extends u and v' a choice that extends v. Finally, let $x = (a_1)^\omega$. It follows that $u'x$ and $v'x$ are elements of Δ. Since $u \neq v$, it follows that there is an $n' \leq n$, such that $\pi_{n'}(u'x) \neq \pi_{n'}(v'x)$. By Lemma 3, there is an i such that

$$\pi_{|u'|+i}(\Phi^{g,c}(u'x)) \neq \pi_{|v'|+i}(\Phi^{g,c}(v'x)) \ . \tag{17}$$

See Fig. 3 for a visual explanation.

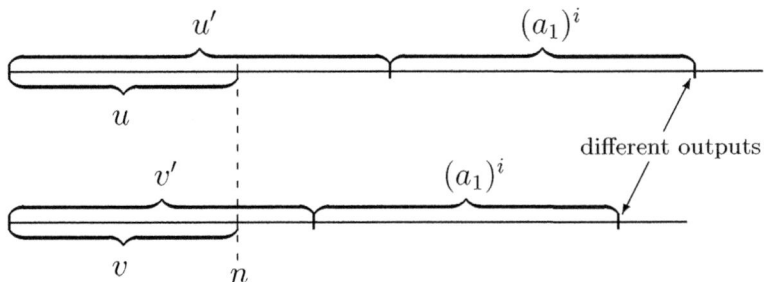

Fig. 3. Two finite choices u and v, such that $u \neq v$, are extended to choices u' and v' respectively. These choices are extended with the infinite string $x = (a_1)^\omega$. Then there is an integer i such that $\pi_{|u'|+i}(\Phi^{g,c}(u'x)) \neq \pi_{|v'|+i}(\Phi^{g,c}(v'x))$

Proposition 5. *Let A be a lineage that non-uniformly realizes $\Phi^{g,c}$. Suppose A_m is able to process all strings of length n. Then A_m has at least $g_c(n)$ states.*

Proof. Suppose A_m has less than $g_c(n)$ states. Then there are two different choices u and v of length n, with choices u' and v' that extend u and v respectively, such that A_m enters the same state r after processing either u' or v', see Fig. 4.

Then there is an i that satisfies (17). Suppose $\mathcal{A}$ is in the m-th update[5], in state r. Now we give $x = (a_1)^\omega$ as further input to $\mathcal{A}$. After i steps, $\mathcal{A}$ enters a state r' with a certain output b. These last i steps are independent of the steps that $\mathcal{A}$ took to reach r. In other words,

$$\pi_{|u'|+i}(\Phi^{g,c}(u'x)) = \pi_{|v'|+i}(\Phi^{g,c}(v'x)) = b , \tag{18}$$

which contradicts (17). It follows that A_m must have at least $g_c(n)$ states.

□

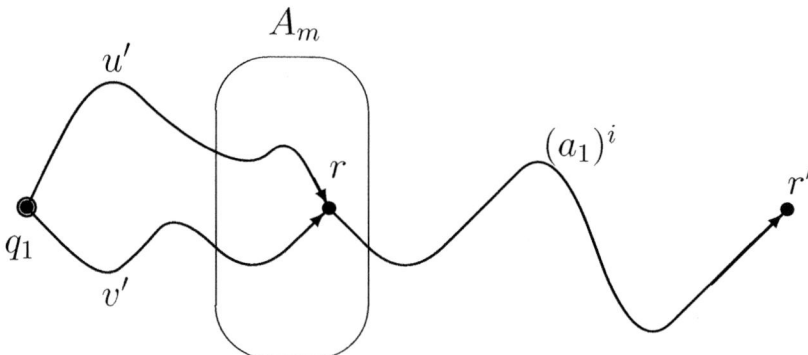

Fig. 4. The paths of two valid input-prefixes u' and v'. After processing u' or v', A_m enters the state r. Then the remainder of the input is processed, which equals $(a_1)^\omega$ in both cases. The rest of the path only depends on r and $(a_1)^\omega$, so after i steps, both paths enter r' and the same output symbol is generated

Corollary 1. *Let $\mathcal{A}$ be a lineage that non-uniformly realizes $\Phi^{g,c}$. Then A_n has at least $g_c(n)$ states.*

Proof. Since each active automaton must read at least one symbol before $\mathcal{A}$ can update, by the time $\mathcal{A}$ is ready to update to the $n + 1$-st automaton, at least n symbols have been read.

□

For any non-decreasing function g and any integer $c > 1$, the complexity class $\mathrm{SIZE}(g_c)$ contains the translation $\Phi^{g,c}$. Furthermore, $\mathrm{SIZE}(g_c)$ is the smallest complexity class that contains $\Phi^{g,c}$.

---

[5] Or the $m + 1$-st, if r is an exit state.

3.2   A Hierarchy Result for Complexity Classes

For clarity, we repeat the results of the preceding section in one Proposition.

Proposition 6. *Let c be a positive integer and g a positive, non-decreasing function such that g equals g_c. Let h be a function such that $h(m) < g(m)$ for a certain m. Then $SIZE(g) - SIZE(h)$ is non-empty.*

Proof. Combine Proposition 4 and Corollary 1.

□

When we are free to choose c, we can show that for any positive non-decreasing function g, translations exist that cannot be non-uniformly realized by any lineage that has less than $g(n)$ states in its n-th automaton, for any n. Observe that this is a stronger claim than before; we no longer require the "growth rate" to be bounded.

Theorem 1. *Let g be a positive, non-decreasing function and let h be a function such that $h(m) < g(m)$ for a certain m. Then $SIZE(g) - SIZE(h)$ is non-empty.*

Proof. Let $c \geq g(1)$ be an integer such that $g(n+1) \leq c \cdot g(n)$ for every n smaller than m. Then $g_c(n) = g(n)$ for all $n \leq m$. It follows that $h(m) < g_c(m)$. The translation $\Phi^{g,c}$ can be non-uniformly realized by a lineage of size g_c (Proposition 4). Hence $\Phi^{g,c} \in SIZE(g)$.

Any lineage $\mathcal{A}$ that non-uniformly realizes $\Phi^{g,c}$ must have at least $g_c(n)$ states in its n-th automaton (Corollary 1). Since $h(m) < g_c(m)$, it follows that $\mathcal{A}$ is not of size h. Hence $\Phi^{g,c} \notin SIZE(h)$.

□

Corollary 2. *Let g and h be positive non-decreasing functions such that $h(n) \leq g(n)$ for all n. If the inequality is strict for a certain n, then $SIZE(h)$ is a proper subset of $SIZE(g)$.*

Proof. By definition, any translation of complexity h is in $SIZE(g)$. By Theorem 1, not every translation of complexity g is in $SIZE(h)$.

□

This means that every extra state of a lineage can be used to gain more potential computing power.

Corollary 3. *Let g and h be positive non-decreasing functions such that $h(n) < g(n)$ for a certain n and $g(m) < h(m)$ for a certain m. Then the classes $SIZE(g)$ and $SIZE(h)$ are incomparable: both contain translations that do not occur in the other.*

4 Conclusion

In this paper, we have developed the theory of lineages as a new model of computation. A lineage is a sequence of finite automata, where each automaton in the sequence is viewed as the next instantiation or incarnation of the evolving system that it models. By using lineages, one can immediately single out the evolutionary aspects of the system. The development of a lineage is modelled by looking at the automata in the sequence, and the relation between each automaton and its immediate successor.

An important characteristic of an automaton is its size. For a lineage, we defined a complexity measure based on the size of the automata in the sequence. We proved in Theorem 1 that lineages of higher complexity are able to non-uniformly realize more translations than lineages of lower complexity. Specifically, for each non-decreasing function g, there is a translation that can be non-uniformly realized by a lineage of complexity g, but not by any lineage that has fewer than $g(n)$ states available for its n-th automaton, for a certain n. On the other hand, once a translation (over an input alphabet of size c) is fixed, we know that it can be non-uniformly realized by a lineage of complexity at most c^n.

We conclude that lineages of automata present an attractive model for evolving interactive systems, with a basic mechanism for the underlying mode of computation. The attractiveness is due to the mathematical elegance of the model which, in spite of its apparent simplicity, still captures the important aspects of many other models. The close relationship of lineages to finite automata makes the model even more interesting since techniques and proofs from automata theory can be adapted to the theory of lineages.

References

1. J.L. Balcázar, J. Díaz, and J. Gabarró. *Structural Complexity I*, 2nd Edition, Springer-Verlag, Berlin, 1995.
2. L.H. Landweber. Decision Problems for ω-Automata, *Math. Syst. Theory*, Vol. 3 : 4, 1969, pp. 376-384.
3. J. van Leeuwen, J. Wiedermann. On Algorithms and Interaction, in: M. Nielsen, B. Rovan (Eds.), *Mathematical Foundations of Computer Science 2000*, Lecture Notes in Computer Science Vol. 1893, Springer-Verlag, Berlin, 2000, pp. 99-113.
4. J. van Leeuwen, J. Wiedermann. The Turing Machine Paradigm in Contemporary Computing, in: B. Enquist, W. Schmidt (Eds.), *Mathematics Unlimited - 2001 and Beyond*, Springer-Verlag, Berlin, 2001, pp. 1139-1156.
5. J. van Leeuwen, J, Wiedermann. A Computational Model of Interaction in Embedded Systems, in: Technical Report UU-CS-2001-02, Institute of Information and Computing Sciences, Utrecht University, 2001.
6. J. van Leeuwen, J. Wiedermann. Beyond the Turing Limit: Evolving Interactive Systems, in: L. Pacholski, P. Ružička (Eds.), *SOFSEM 2001: Theory and Practice of Informatics*, 28th Conference on Current Trends in Theory and Practice of Informatics, Lecture Notes in Computer Science Vol. 2234, Springer-Verlag, Berlin, 2001, pp. 90-109.

7. M. Li, P. Vitányi. *An Introduction to Kolmogorov Complexity and Its Applications*, 2nd Edition, Springer-Verlag, New York, 1997.
8. L. Staiger. ω-Languages, in: G. Rozenberg, A. Salomaa (Eds.), *Handbook of Formal Languages*, Vol. 3, *Beyond Words*, Springer-Verlag, Berlin, 1997, pp. 339-387.
9. W. Thomas. Automata on Infinite Objects, in: J. van Leeuwen (Ed.), *Handbook of Theoretical Computer Science*, Vol. B, Elsevier Science, Amsterdam, 1990, pp. 134-191.
10. P. Verbaan, J, van Leeuwen, J. Wiedermann. Lineages of Automata, Technical Report UU-CS-2004-018, Institute of Information and Computing Sciences, Utrecht University, 2004.
11. P. Wegner. Why Interaction is more Powerful than Algorithms, *Comm. of the ACM*, Vol. 40:5, 1997, pp. 81-91.
12. P. Wegner, E. Eberbach. New Models of Computation, *The Computer Journal*, Vol. 47:1, 2004, pp. 4-9.
13. P. Wegner, D. Goldin. Computations beyond Turing Machines, *Comm. of the ACM*, Vol. 46, 2003, pp. 100-102.
14. J. Wiedermann, J. van Leeuwen. Emergence of a Super-Turing Computational Potential in Artificial Living Systems, in: J. Kelemen, P. Sosík (Eds.), *Advances in Artificial Life*, 6th European Conference (ECAL 2001), Lecture Notes in Artificial Intelligence Vol. 2159, Springer-Verlag, Berlin, 2001, pp. 55-65.
15. J. Wiedermann, J. van Leeuwen. The Emergent Computational Potential of Evolving Artificial Living Systems, *AI Communications*, Vol. 15 No. 4, IOS Press, Amsterdam, 2002, pp. 205-215.

Author Index

Lecture Notes in Computer Science

For information about Vols. 1–3019

please contact your bookseller or Springer-Verlag